WENNER–GREN CENTER
INTERNATIONAL SYMPOSIUM SERIES

VOLUME 47

DUALITY AND UNITY OF THE BRAIN

W0036554

DUALITY AND UNITY OF THE BRAIN

Unified Functioning and Specialisation
of the Hemispheres

Proceedings of an International Symposium held at
The Wenner-Gren Center, Stockholm, May 29 – 31, 1986

Edited by

David Ottoson
Wenner-Gren Center Foundation
Stockholm, Sweden

PLENUM PRESS • NEW YORK AND LONDON

First published 1987

Published in Great Britain by
THE MACMILLAN PRESS LTD
Houndmills, Basingstoke, Hampshire RG21 2XS
and London
Companies and representatives
throughout the world

Published in the United States of America by
PLENUM PUBLISHING CORPORATION
233 Spring Street, New York, NY 10013, USA

ISBN-13: 978-1-4612-9081-0   e-ISBN-13: 978-1-4613-1949-8
DOI: 10.1007/ 978-1-4613-1949-8
LCCN 87–042720

Contents

CONTENTS

Session V
Integration of Hemispheric Functions
Chairman: C. Trevarthen

The Participants

Aniko Bartfai
Department of Psychiatry
Karolinska Hospital
S-104 01 STOCKHOLM
Sweden

Camilla Benbow
Department of Psychology
Iowa State University
AMES
Iowa 50011
USA

Giovanni Berlucchi
Department of Physiology
University of Verona
Strada Le Grazie
I-37134 VERONA
Italy

Erik Borg
Department of Audiology
Karolinska Hospital
S-104 01 STOCKHOLM
Sweden

Nathan Brody
Department of Psychology
Wesleyan University
MIDDLETOWN
Connecticut 06457
USA

Philip Bryden
Department of Psychology
University of Waterloo
WATERLOO
Ontario
Canada N2L 3G1

Björn Brådvik
Department of Neurology
University Hospital
S-221 85 LUND
Sweden

Aila Collins
Olof af Acrels väg 1
S-171 64 SOLNA
Sweden

Otto-Detlev Creutzfeldt
Department of Neurobiology
Max-Planck-Institute for
Biophysical Chemistry
P O Box 2841
D-3400 GÖTTINGEN-NIKOLAUS-
BERG
FRG

Ennio de Renzi
Department of Neurology
University of Modena
Via del Pozzo, 71
I-41100 MODENA
Italy

Marian Diamond
2549 Life Sciences Bldg.
University of California
BERKELEY
California 94720
USA

Gunnar Edman
Department of Psychiatry
Karolinska Hospital
S-104 01 STOCKHOLM
Sweden

Curt von Euler
Nobel Institute of Neuro-
  physiology
Karolinska Institute
S-104 01 STOCKHOLM
Sweden

Howard Ehrlichman
44-59 Kissena Blvd
FLUSHING
New York 113 55
USA

Lars Farde
Department of Psychiatry
Karolinska Hospital
S-104 01 STOCKHOLM
Sweden

Marianne Frankenhaeuser
Department of Psychology
University of Stockholm
S-106 91 STOCKHOLM
Sweden

Guido Gainotti
Department of Psychiatry
Università Cattolica del
  Sacro Cuore
Largo Agostino Gemelli, 8
I-00168 ROMA
Italy

Ida Gerendai
2nd Department of Anatomy
Semmelweis University
  Medical School
Tüzolto 58
H-1094 BUDAPEST
Hungary

Stanley Glick
Department of Pharmacology
  and Toxicology
Albany Medical College
47 New Scotland Ave.
ALBANY
New York 12208
USA

Gunnar Grant
department of Anatomy
Karolinska Institute
S-104 01 STOCKHOLM
Sweden

Stefan Hagstadius
Department of Psychiatry
University Hospital
S-221 85 LUND
Sweden

Richard Harshman
Department of Psychology
University of Western
  Ontario
LONDON
Ontario
Canada N6A 6C2

Joseph Hellige
Department of Psychology
Seeley G. Mudd Building
University of Southern
  California
University Park
LOS ANGELES
California 90089
USA

Eric Hellstrand
Department of Clinical
  Neurophysiology
Karolinska Hospital
S-104 01 STOCKHOLM
Sweden

Kenneth Hugdahl
Department of Psychology
University of Bergen
Årstadveien 21
N-5000 BERGEN
Norway

Tomas Hökfelt
Department of Histology
Karolinska Institute
S-104 01 STOCKHOLM
Sweden

David Ingvar
Department of Clinical
  Neurophysiology
University Hospital
S-221 85 LUND
Sweden

Barbro Johansson
Department of Neurology
University Hospital
S-221 85 LUND
Sweden

Birgitta Johnsen
Department of Phoniatry
Akademiska Hospital
S-751 85 UPPSALA
Sweden

Doreen Kimura
Department of Psychology
University of Western
  Ontario
LONDON
Ontario
Canada N6A 5C2

Britt af Klinteberg
Askrikevägen 21
S-181 46 LIDINGÖ
Sweden

Sven Levander
Department of Psychiatry
Pb 3008
N-7001 TRONDHEIM
Norway

Susan Levine
Department of Behavioural
  Sciences
University of Chicago
5848 University Ave.
CHICAGO
Illinois 60637
USA

Jerre Levy
Department of Behavioural
  Sciences
University of Chicago
5848 University Ave.
CHICAGO
Illinois 60637
USA

Sivert Lindström
Department of Physiology
University of Göteborg
S-400 33 GÖTEBORG
Sweden

Thomas Lundeberg
Department of Physiology
Karolinska Institute
S-104 01 STOCKHOLM
Sweden

Peter MacNeilage
Department of Linguistics
Calhoun 501
University of Texas
AUSTIN
Texas 78712
USA

David Magnusson
Department of Psychology
University of Stockholm
S-106 91 STOCKHOLM
Sweden

Bengt Meyerson
Department of Pharmacology
Biomedicum
S-751 24 UPPSALA
Sweden

Björn Meyerson
Department of Neurological
  Surgery
Karolinska Hospital
S-104 01 STOCKHOLM
Sweden

David Milner
MRC Cognitive Neuroscience
  Research Group
Psychological Laboratory
University of St. Andrews
FIFE KY16 9JU
Scotland

Dolores Montero
Department of Psychiatry
Karolinska Hospital
S-104 01 STOCKHOLM
Sweden

Morris Moscovitch
The Institute for Advanced
  Studies
Hewbrew University of
  Jerusalem
GIVAT RAM 91904
Israel

Michael Myslobodsky
Department of Psychology
National Institute of
  Mental Health
BETHESDA
Maryland 20205
USA

Björn Mårtensson
Department of Psychiatry
Karolinska Hospital
S-104 01 STOCKHOLM
Sweden

Margaret Naeser
Veterans Administration
  Medical Center
150 South Ave.
BOSTON
Massachusetts 02130
USA

Peter Nordström
Department of Psychiatry
Karolinska Hospital
S-104 01 STOCKHOLM
Sweden

Ulf Norrsell
Department of Physiology
University of Göteborg
S-400 33 GÖTEBORG
Sweden

Henrik Nybäck
Department of Psychiatry
Karolinska Hospital
S-104 01 STOCKHOLM
Sweden

Håkan Nyman
Department of Psychiatry
Karolinska Hospital
S-104 01 STOCKHOLM
Sweden

George Ojemann
Department of Neurological
  Surgery
University of Washington
SEATTLE
Washington 98195
USA

Nancy Pedersen
Department of Hygiene
Karolinska Institute
S-104 01 STOCKHOLM
Sweden

Jarl Risberg
Department of Psychiatry
University Hospital
S-221 85 LUND
Sweden

Glenn Rosen
Neurological Unit
Beth Israel Hospital
330 Brookline Ave.
BOSTON
Massachusetts 02215
USA

Eva Rydin
Karolinska Hospital
Department of Psychiatry
Karolinska Hospital
S-104 01 STOCKHOLM
Sweden

Daisy Schalling
Department of Psychiatry
Karolinska Hospital
S-104 01 STOCKHOLM
Sweden

Hans Sjöberg
Department of Psychiatry
Karolinska Hospital
S-104 01 STOCKHOLM
Sweden

Ann-Charlotte Smedler
Department of Psychology
University of Stockholm
S-106 91 STOCKHOLM
Sweden

Karl-Erik Spens
Department of Audiology
Karolinska Hospital
S-104 01 STOCKHOLM
Sweden

Ulf Söderberg
Department of Neurophysio-
   logy
Ulleråker Hospital
S-750 15 UPPSALA
Sweden

Lars Terenius
Department of· Pharm.
   Pharmacology
Biomedicum
S-751 24 UPPSALA
Sweden

Colwyn Trevarthen
Department of Psychology
University of Edinburgh
7 George Square
EDINBURGH EH8 9JZ
Scotland

Lil Träskman-Bendz
Sveavägen 109
S-113 50 STOCKHOLM
Sweden

Don Tucker
Department of Psychology
University of Oregon
EUGENE
Oregon 97403
USA

Per Uddén
Hofstrasse 1
CH-6064 KERNS
Switzerland

Siegbert Warkentin
CBF-lab.
S:t Lars Hospital
Box 638
S-220 06 LUND
Sweden

Frits-Axel Wiesel
Department of Psychiatry
Karolinska Hospital
S-104 01 STOCKHOLM
Sweden

Ann Wirsén
Department of Psychiatry
Karolinska Hospital
S-104 01 STOCKHOLM
Sweden

Sandra Witelson
Department of Psychiatry
McMaster University
1200 Main street West
HAMILTON
Ontario L8N 3Z5
Canada L8N 3Z5

Anna Wägner
Department of Psychiatry
Karolinska Hospital
S-104 01 STOCKHOLM
Sweden

Eran Zaidel
Department of Psychology
University of California
LOS ANGELES
Ca 90024
USA

Rolf Öhman
Department of Psychiatry
University of Lund
S-221 00 LUND
Sweden

Opening Address*

Roger Sperry

(Read by David Ottoson)

Rather than address select aspects of our topic, or particular researches in which I have been involved, or to personally reminisce about golden "days that are no more" I plan to respond to Professor Ottoson's kind invitation to open this conference by reviewing briefly some related conceptual developments that, as much as the empirical findings themselves perhaps, are of general ongoing concern to all of us in science.

Back in the mid-1960s a number of things came together in psychobiology to cause a shift in theory away from behaviorism with its strictly objective, physical explanations, over to a new mentalist (for want of a better term) position accepting the causal efficacy of subjective or introspective mental events, in a supervening sense. Earlier arguments in neuroscience that had excluded any causal influence of consciousness and which seemingly had been closed, airtight and irrefutable, were discovered to have a flaw, a loophole or shortcoming, and to be outweighed by a new 'emergent interactionist' reasoning.

A new formula for mind-brain interaction was perceived that involved a revised philosophic stance described initially as a compromise between materialism and dualism because it integrated formerly contradictory features from opposite sides of the old dichotomy into a single consistent worldview framework.

Instead of excluding mind and spirit from science, as had been the dominant practice for more than half a century, the new outlook puts subjective, mental forces near the top of the brain's causal control hierarchy and gives them primacy in determining what a person is and does. These revisions allow one to retain belief in science and the scientific method and at the same time they reinforce traditional humanist values – and support also mentalist (rather than behaviorist) concepts of the conscious self, freedom of will and personhood.

Cognitive psychology, humanistic psychology, clinical, abnormal and related psychologies and psychiatries, as well as the social and other sciences that depend on introspective mental phenomena and subjective explanations gain a more prestigious scientific status in the new outlook. Cognitive states had formerly been conceived to be either in parallel with or identical to their neural correlates. Either way it was generally supposed that science could provide, in principle, a complete explanation of brain function and behavior strictly in terms of the neural correlates themselves without reference to the introspective phenomena, and this was taken to be the more rigorous

---

* Based on edited excerpts from recent papers of Sperry, this address was compiled and read by Professor David Ottoson when it was learned that Dr. Sperry would be unable to present opening remarks in person.

and more truly scientific way to go. Accordingly, reliance on subjective mental explanations tended to be put down as something short of true science.

In the new interactionist reasoning the cognitive properties are recognized to be different from, and more than, their collective neural correlates, to have their own dynamics and laws of interaction, and to exert downward emergent control over their physiological constituents. On these revised terms psychology, cognitive science and the social sciences represent distinct sciences in their own right at their own level with their own laws and principles that are not, even in theory, reducible to the kinds of laws which identify brain physiology or biophysics – though advancements in brain physiology should enormously enhance their understanding. Modified concepts incorporating 'downward causation' give the higher cognitive forces in brain processing control over the lower forces of neurophysiology.

The new outlook in behavioral science is found on analysis to rest on a shift to a revised form of causal determinism. Specifically, it involves a switch from traditional microdeterminism, in which everything is determined exclusively from below upward following the course of evolution, to a paradigm that gives due emphasis to macrodeterminism in which higher emergent properties of organizational hierarchies at all levels interact autonomously at their own level and also exert downward supervenient control over the subelements at lower levels. The bottom-up controls are retained but are supplemented with top-down controls. The recent view is believed to represent a more valid paradigm for all science, not just behavioral science.

This new 'macro' mentalist paradigm profoundly transforms the traditional scientific descriptions of ourselves and the world. Among other things it turns around the science-values relation and largely resolves the dilemma of free-will and determinism. More relevant to our present conference, it also changes our conceptions of the causal controls in physical brain processing and provides an explanation of why the left and right hemispheres, though they may each, when surgically separated, function differentially each as a separate conscious unit, may nevertheless when normally joined together, function together as a single conscious self.

Session I
Neuroanatomical and Neurochemical Asymmetries of the Brain

Chairman: J. Levy

1
Functional and Neurochemical Asymmetry in the Corpus Striatum

S.D. Glick, J.N. Carlson, K.L. Drew and R.M. Shapiro

INTRODUCTION

Although it is well documented that the human brain is asymmetric with regard to a variety of functions, most of these functions, notably language and affect, have been ascribed to the cortex. There has been relatively little interest, and consequently relatively little data, bearing on the issue of subcortical asymmetry. However, asymmetry in one subcortical structure, the corpus striatum, has been investigated extensively in rats, and recently to some extent in humans. Our overall understanding of the corpus striatum, in terms of its anatomy, neurochemistry and functions, is certainly considerable vis-a-vis our understanding of the cortex and other portions of the neuraxis. Study of the corpus striatum may therefore elucidate basic mechanisms involved in laterally organized neural systems.

The functional significance of the dopamine-containing nigrostriatal pathway, the major input to the corpus striatum, has been intensively studied during the past two decades. The discovery of L-Dopa's efficacy in treating the symptoms of Parkinsonism (Cotzias et al., 1967) led to a search for animal models that could be used to evaluate new and potentially more useful agents for this disorder. Ungerstedt (1971) demonstrated that rats with unilateral lesions of the nigrostriatal pathways would turn in circles (rotate) in response to a variety of dopaminergic agents (e.g., d-d-amphetamine, L-Dopa, apomorphine). Soon thereafter, in this laboratory, it was found that such drugs also induced rotation, although at lower rates, in normal intact rats (e.g., Jerussi and Glick, 1976). Subsequent studies demonstrated that, as with lesioned rats, the rotation in normal rats was consistent in direction and correlated in magnitude upon repeated testing. Since the rotational behavior in lesioned rats was attributed to a functional asymmetry resulting from the unilateral loss of dopaminergic neurons, it was hypothesized that this behavior in normal rats was a manifestation of an endogenous asymmetry of these same neurons.

In this chapter we will review evidence bearing on the neuronal mechanisms responsible for the consistent direction of circling behavior in normal rats. The plasticity of the strength and direction of this behavior with respect to heritability, learning, and the effects of stress will also be discussed as will some recent data in humans.

DUAL ASYMMETRIES IN NIGROSTRIATAL FUNCTION: A NEW MODEL

When tested at night, during the active part of their circadian cycle, normal untreated rats will rotate spontaneously, and the direction of nocturnal rotation is usually the same as that induced by d-amphetamine (Glick and Cox, 1978). Correlations been rats' left-right preferences in other contexts and their preferred directions of d-amphetamine-induced circling led to the conclusion that normal circling is "simply the logical result of an intrinsic and persistent side preference" (Glick et al., 1977). In the first studies aimed at finding a neurochemical basis of such preferences, the levels of dopamine in the two striata differed by about 15% and the side containing more dopamine was contralateral to rats' side preferences (Zimmerberg et al., 1974). A high dose (20 mg/kg) of d-amphetamine increased this asymmetry to about 25% (Glick et al., 1974). It is important to note that these initial experiments were conducted with female rats as later experiments in this laboratory and elsehwere (Robinson et al., 1980) showed that dopamine levels in the contralateral and ipsilateral striata of male rats were not significantly different. Similarly, DOPAC, the major metabolite of dopamine in rats, was found to be asymmetrically distributed in the striata of female (Jerussi et al., 1977) but not of male (Yamamoto and Freed, 1984) rats. Although both nocturnal and d-amphetamine-induced rotation occur at lower rates in male than in female rats (Brass and Glick, 1981; Becker et al., 1982; Glick, 1985), the lack of an apparent dopaminergic asymmetry in the striata of male rats questioned the role of the nigrostriatal system in mediating the lateralized behavioral phenomena. Recent work (Shapiro et al., 1986) has focused on this issue.

Side-to-side differences in striatal dopamine content and metabolites could result from either a functional or anatomical asymmetry. That is, the same numbers of nerve terminals could contain or release different amounts of dopamine, or there might simply be different numbers of terminals. With the goal of distinguishing these possibilities, dopamine uptake in vitro was measured in crude mitochondrial pellets (Gray and Whittaker, 1962; Kuhar, 1973) obtained from the striata of rats previously tested for nocturnal rotation. Side differences in Vmax should reflect an asymmetry in the density of dopamine terminals. The results indicated that there are indeed asymmetries in dopamine uptake and, most importantly, in both sexes. But the relationships between these Vmax asymmetries and rats' circling behavior were dichotomous suggesting that, in both sexes, there are two populations of rats. For females, but not for males, the contralateral Vmax was significantly greater than the ipsilateral Vmax. These data appeared consistent with the sex differences in dopamine and DOPAC asymmetries noted above. However, because multiple determinations from each striatum were obtained, it was possible to calculate for each individual rat whether or not there was a significant difference in Vmax. Many males did indeed have a significant Vmax asymmetry. In both sexes there were some rats having a greater Vmax in the contralateral striatum and other rats having a greater Vmax in the ipsilateral striatum. Most females (19/26) had a higher contralateral Vmax-- hence the overall group difference between sides. Most males (13/19) had a higher ipsilateral Vmax-- but the prevalence was less than in females so that the overall group difference was not significant. In other words, males and females were differentially distributed in the two populations, and males were more equally represented than females (see Glick and Shapiro, 1985).

The magnitude of the Vmax asymmetry for dopamine uptake was positively correlated with the strength of a rat's rotational bias. This was true for both sexes and for both populations of rats (for all rats combined: r=.50, p<0.001, n=45). That is, the larger the Vmax asymmetry, the stronger was a rat's rotational bias, regardless of which side Vmax was higher on and which way the rat turned. This might be interpreted to indicate that the striatal dopaminergic asymmetry is a determinant of the strength but not of the direction of a rat's turning. Alternatively, there might really be two populations of rats related to whether the net influence of striatal function is excitatory or inhibitory in nature. Evidence of functionally distinct and antagonistic systems has been reported with respect to dopaminergic nigrostriatal projections (Vaccarino and Franklin, 1982), intrinsic striatal neurons (Cools and van Rossum, 1980), and striatal outputs (Groves, 1983). The present "two-population" hypothesis is more directly supported by the fact that a neurochemical difference distinguished the two populations: whereas the mean ipsilateral Vmax values were virtually identical in the two populations, the contralateral Vmax values were significantly different (Shapiro et al., 1986).

One important prediction resulting from the dopamine uptake data is that the effect of a unilateral nigrostriatal lesion would differ in the two populations. Although d-amphetamine commonly induces ipsilateral rotation in rats with such lesions, the relationship between the intensity of rotation and the size of the lesion is not always clear (e.g., Heikkila et al., 1981). Indeed the direction of preoperative rotation with regard to the side of a subsequent lesion has been shown to be an important determinant of the intensity of postoperative rotation (Glick and Cox, 1978; Robinson and Becker, 1983). We recently examined the relationship between lesion size and effect more closely. 6-Hydroxydopamine

6

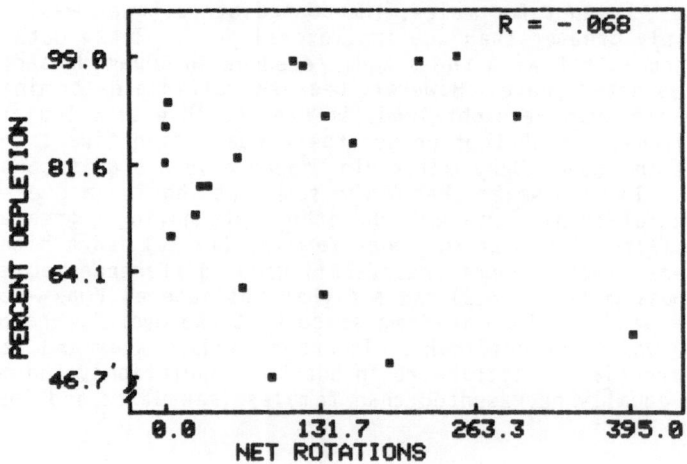

Figure 1. Scatter diagram showing lack of relationship between the extent of 6-hydroxydopamine-induced dopamine depletion in striatum and the intensity of ipsilateral rotation elicited by d-amphetamine (1.56 mg/kg) in male rats.

lesions were placed unilaterally in the substantia nigra of 20 male rats. All rats were tested the day before and eight days after surgery for rotation induced by d-amphetamine (1.56 mg/kg). All lesions were made ipsilateral to the preoperative direction of rotation so that side of the lesion could not be a confounding variable. Figure 1 shows that there is no apparent correlation between striatal dopamine depletion and intensity of postoperative rotation. Moreover, compared to their preoperative behavior, 10 of the 20 rats either showed little change or actually rotated less than they had initially. These data are seemingly inconsistent with the results of a large number of studies conducted over the last 15 years. However, the assumption that only high rates of rotation are associated with large lesions has been so prevalent that one wonders whether "exceptional" rats have been innocently excluded from consideration. For example, in one recent study (Sakurai et al., 1985), it was stated that lesions of the substantia nigra with 6-OHDA were regarded as successful only "if rats turned more than 20 times in at least one period of 5 minutes..." Figure 1 shows that low rates of rotation can in fact be associated with large lesions; and high rates can be associated with small lesions (or no lesion, as we have observed rates in excess of 300 turns per hour in many intact rats).

The results of the preceding experiments are not consistent

with there being a homogeneous population of rats. Figure 2 shows
one possible model of nigrostriatal inputs and outputs that may
underlie the postulated two populations of rats. In this model,
dopamine would be uniformly inhibitory and would function to
modulate the intensity rather than the kind of striatal output. The
differences between rats would be mostly a function of whether the
striatal outputs are predominantly excitatory or inhibitory. As
suggested by the uptake data, rats would also be distinguishable by
virtue of having different numbers of neurons on the contralateral
side. The model shows the clearest or extreme cases. That is, the
two populations are probably not totally discrete. Rather, there
probably exists a continuum of rats having varying ratios of
excitatory and inhibitory outputs-- and the proposed two populations
really represent rats having a preponderance of one versus the other
kind of output. Figure 3 illustrates how rats having a mixture of
excitatory and inhibitory outputs could be responsible for a lack of
correlation between the dopaminergic asymmetry and rotation. The
fact that normally, in intact rats, there is such a correlation (see

TWO POPULATIONS OF RATS:

POSSIBLE MECHANISMS OF STRIATAL ASYMMETRY

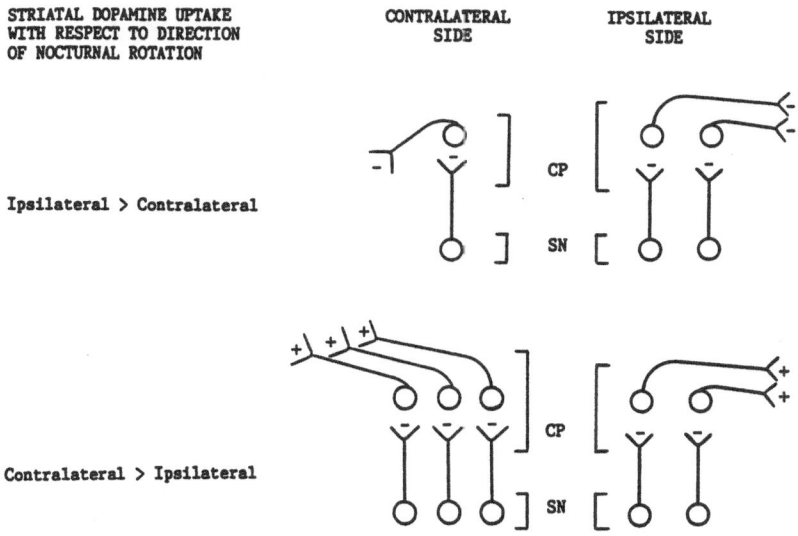

Figure 2. A model of nigrostriatal asymmetry in which "two
populations" of rats result from a preponderance of excitatory vs.
inhibitory striatal outputs.

8

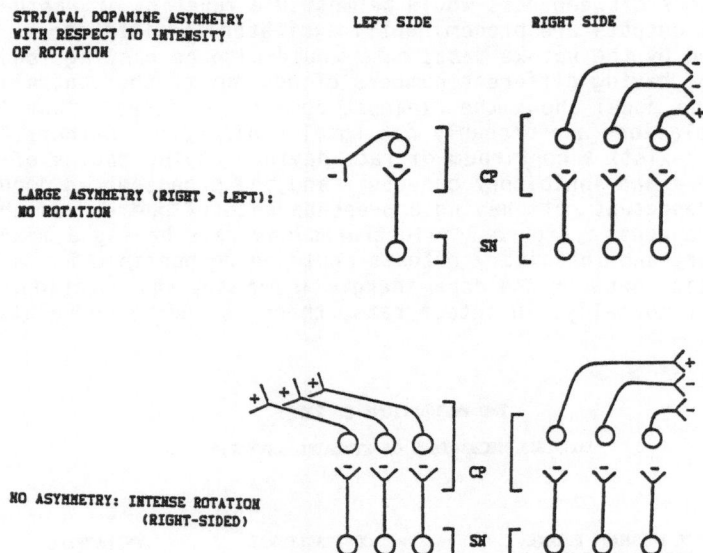

Figure 3. Variations in striatal outputs that may result in a lack of correlation between asymmetries in striatal dopamine levels and rates of rotation.

above; also Shapiro et al., 1986) suggests that most rats have, in both striata, a preponderance or net balance of outputs of one kind or the other (either excitatory or inhibitory). However, a lesion may upset this balance and, although seeming paradoxical, this may result in a lack of correlation between the asymmetry and the behavior (Figure 1).

PLASTICITY OF LATERALIZED BEHAVIOR

A vast literature attests to the fact that most rats (cf. Zimmerberg et al., 1978) can learn to exhibit left or right side preferences in response to an appetitive or aversive stimulus. The demonstration of an endogenous side preference led to the question of how endogenous and learned preferences interact. Early studies using a T-maze showed that rats trained to reverse their initial intrinsic preferences gradually reverted back to their initial preferences when subsequently allowed to make spontaneous choices (Glick et al., 1977). More recent work has employed operant and

classical conditioning techniques to examine this question more
fully as well as to determine the extent to which conditioned
changes in laterality can be used as a model to explore the
plasticity of a well-defined neuronal system.

An operant procedure has been used to train rats to turn in
circles for water reinforcement during the daytime (Glick, 1982).
Using a continuous reinforcement schedule in which each 360° turn
results in reward, rats typically make between 200 and 300 net
rotations per 1-hr daily test session. In one experiment (Glick and
Hinds, 1984) different groups of rats were reinforced for turning
either in the same or opposite direction as that elicited in a
previous test with d-amphetamine. All of 14 rats trained in the
"same" direction readily acquired the task whereas only 13 of 33
rats trained in the "opposite" direction showed evidence of
learning. Rats were retested with d-amphetamine at two and nine
days after cessation of training. Initially, at two days, the
effect of d-amphetamine was greater in rats trained in the "same"
direction and decreased or reversed in rats successfully trained in
the "opposite" direction (i.e., for the latter rats, d-amphetamine
now elicited more turning in the trained than in the endogenous
direction). A week later, however, these changes mostly
disappeared-- the reversed or "oppositely" trained rats reverted to
their endogenous directions. This would suggest that a mechanism to
reset the asymmetry is intrinsic to the organization of the
nigrostriatal system. This might be predicated on an anatomical
asymmetry; that is, if the normally more active side of the system
contained more neurons than the other side. Training might reverse
the functional asymmetry by selectively activating the side with
fewer neurons; afterwards, when normal inputs were restored, the
system would gradually return to normal.

The operant turning procedure has recently been modified in an
interesting way: rats are now being trained to turn in both
preferred and non-preferred directions within the same 1-hr test
session. The session is divided into four consecutive parts, each
signaled by one of two discriminative stimuli (light or dark). In
the first 15 minutes of the session, a stimulus (e.g., light)
signals that only turns in one direction are being reinforced;
during the next 15 minutes, the other stimulus (dark) signals that
only turns in the opposite direction are being reinforced. This
sequence is repeated during the remainder of the session such that
rats learn to switch directions of turning every 15 minutes.
Although, using this procedure, rats exhibit equivalent rates of
turning in each direction, the existence of an endogenous preference
can still be demonstrated: low doses (0.125-0.5 mg/kg) of d-
amphetamine enhance rates of turning in the initially preferred
direction and diminish rates of turning in the initially non-
preferred direction. It is important to note that these doses are
below those that are commonly required to elicit turning in either
intact or unilaterally lesioned rats. Higher doses (1.0-2.0 mg/kg)
of d-amphetamine, comparable to those usually used in other

preparations, override the training-induced preference; that is, rats turn in the initially preferred direction regardless of which stimulus is present and which direction is being reinforced. Although the mechanism of asymmetry clearly shows plasticity, in terms of its functional responsiveness to experiential contingencies, it also shows a high degree of stability that is likely to have an anatomical basis.

Another aspect of plasticity is being explored within the context of a classical conditioning paradigm (Drew and Glick, 1986). The objective here has been to determine if environmental cues will, as a consequence of being associated with drug-induced rotation, subsequently elicit rotation in the absence of the drug. Using the test environment as the conditioned stimulus and 1.25 mg/kg of d-amphetamine as the unconditioned stimulus, both lateralized activity (rotation) and non-lateralized activity (extra quarter turns, cf. Greenstein and Glick, 1975) were successfully conditioned in female rats after only two conditioning trials (one day between trials). Lateralized and non-lateralized responses were dissociable: conditioned lateralized activity extinguished more rapidly than conditioned non-lateralized activity. Conditioned rotation, like the drug-induced response (Jerussi and Glick, 1976), appears to be mediated by dopamine: a low dose (0.1 mg/kg) of haloperidol, a dopamine antagonist, blocked the expression of the conditioned response. Further studies are attempting to determine if conditioned rotation is in fact directly attributable to the conditioning of nigrostriatal neurons. If this can be demonstrated, then classically conditioned changes in nigrostriatal asymmetry might be employed as a therapeutic regimen in situations where enhanced asymmetry is beneficial (cf. Glick et al., 1977).

LEFT- vs. RIGHT-SIDEDNESS: MODULATION BY STRESS

For a long time it was thought that left- and right-sided rats differed in only subtle ways (Glick and Ross, 1981). This premise was reconsidered following the discovery (Glick et al., 1983) that an acute dose (20 mg/kg) of cocaine elicited pronounced differences in rotational behavior: right-rotating female rats rotated much more than left rotators while left-rotating male rats rotated much more than right rotators. These findings were in sharp contrast to those obtained with d-amphetamine where no significant rotational differences between left- and right-biased rats of either sex were found. While cocaine and d-amphetamine are known to be similar in many of their neurochemical and behavioral effects, some difference(s) must obviously exist. For example, although both drugs activate dopaminergic pathways in brain, d-amphetamine preferentially affects nigrostriatal neurons (Bowers and Hoffman, 1984) and cocaine preferentially affects mesocortical neurons (Hadfield and Nugent, 1983). Mesocortical dopamine neurons have also been shown to be activated selectively by inescapable footshock stress (Deutch et al., 1985; Herman et al., 1983; Thierry et al.,

1976). It was therefore postulated that cocaine-like effects on rotational behavior might be produced by a combination of inescapable stress and d-amphetamine. This expectation was verified when male and female rats, selected on the basis of their rotational behavior in response to d-amphetamine (1.56 mg/kg, males; 1.25 mg/kg females), were exposed to inescapable footshock stress and then retested with d-amphetamine. Right-rotating males and left-rotating females shifted their directional bias toward the opposite side, while left-rotating males and right-rotating females exhibited increased rotation in their pre-stress direction (Carlson and Glick, 1985). We have recently found that food deprivation (24 hours) is another stressor that will selectively activate mesocortical dopamine neurons; and, like footshock stress, a combination of food deprivation and d-amphetamine mimicked the left versus right rotational effects of cocaine (Carlson and Glick, 1986). These results suggest that individual differences in reaction to stress may in part be dependent upon endogenous asymmetries in brain.

INDIVIDUAL AND HERITABLE DIFFERENCES IN ROTATION

The intensity of circling behavior as well as the relative incidence of left- and right-sidedness varies among strains (Brass and Glick, 1981; Glick and Ross, 1981; Glick, 1985) as well as between sexes (Becker et al., 1982; Brass and Glick, 1981). These observations as well as the evidence that striatal asymmetry and side preferences are present at birth (Ross et al., 1981) suggested that some aspects of this lateralized behavior would be under heritable control. An earlier study from this laboratory indicated that whether a particular rat develops a left or right circling preference is complexly determined by an interaction between heritable and sex-related hormonal factors (Glick, 1983). An attempt to breed strains of strongly and weakly rotating rats indicated that the strength of left or right preferences is under similar controls. The objective was to breed one strain that would exhibit intense circling behavior (nocturnal) in a preferred direction and another strain that would exhibit little or no directional preference in circling behavior. A strain difference (approximately twofold) in rotation parameters was limited to females and developed gradually over eight generations before asymptoting (Glick, 1985). A totally unexpected finding was that in both males and females of the weakly rotating strain there developed a left-sided population bias. There were no strain differences in locomotor activity (measured in a photocell box). The results of this work indicated that the strength of rotational preferences is separable from the level of locomotor activity and to some extent heritable. A modulatory role of testosterone was postulated to account for the complexity of results with respect to sex dependence. Neonatal or perinatal exposure to testosterone is known to masculinize the central nervous system (e.g., MacLusky and Naftolin, 1981). In this laboratory neonatal administration of testosterone propionate (1.25 mg, s.c.) reduced adult rotation in female rats. Thus the absence of a breeding effect on the rates of male rotation might be attributable to a dampening effect of the obviously stronger and longer exposure to testosterone in males (females are exposed to some extent in utero, cf. Glick, 1983). Testosterone has been hypothesized to delay preferentially the development of the left side of the brain (Geschwind and Behan, 1982) and, as one result in humans, to be responsible for the greater incidence of left-handedness in males versus females. The left bias in the weakly rotating strain may have developed because the parents were selected for having very weak or negligible left-right preferences and the effect of testosterone on brain maturation could be exerted independently of a parental directional influence (Glick, 1983).

Another resource for comparing lateralized and non-lateralized rats was discovered as a consequence of this laboratory moving from Mount Sinai School of Medicine to Albany Medical College in July of 1984. For the ten years preceding the move, Sprague-Dawley derived rats had been obtained from Perfection Breeders, Douglassville, PA. Because shipping the same rats to Albany would have increased their costs considerably, it was decided to switch breeders and purchase naive rats from a closer breeder-- and there were two nearby, Taconic Farms and Blue Spruce. Quite surprisingly, both male and female rats from both sources rotated (nocturnal) much less than had the rats from Perfection Breeders previously tested in New York. Rats from Perfection Breeders were then purchased and tested in Albany-- and the data very closely resembled data obtained a couple of months earlier in New York. Rats from two other breeders-- Harlan and Charles River-- were then purchased; rates of nocturnal rotation exhibited by rats from each of these latter sources were intermediate between Taconic/Blue Spruce and Perfection. It was then decided that, despite the added cost, to return to Perfection rats. This decision, made more than two months after the initial Albany shipment of Perfection rats was tested, resulted in the unaccountable and frustrating finding that Perfection rats now also rotated at relatively low rates. Several dozen Perfection rats were tested with increasing bewilderment. An inquiry to the breeder resolved this puzzle: unlike the other breeders, Perfection's colony of rats was not "closed"-- new breeding rats from an outside source were periodically introduced and new lines established every few years. The Perfection rats initially tested in Albany were from the same "old" line used in New York whereas the subsequent shipments only contained rats from a "new" line, the "old" line having been discontinued. This disconcerting news prompted a decision to order rats from a "closed" colony breeder. Because it was also desirable to use rats that had a behavioral baseline similar to the rats used for several years in New York, Zivic-Miller rats were purchased and tested-- these rats had been used in New York from 1971 to 1974 and were indistinguishable, in 1974, from Perfection rats; the switch to Perfection in 1974 was made because of a much lower cost per rat at the time. The newly arrived Zivic-Miller rats indeed rotated at rates comparable to those of the "old" line Perfection and formerly used Zivic-Miller rats. Although this process entailed considerable time and expense, it was not unrewarded. As all rats tested were Sprague-Dawley derived, the foregoing search had essentially characterized a six- to seven-fold continuum of rats' rotational behavior (net rotations per night: males ranged form 5.0 to 31.2 and females from 6.2 to 42.8) and located sources for providing rats at quantitative steps along this continuum. An intriguing finding was that Taconic and Blue Spruce rats clearly rotated less than the weakly rotating strain bred in this laboratory. It is therefore possible for weak rotation to be bred in males as well as in females-- perhaps continued inbreeding in the laboratory would have yet produced this effect. Furthermore, a left population bias was observed in both male and female rats from each of five low-rotating sources (but not from two high rotating sources). Consistent with the laboratory bred rats, it appears that a left bias is always a consequence of breeding for weak rotation.

DOPAMINE ASYMMETRY AND CIRCLING IN HUMANS

A reanalysis of postmortem data (Rossor et al., 1980) revealed that the human brain has neurochemical asymmetries in several structures and neurotransmitter systems (Glick et al., 1982). Most relevant to the foregoing animal work was the finding of a left-biased dopamine asymmetry in human globus pallidus. Pallidal dopamine is apparently higher on the side contralateral to hand preferences-- analogous, perhaps, to striatal dopamine levels being higher on the side contralateral to side preferences in rats (Zimmerberg et al., 1974). The similar dopamine asymmetries in rats and humans suggested, however, that a more direct analogy might be possible. In collaboration with Dr. Stefan Bracha (University of California at San Diego), an attempt was made to determine whether "normal" humans would exhibit circling during the course of a "normal" day's activities. For this purpose, a new electronic device was developed for measuring in humans the same kinds of rotational movements observed in rats. The device consists of a position sensor and an electronic processing circuit. The position sensor monitors changes in the orientation of the dorsal-ventral axis of the subject. Magnetic north is used an an external reference and a compass is used to track this reference. As in rats (Greenstein and Glick, 1975), the output provides left and right full (360 degree) turns.

An initial study with 22 males showed that, unbeknown to themselves, "normal" humans preferentially rotate to the left or to the right; only one subject had no directional preference. There were obvious relationships between the direction of turning preferences and other indices of sidedness. All 12 subjects having right turning preferences were also right-sided for hand, foot and eye preference. Seven of the nine subjects having left turning preferences were also left-sided for at least one of the other indices. Subjects found to be consistently "left hemisphere dominant" made significantly more turns to the right than to the left, whereas "mixed dominance" subjects made significantly more turns to the left than to the right. Considered separately, eye preference was more clearly associated with turning preference than either hand or foot preference-- this demonstrates the importance of the visual modality in guiding human movement and might be indicative of the extensive input into the basal ganglia from the frontal eye fields (Kunzle and Akert, 1977). Further studies with larger numbers of normal subjects as well as with neurological and psychiatric patients are in progress (Bracha, Seitz, Judd and Glick, in preparation). It is hoped that the device developed here will become a useful tool for obtaining quantitative information regarding the status of basal ganglia function in humans.

ACKNOWLEDGMENTS

This research was supported by grants DA01044 and DA03817 from the National Institute on Drug Abuse. The authors thank P.A. Hinds, N. Camarota, J. Baird, M. Bryda, and D. Hackley for technical assistance.

REFERENCES

Becker, J.B., Robinson, T.E. and Lorenz, K.A. (1982). Sex differences and estrous cycle variations in amphetamine-elicited rotational behavior. Europ. J. Pharmacol., 80, 65-72.

Bowers, M.B. and Hoffman, F.J. (1984). Homovanillic acid in rat caudate and prefrontal cortex following phencyclidine and amphetamine. Psychopharmacology, 84, 136-137.

Brass, C.A. and Glick, S.D. (1981). Sex differences in drug-induced rotation in two strains of rats. Brain Res., 223, 229-234.

Carlson, J.N. and Glick, S.D. (1985). Stress-induced changes in rotational behavior. Neurosci. Abstr., 11, 870.

Carlson, J.N. and Glick, S.D. (1986). Changes in rotational behavior induced by food deprivation in the Long-Evans rat. Neurosci. Abstr., in press.

Cools, A.R. and van Rossum, J.M. (1980). Multiple receptors for brain dopamine in behavior regulation: concept of dopamine-e and dopamine-i receptors. Life Sci., 27, 1237-1253.

Cotzias, G.C., Van Woert, M.H. and Shiffer, L.M. (1967). Aromatic amino acids and modification of Parkinsonism. New Engl. J. Med., 276, 374-379.

Deutch, A.Y., Tam, S.-Y. and Roth, R.H. (1985). Footshock and conditioned stress increase 3,4-dihydroxyphenylacetic acid (DOPAC) in the ventral tegmental area but not the substantia nigra. Brain Res., 333, 143-146.

Drew, K.L. and Glick, S.D. (1986). Classical conditioning of amphetamine-induced rotation in unlesioned rats. Neurosci. Abstr., in press

Geschwind, N. and Behan, P.O. (1982). Left-handedness: Association

with immune disease, migraine. and developmental learning
disorders. Proc. Natl. Acad. Sci. USA, 79, 5097-5100.

Glick, S.D. (1982). Operant control of turning in circles: A new
model of dopaminergic drug action. Brain Res., 245, 394-397.

Glick, S.D. (1983). Heritable determinants of left-right bias in the
rat. Life Sci., 32, 2215-2221.

Glick, S.D. (1985). Heritable differences in turning behavior of
rats. Life Sci., 36, 499-503.

Glick, S.D. and Cox, R.D. (1978). Nocturnal rotation in normal rats:
Correlation with amphetamine-induced rotation and effects of
nigrostriatal lesions. Brain Res., 150, 149-161.

Glick, S.D. and Hinds, P.A. (1984). Modulation of turning
preferences by learning. Behav. Brain Res., 12, 335-337.

Glick, S.D., Hinds, P.A. and Shapiro, R.M. (1983). Cocaine-induced
rotation: sex-dependent differences between left- and right-
sided rats. Science, 221, 775-777.

Glick, S.D., Jerussi, T.P., Waters, D.H. and Green, J.P. (1974).
Amphetamine-induced changes in striatal dopamine and
acetylcholine levels and relationship to rotation (circling) in
rats. Biochem. Pharmacol., 23, 3223-3225.

Glick, S.D. and Ross, D.A. (1981). Right-sided population bias and
lateralization of activity in normal rats. Brain Res., 205, 222-
225.

Glick, S.D. and Shaprio, R.M. (1985). Functional and neurochemical
mechanisms of cerebral lateralization in rats. In Cerebral
Lateralization in Nonhuman Species. (ed. S.D. Glick). Academic
Press, Orlando.

Glick, S.D., Zimmerberg, B. and Jerussi, T.P. (1977). Adaptive
significance of laterality in the rat. Ann. N.Y. Acad. Sci.,
299, 180-185.

Gray, E.G. and Whittaker, V.P. (1962). The isolation of nerve
endings from brain: An electron-microscopic study of cell
fragments derived by homogenization and centrifugation. J. Anat.,
96, 79-88.

Greenstein, S. and Glick, S.D. (1975). Improved automated apparatus
for recording rotation (circling behavior) in rats or mice.
Pharmacol. Biochem. Behav., 3, 507-510.

Groves, P.M. (1983). A theory of the functional organization of the
neostriatum and the neostriatal control of voluntary movement.
Brain Res., 5, 109-132.

Hadfield, G.M. and Nugent, E.A. (1983). Cocaine: Comparative effect
on dopamine uptake in extrapyramidal and limbic systems. Biochem.
Pharmacol., 32, 744-746.

Heikkila, R.E., Shapiro, B.S. and Duvoisin, R.C. (1981). The
relationship between loss of dopamine nerve terminals, striatal
[^{3}H]spiroperidol binding and rotational behavior in unilaterally
6-hydroxydopamine-lesioned rats. Brain Res., 211, 285-292.

Herman, J.P., Guillonneau, D., Dantzer, R., Scatton, B.,
Smerdjian-Rouquier, L. and LeMoal, M. (1982). Differential
effects of inescapable footshocks and stimuli previously paired
with footshocks on dopamine turnover in cortical and limbic areas
of the rat. Life Sci., 30, 2207-2214.

Jerussi, T.P. and Glick, S.D. (1976). Drug-induced rotation in rats without lesions: Behavioral and neurochemical indices of a normal asymmetry in nigrostriatal function. Psychopharmacology, 47, 249-260.

Jerussi, T.P., Glick, S.D. and Johnson, C.L. (1977). Reciprocity of pre- and postsynaptic mechanisms involved in rotation as revealed by dopamine metabolism and adenylate cyclase stimulation. Brain Res., 129, 385-388.

Kuhar, M.J. (1973). Neurotransmitter uptake: A tool in identifying neurotransmitter-specific pathways. Life Sci., 13, 1623-1634.

Kunzle, H. and Akert, K.J. (1977). Efferent connections of cortical area 8 (frontal eye field) in Macaca fascicularis. A reinvesigation using the autoradiographic technique. J. Comp. Neurol., 173, 147-164.

MacLusky, N.J. and Naftolin, F. (1981). Sexual differentiation of the central nervous system. Science, 211, 1294-1303.

Robinson, T.E. and Becker, J.B. (1983). The rotational behavior model: Asymmetry in the effects of unilateral 6-OHDA lesions of the substantia nigra in rats. Brain Res., 264, 127-131.

Robinson, T.E., Becker, J.B. and Ramirez, V.D. (1980). Sex differences in amphetamine-elicited rotational behavior and the lateralization of striatal dopamine in rats. Brain Res. Bull., 5, 539-545.

Ross, D.A., Glick, S.D. and Meibach, R.C. (1981). Sexually dimorphic brain and behavioral asymmetries in the neonatal rat. Proc. Natl. Acad. Sci. USA, 78, 1958-1961.

Sakurai, Y., Ohta, H., Shimazoe, T., Kataoka, Y., Fujiwara, M. and Ueki, S. (1985). Tetrahydrocannabinol elicited ipsilateral circling behavior in rats with unilateral nigral lesion. Life Sci., 37, 2181-2185.

Shapiro, R.M., Glick, S.D. and Hough, L.B. (1986). Striatal dopamine uptake asymmetries and rotational behavior in unlesioned rats: Revising the model? Psychopharmacology, in press.

Thierry, A.M., Tassin, J.P., Blanc, G. and Glowinski, J. (1976). Selective activation of the mesocortical DA system by stress. Nature, 263, 242-244.

Ungerstedt. U. (1971). Striatal dopamine release after amphetamine or nerve degeneration revealed by rotational behaviour. Acta Physiol. Scand. Suppl. 367, 49-68.

Vaccarino, F.J. and Franklin, K.B.J. (1982b). Self-stimulation and circling reveal functional differences between medial and lateral substantia nigra. Behav. Brain Res., 5, 281-295.

Yamamoto, B.K. and Freed, C.R. (1984). Reversal of amphetamine-induced circling preference in trained circling rats. Life Sci., 34, 675-682.

Zimmerberg, B., Glick, S.D. and Jerussi, T.P. (1974). Neurochemical correlate of a spatial preference in rats. Science, 185, 623-625.

Zimmerberg, B., Strumpf, A.J. and Glick, S.D. (1978). Cerebral asymmetry and left-right discrimination. Brain Res., 140, 194-196.

2
Laterality in the Neuroendocrine System

Ida Gerendai

Neuroendocrinology grew out from two disciplines;
endocrinology and neurology. The discovery of reciprocal regulation between the two main control systems of
the body, the hormonal and the neural one, led to studies that revealed several new aspects of regulation.
Our investigations demonstrating lateralization of neuroendocrine control described in this chapter seem to
contribute to developments in further elucidation of
fundamental biological mechanisms

The primary goal of studies which led us to the
first demonstration of asymmetry in the neuroendocrine
system was to explore the existence of a possible direct neural connection between the brain and the endocrine glands. We supposed that the brain, in addition
to regulating endocrine functions with neurohormones
that are released by the hypothalamus might control
peripheral endocrine gland functions by a direct neural mechanism. The model which we have used concerned
the effects of compensatory hypertrophy of the endocrine organs that follows the removal of one of the
paired endocrine glands

LATERALITY IN GONADAL CONTROL

Several authors reported increased gonadotrophin
(primarily FSH) secretion following hemiovariectomy.
It was assumed that this is the consequence of increased synthesis and/or release of hypothalamic luteinizing hormone-releasing hormone (LHRH) Therefore, we studied the LHRH content of the two sides of the hypothalamus following unilateral ovariectomy (Gerendai et al
1978). Unexpectedly, we found that in intact female
rats the LHRH content of the two sides of the hypothalamus differs significantly; the right-side

hypothalamus contains significantly more LHRH than the
left-side one. The biochemical asymmetry was remarkably
stable, and there were no instances of lack of asymmetry or reversal in the direction of asymmetry. Unilateral ovariectomy induced significant increase in the
LHRH content of the hypothalamus-half ipsilateral to
hemispaying. Thus, right-side ovariectomy enhanced the
LHRH asymmetry while left-side intervention diminished
it. In bilaterally ovariectomized animals, consistent
with earlier reports, there was a decrease in total hypothalamic LHRH content. This reduction, however, occur
red exclusively on the right side. Further studies have
revealed that the shifts in hypothalamic LHRH asymmetry
pattern are not the consequences of the removal of the
specific hormone producing tissue but rather of the in-
terruption of the postulated neural pathway between the
hypothalamus and the ovary. Pharmacological sympathec-
tomy of the remaining ovary (local treatment of the gonad with 6-OHDA) in unilaterally ovariectomized rats
resulted in LHRH content changes similar to those observed in bilaterally ovariectomized animals (Gerendai
et al., 1979). These studies indicated that the distribution of LHRH in the hypothalamus is asymmetrical, and
that unilateral ovariectomy influences the neurohormone
content of the hypothalamus strictly monolaterally,-an
observation that strongly suggests the existence of a
direct hypothalamo-ovarian neural feed-back.

In recent studies the functional significance of hypothalamic LHRH asymmetry was confirmed (Fukuda et al.,
1984). Electrolytic lesion of the anterior hypothalamus
the region where the majority of LHRH perikarya is located in rats, prevented the development of compensatory ovarian hypertrophy regardless the side of hemiovariectomy. Furthermore, in female rats the development
of sex-specific reproductive functions could be disturbed differentially, depending on the side of early
postnatal implantation of estradiol in the hypothalamus
(Nordeen and Yahr, 1982). Implantation of the steroid
in the right hypothalamus led to masculinization, while
implantation on the left resulted in the loss of oestrus cycle (defeminization).

The hypothalamic laterality in gonadal control raised the question as to whether extrahypothalamic neural
structures also involved in gonadal regulatory processes are lateralized or not.

It is generally accepted that norepinephrinergic
input to the medial preoptic are arising mainly from
the locus coeruleus is essential for the control of
cyclic gonadotrophin secretion. Injection of the

excitotoxin kainic acid into the right locus coeruleus prevented the development of compensatory hypertrophy regardless the side of hemicastration. Destroying the left locus coeruleus did not interfere with the response (Gerendai, 1984). This observation indicates that certain afferents to the hypothalamus involved in the control of gonadal functions are also lateralized

In further studies we tested the possibility as to whether neural fibers of vagal origin, that besides the sympathetic nerve provide the final neural supply of the ovary (Burden, 1978), are lateralized or not (Gerendai and Nemeskéri, 1983). Left- or right-sided vagotomy in unilaterally ovariectomized adult rats reduced the rate of compensatory ovarian hypertrophy in each experimental group. Similar interventions in immature rats (on day 15 or 19 of age) resulted in changes different from those found in adults Left-sided vagotomy with left- or right-sided ovariectomy (on day 15 of age) did not lead to compensatory ovarian hypertrophy as late as day 53. In addition, left-sided vagotomy with or without hemicastration delayed the onset of puberty. Right-sided vagotomy was ineffective on the parameters studied. When the same surgery was performed on day 19 of life the time of vaginal opening and the rate of hypertrophy were similar to those of control animals, except the group with right vagotomy+right ovariectomy. This combination resulted in precocious puberty and enhanced ovarian hypertrophy.

These results indicate·the involvement of the vagus in the control of certain ovarian functions. Furthermore, they also suggest that the bilateral vagal control over ovarian functions develops only after puberty In immature animals there is vagal predominance and the asymmetrical vagal control shifts from one side to the other. Consistent with this observation, changes in hypothalamic dominance have been reported by Nance et al. (1983, 1984).

The principle of asymmetric neural control over female gonadal functions could be extended to male repro ductive processes. Unilateral orchidectomy has been reported to induce unilateral LHRH content changes in the hypothalamus (Mizunuma et al , 1983; Bakalkin et al., 1984). Deafferentation of the hypothalamus in unilaterally orchidectomized rats prevented the hemicastration-induced FSH rise only in the case if right-sided hypothalamic surgery was combined with right hemicastration (Nance and Moger, 1982) .In our recent studies we found that in immature rats Sertoli cell function is under asymmetric vagal control (in preparation).

From the available evidences it can be concluded that the neural control of gonadal functions is asymmetrical. The structures involved in the asymmetric control are; the hypothalamus, the locus coeruleus,and the vagus, but it is reasonable to suppose that further cerebral structures will be demonstrated to exert an asymmetrical control over gonadal functions. The program generated in the asymmetrical cerebral structures can affect the gonads through the pituitary by the common hypothalamo-pitutary neuroendocrine regulatory mechanism and via direct hypothalamo-gonadal neural pathway Interestingly enough, in the majority of mammals no sided predominance is present in gonadal functions. Some exceptions are known to exist, for instance in bats ovulation occurs predominantly from one ovary. The contralateral gland has the potential to function but starts to work only if the dominant ovary is removed (Bleiber and Ehteshami, 1981). Recently Rao and Edgerton (1984) have reported biochemical differences between right and left ovaries of pigs. There is another aspect of gonadal asymmetry; the right gonad develops earlier than the left one (Mittwoch, 1975). Chang et al. (1960) have reported that in human the right testis is heavier than the left gonad. In humans varicocele affects left testis preferentially (Olson and Stone, 1949). Similarly, greater vulnerability of the left testis has been observed in hemivasectomized neonatal (Gerendai et al., in press) and adult (Plaut,1973) rats.

Overall on the basis of the available data it can be supposed on one hand, that the full control of gonadal functions requires asymmetry of certain regulatory nervous structures, but on the other hand, the pattern of cerebral asymmetry can be modified by the gonads. The latter phenomenon can be realized by hormonal (sex steoids) and/or by neural mechanism. The effects of sex hormones on asymmetry of cerebral cortex have been demonstrated in both humans (Geschwind and Galaburda, 1985) and rats (Diamond et al., 1981). The possibility of modification of asymmetry pattern by neural inputs arising from the gonads is suggested by the finding that removal of one gonad leads to unilateral changes in hypothalamic LHRH content (Gerendai et al., 1978; Mizunuma et al., 1983; Bakalkin et al., 1984) Similarly even our early studies on the neural control of the gonads revealed that hemicastration can induce metabolic asymmetry in the hypothalamus (Gerendai and Rácz, 1975; Gerendai and Halász, 1976).

LATERALITY IN THYROID CONTROL

In further experiments we studied the possibility of asymmetric control of the thyroid gland. Unilateral or bilateral thyroidectomy was performed in adult male rats and the thyrotropin-releasing hormone (TRH) content of the hypothalamus halves was measured. In contrast to the asymmetrical hypothalamic distribution of LHRH no sided difference in the TRH content of the two halves of the hypothalamus occurred. Right- or left-sided thyroidectomy resulted in a significant increase in the neurohormone content on both sides. These data indicate the lack of TRH asymmetry in the hypothalamus, and suggest that hypothalamic TRH is mainly controlled by hormonal and not by neural mechanisms (Gerendai et al., 1985). In an other study, however, we found functional asymmetry in the neural control of the mitotic activity of the thyroid follicular cells. Left-sided posterior deafferentation of the hypothalamus reduced the basal mitotic index of the follicular cells in both sides of the thyroid gland, and prevented the hemithyroidectomy-induced increase in mitotic activity (Lewinski et al., 1982) It can be concluded from the above described studies and from those indicating asymmetric control of gonadal functions that the pattern of asymmetric cerebral control of different endocrine glands is not uniform.

LATERALITY IN THE CONTROL OF PROLACTIN SECRTION

Early clinical observations indicated that galactorrhoea can occur after chest wall injury, such as mechanical trauma, burns or herpes zoster of thoracic dermatomes (Grimm, 1955; Berger et al., 1966; Barnes, 1966). Morley et al (1977) have reported that in a patient with galactorrhoea and hyperprolactinaemia observed after burns to the chest wall, intercostal nerve block normalized serum prolactin level Patients after mastectomy operation have been reported to have in several cases hyperprolactinaemia (Jillis et al, 1977) From similar observations Herman et al. (1981) concluded that the mastectomy-induced hyperprolactinaemia may be the result of neurogenic stimuli arising from the damaged nerve supply to the breast In the rat electrical stimulation of the mammary nerve leads to changes in pituitary and plasma prolactin levels similar to those seen during suckling (Mena et al., 1980) On the basis of these observations we aimed to study whether uni- or bilateral mastectomy could influence prolactin secretion in male rats (Gerendai et al, 1985).

Unexpectedly, we could demonstrate a differential

prolactin response to right- and left-sided mastectomy.
One week after right-sided mastectomy animals had hyper
prolactinaemia while those with left-sided ablation exhibited hypoprolactinaemia. Interestingly, the right-
sided mastectomy-induced enhanced serum prolactin level
could be observed as late as two weeks postsurgery,
while the effect of left-sided mastectomy on serum prolactin level was not already apparent at this postoperative day. In bilaterally mastectomized animals serum
prolactin levels varied widely, but the mean value was
higher than that in sham operated controls These data
indicate the involvement of a neural pathway arising
from the mammary gland in the control of prolactin secretion. Data also indicate the asymmetry of the mammary
gland and/or the neural pathway involved in the mastectomy-induced prolactin response. Furthermore,the facts
that the right-sided mastectomy-induced hyperprolactinaemia lasted longer, and that following bilateral
mastectomy the number of animals with hyperprolactinaemia exceeded the number of rats with hypoprolactinaemia
suggest that right mastectomy is more effective to influence prolactin secretion than left-sided surgery. We
know of no studies of similar sided differences in human, but it is widely known that breast cancer is more
common on the left than on the right side

LATERALITY IN THE CONTROL OF GROOMING BEHAVIOR

In the previous section we described that unilateral mastectomy is male rats causes differential prolactin changes depending on the side of surgery. In addition, our data also showed that unilateral vagotomy is
followed by hyperprolactinaemia regardless the side of
nerve transection (Gerendai et al , 1983) Since hyperprolactinaemia is known to affect several types of behavior (Drago et al , 1980; Drago et al , 1981; Drago
et al., 1982) we aimed to study whether the changes in
plasma prolactin levels induced by right- or left-sided
mastectomy and right- or left-sided vagotomy influence
grooming behavior of the rat (Gerendai et al , 1984)
Six days after unilateral or bilateral mastectomy,
grooming activity of each operated group differed from
that of sham-operated rats Right-sided mastectomy induced excessive grooming while left-sided mastectomy
resulted in a slight not significant decrease in grooming activity. Animals with bilateral mastectomy exhibited enhanced grooming behavior The correlation between plasma prolactin levels and grooming activity appeared to be positive and highly significant

In right-sided vagotomized rats no difference in
grooming activity was found between vagotomized and

control animals. Left-sided vagotomy, however, resulted in a significant increase in the occurrence of grooming. Since both right- and left-sided vagotomy was found to increase plasma prolactin levels no correlation could be demonstrated between the hormone levels and grooming activity.

Mechanism of action of prolactin in enhancing grooming behavior is still unclear, but several data indicate that prolactin influences concentration and turnover of dopamin in various brain regions. It seems that unilateral or bilateral mastectomy-induced changes in grooming behavior might be due to changes in prolactin levels. In contrast, left vagotomy is effective to induce enhanced grooming behavior with high plasma prolactin concentrations while right-sided vagotomy in spite of the presence of hyperprolactinaemia failed to modify grooming behavior. This discrepancy might indicate an asymmetric vagal control in central mechanisms responsible for the expression of grooming behavior. The lateralized vagal control in grooming behavior is supported by the finding of asymmetry in paw-licking response in unilaterally vagotomized rats (Gerendai, 1984). The latency of paw-licking response on a hot-plate test did not differ in right-sided vagotomized animals from that of controls. However, left vagotomy induced a higher level of paw-licking responses compared to sham-operated or intact controls. Vagal fibers, or structures receiving vagal inputs, which are involved in the perception of pain, apparently are left-biased.

Taken together, the data on functional vagal asymmetry indicate that each vagus is superior in certain functions.

MORTALITY RATES AFTER UNILATERAL BRAIN SURGERIES

Our empirical observation concerning the higher mortality rate of animals with right-sided cerebral interventions suggests cerebral laterality in vital functions (Gerendai, 1984). At present it is difficult to provide an explanation for this unexpected finding Consistent with our observation, Robinson and Coyle (1980) reported that ligation of the right middle cerebral artery caused bilateral decrease of catecholamine content of the brain and spontaneous hyperactivity. Heilman et al. (1977) demonstrated that the right hemisphere controls certain important autonomic functions. Geschwind and Galaburda (1985) have pointed out that the role of the right hemisphere in vital functions might be the explanation of the earlier development

of right hemisphere both in the course of phylogenesis
and ontogenesis. The "conservatism" of right hemisphere function further suggests the important controls by
the right hemisphere over vital functions.

CONCLUSION

The above described findings suggest lateralization in the control of endocrine processes in the rat.
It should be emphasized, however, that the studies on
neuroendocrine asymmetry in isolation are of limited
value in our understanding of the nature of asymmetry.
Increasing number of evidences on morphological, pharmacological, and behavioral asymmetry of the brain of
the most commonly used experimental animal, the laboratory rat suggests that laterality of the cerebral
control mechanisms in rodents is the rule rather than
the exception. We are aware that according to the common view cerebral dominance is limited to the asymmetry of certain human cortical functions The skepticism
concerning cerebral asymmetry in animals and asymmetry
of subcortical structures both in humans and animals
is hardly understandable. It is unreasonable to suppose that the asymmetries in different animal species
such as in the unicellular organism, in the lamprey,
in song birds, and in great apes -just to mention a
few, have evolved by chance. Further systematic studies are required to fill the gap in our understanding
of the development of asymmetry of living organisms in
general, and of the brain, in particular

It is beyond the scope of the present review to
discuss asymmetries of the non-living matter. But just
considering the fact that living systems contain L-ami-
no acids and D-sugars (with only some exceptions) it
is difficult to avoid speculating that asymmetries existing in physics and chemistry might be the ultimate
origin of biological asymmetry. Starting from the theory of parity non-conservation based on the asymmetry
of weak interactions during β-decay (Lee and Yang,
1956) the possible determining role of physical asymmetry in the origin of optical asymmetry of biomolecu-
les during chemical evolution has been suggested
(Keszthelyi, 1976). No data are available whether chemical asymmetry through complex biochemical processes
might generate biological asymmetries or not. Perhaps
it is not too far-fetching to speculate that physical
and chemical asymmetries could be amplified by biological processes, which during the course of evolution
might lead to cerebral dominance.

REFERENCES

Bakalkin, G.Y., Tsibezov, V.V., Sjutkin, J.A, Veselova, S.P., Novikov, I.D and Krivosheev, O G (1984). Lateralization of LH-RH in rat hypothalamus Brain Res 296, 361-364.

Barnes, A.B. (1966). Diagnosis and treatment of abnormal breast secretions. N Engl. J. Med , 275, 1184-1187.

Berger, R.L., Joison, J.and Braverman, L (1966) Lactation after incision of the thoracic cage N Engl J. Med., 274, 1493-1497.

Bleier, N.J. and Ehteshami, M. (1981). Ovulation following unilateral ovariectomy in the California leaf-nosed bat (Macrotus californicus) J. Reprod Fert., 63, 181-183.

Burden, H.W (1978) Neural modulation of ovarian function. Trends in Neurosci., 1, 85-86.

Chang, K.S.F, Hsu, F.K. and Chan, S.T (1960) Scrotal asymmetry and handedness. J Anat., 94, 543-548

Diamond, M.C., Dowling, G A. and Johnson, R.E (1981) Morphologic cerebral cortical asymmetry in male and female rats. Exp. Neurol., 71, 261-268

Drago, F., Bohus, B. and Mattheij, J A M (1982) Endogenous hyperprolactinaemia and avoidance behaviors of the rat Physiol Behav 28, 1-4.

Drago, F., Canonico, P-L., Bitetti, R and Scapagnini, U. (1980). Systemic and intraventricular prolactin induces excessive grooming. Eur. J Pharmacol., 65, 457-458.

Drago, F., Pellegrini-Quarantotti, B, Scapagnini, U and Gessa, G.L. (1981). Short-term endogenous hyperprolactinaemia and sexual behavior of male rats. Physiol. Behav., 26, 277-279.

Fukuda, M., Yamanouchi, K, Nakano, Y, Furuya, H.and Arai, Y. (1984). Hypothalamic laterality in regulating gonadotropic function: unilateral hypothalamic lesion and ovarian compensatory hypertrophy Neurosci. Lett., 51, 365-370.

Gerendai, I. (1984). Lateralization of neuroendocrine control. In Cerebral Dominance. The Biological Foundations. (eds. N. Geschwind and A.M. Galaburda). Harvard Univ. Press, Cambridge, Mass. pp. 167-178

Gerendai, I. and Halász, B (1976). Hemigonadectomy-induced unilateral changes in the protein-synthesizing activity of the rat hypothalamic arcuate nucleus Neuroendocrinology, 21, 331-337.

Gerendai, I. and Nemskéri, Á (1983). The effect of unilateral vagotomy on compensatory ovarian hypertrophy and on the onset of puberty. In Neuropeptides,Neurotransmitters, and Regulation of Endocrine Processes. (eds. E. Endrőczi, L. Angelucci, U Scapagnini and D. deWied). Akadémiai Kiadó, Budapest, pp. 191-198.

Gerendai, I. and Rácz, K. (1975). Differences in the RNA-synthesizing activity between the two sides of the hypothalamic arcuate nucleus following unilateral orchidectomy. Acta biol. Acad. Sci hung.,26, 229-231.

Gerendai, I , Nemeskéri, Á. and Csernus, V The effect of neonatal vasectomy on testicular function. Andrologia (in press).

Gerendai, I., Clementi, G., Prato, A. and Scapagnini, U. (1983). Unilateral vagotomy induces hyperprolactinaemia in male rats. Neuroendocrinol. Lett., 5, 41-45.

Gerendai, I., Drago, F , Continella, G and Scapagnini, U. (1984). Effects of mastectomy and vagotomy on grooming behavior of the rat; possible involvement of prolactin. Physiol. Behav., 33, 1-4.

Gerendai, I., Prato, A., Clementi, G. and Scapagnini, U. (1985). Effect of unilateral or bilateral mastectomy on prolactin secretion in male rats. Neuroendocrinol. Lett. 7, 31-36.

Gerendai, I., Rotsztejn, W., Marchetti, B and Scapagnini, U. (1979). LH-RH content changes in the mediobasal hypothalamus after unilateral ovariectomy. In Neuroendocrinology: Biological and Clinical Aspects (eds. A. Polleri and R. MacLeod). Academic Press, New York, pp. 97-102.

Gerendai, I., Nemeskéri, Á., Faivre-Bauman, A., Grouselle, D. and Tixier-Vidal, A. (1985) Effects of unilateral or bilateral thyroidectomy on TRH content of hypothalamus halves. J. Endocrinol. Invest., 8, 321-323.

Gerendai, I , Rotsztejn, W., Marchetti, B , Kordon, C and Scapagnini, U. (1978). Unilateral ovariectomy induced luteinizing hormone-releasing hormone content changes in the two halves of the mediobasal hypothalamus. Neurosci. Lett., $\underline{9}$, 333-336.

Geschwind, N. and Galaburda, A.M. (1985). Cerebral lateralization. Biological mechanisms, associations, and pathology: I. A hypothesis and a program for research Arch. Neurol., $\underline{42}$, 428-459.

Geschwind, N. and Galaburda, A.M (1985). Cerebral lateralization. Biological mechanisms, associations, and pathology: II. A hypothesis and a program for research. Arch. Neurol., $\underline{42}$, 521-552

Grimm, E.G. (1955). Non-puerperal galactorrhoea with case reports, G. Bull Northwestern Univ. Med. Sch.,$\underline{29}$, 350-353.

Heilman, K.M., Schwartz, H. and Watson, R.T (1977). Hypoarousal in patients with the neglect syndrome and emotional indifference, Neurology, $\underline{28}$, 229-232.

Herman, V , Kalk, W.J., DeMoor, N G and Levin, J. (1981). Serum prolactin after chest wall surgery: elevated levels after mastectomy. J. Clin. Endocrinol. Metab., $\underline{52}$, 148-151

Keszthelyi, L (1976). Chemical evolution: effect of high energy radiation. Origins of Life, $\underline{7}$, 349-354.

Lee, T.D. and Yang, C N (1956). Question of parity conservation in weak interactions Physic. Rev., $\underline{104}$, 254-258.

Lewinski, A., Gerendai, I., Pawlikowski, M. and Halász, B. (1982). Unilateral posterior deafferentation of the hypothalamus and mitotic activity of thyroid follicular cells under normal conditions and after hemithyroidectomy. Endocrinol. Exp.,$\underline{16}$, 75-80.

Mena, F., Pacheco, P. and Grosvenor, C.E (1980) Effect of electrical stimulation of mammary nerve upon pituitary and plasma prolactin concentrations in anesthetized lactating rats. Endocrinology, $\underline{106}$, 458-462

Mittwoch, V. (1975). Lateral asymmetry and gonadal differentiation. Lancet, $\underline{1}$, 401-402.

28

Mizunuma, H., DePalatis, L R. and McCann, S M (1983).
Effect of unilateral orchidectomy on plasma FSH concentration: evidence for a direct neural connection
between testes and CNS. Neuroendocrinology, 37,291-296.

Morley, J.E., Dawson, M , Hodgkinson, J and Kalk, W J
(1977). Galactorrhoea and hyperprolactinaemia associated with chest wall injury J Clin Endocrinol Metab
45, 931-935.

Nance, D.M. and Moger, W H. (1982) Ipsilateral hypothalamic deafferentation blocks the increase in serum
FSH following hemicastration. Brain Res Bull., 8, 299-
302.

Nance, D M., Bhargava, M and Myatt, G A (1984) Further evidence for hypothalamic asymmetry in endocrine
control of the ovary Brain Res Bull , 13,651-655

Nance, D.M , White, J P and Moger, W H (1983) Neural
regulation of the ovary: evidence for hypothalamic a-
symmetry in endocrine control. Brain Res Bull , 10,
353-355.

Nordeen, E.J. and Yahr, P. (1982) Hemispheric asymmetries in the behavioral and hormonal effects of sexually differentiating mammalian brain. Science, 218,
391-394.

Olson, R.O. and Stone, E.P. (1949) Varicocele: symp-
tomatologic and surgical concepts N. Engl J Med ,
240, 877-880

Plaut, S.M. (1973). Testicular morphology in rats vas-
ectomized as adults. Science, N Y , 181, 554-555

Rao, C.V. and Edgerton, L A (1984). Dissimilarity of
corpora lutea within the same ovaries or those from
right and left ovaries of pigs during the oestrus cycle. J. Reprod. Fert., 70, 61-66

Robinson, R G. and Coyle, J.T (1980) The differential effect of right versus left hemispheric cerebral infarction on catecholamines and behavior in the rat
Brain Res., 188, 63-78.

Willis, K.J., London, D.R., Ward, H W C , Butt, W R.,
Lynch, S.S. and Rudd, B T. (1977) Recurrent breast
cancer treated with the antioestrogen tamoxifen: correlation between hormonal changes and clinical course.
Br. Med. J., 1, 425-428

3
Mechanisms of Brain Asymmetry: New Evidence and Hypotheses

Glenn D. Rosen, Albert M. Galaburda and Gordon F. Sherman

INTRODUCTION

Human functional and anatomical asymmetries have been well documented for over 100 years (Galaburda et al., 1978b; Springer and Deutsch, 1985; Sperry, 1974). Thus, the left hemisphere of most individuals is specialized for language functions and the right hemisphere for spatial and nonlinguistic tasks. There are numerous examples of anatomical asymmetries in areas purported to underly language function. The length of the outer border of the planum temporale, an area on the posterior portion of the upper surface of the superior temporal gyrus (and thought to comprise a major portion of Wernicke's area) was shown to be greater on the left in the majority of cases examined (Geschwind and Levitsky, 1968). Thus, out of 100 cases, 65% had larger left plana, 24% were symmetrical, and 11% had plana that were larger on the right. Similar findings of planum asymmetry were reported by others (Wada et al., 1975; Witelson and Pallie, 1973). Asymmetries in cytoarchitectonic area Tpt (which is located mostly within the planum temporale) are correlated strongly with planum temporale asymmetry and can be as much as 620% larger in the left hemisphere (Galaburda et al., 1978a). Cytoarchitectonic area 44 (part of Broca's area) is also asymmetrical in favor of the left (Galaburda, 1980).

The study of lateralization of function and structure has centered on the establishment of a population bias and on gathering information concerning the "standard" pattern of cerebral dominance. Yet there are certain populations of individuals, non-righthanders making up a large percentage of these, that have proved interesting exactly because of their lack of conformity to this standard pattern. Thus, non-right handedness has been associated with anomalous lateralization of function and structure. Non-righthanders score consistently differently on a variety of tests of lateralization of function (see Merron, 1980). Anatomically, the brains of non-righthanders are more likely to be symmetrical, and if they are asymmetrical, they are more likely than right handers to have an asymmetry in the opposite direction from the standard pattern seen in the right-handed population. In addition, asymmetries in the opposite direction are usually smaller in magnitude (Galaburda et al., 1978b).

29

In this paper, we present evidence of anomalous functional and structural asymmetry, and discuss ways by which information gleaned from the study of these cases may yield increasingly important conclusions as to the mechanisms of asymmetry.

DYSLEXIA-ANOMALOUS ANATOMY AND BEHAVIOR

The role of hemispheric specialization in the behavioral manifestations of dyslexia has been extensively investigated; thus, it has been suggested that dyslexia is an example of anomalous functional lateralization (e.g., Orton, 1928; Witelson, 1977). It is equally evident that this example of anomalous functional lateralization is associated with anomalous anatomy. We have studied four consecutive cases of dyslexia and have found that there are striking cortical anomalies associated with this learning disability. These anomalies consist primarily of ectopic collections of neurons in layer I of the cerebral cortex, which implicate problems in neuronal migration to the cortex.

More importantly for this discussion however, is the finding of planum temporale symmetry in all four cases. Recalling that symmetry of the planum temporale is seen in only 25% of the baseline population, it is clear that this pattern appearing in four consecutive cases cannot be explained by chance. We contend, therefore, that anomalous asymmetry and anomalous development of the cortex may underly dyslexia (Galaburda et al., 1985).

Testosterone Hypothesis

An hypothesis to explain the co-occurence of anomalous dominance and problems in cortical migration was derived from Geschwind and Behan's (1982) study of the relationship of left-handedness, autoimmune disease, and dyslexia. They found that lefthanders (as well as their first and second degree relatives) were more likely to suffer from autoimmune disease as well as learning disabilities. These initial findings were confirmed in a patient population with autoimmune-disease in whom it was found that there was a higher degree of left-handedness than in other types of patients (Geschwind and Behan, 1984). Because learning disabilities are more common in the male population and because of this link among left-handedness, immune disorders, and learning disabilities, Geschwind and coworkers (Geschwind and Behan, 1982; Geschwind and Galaburda, 1985) proposed the "testosterone hypothesis" to explain these results. According to this hypothesis excessive amounts of, and/or supersensitivity to, testosterone in utero has multiple effects. First, it affects the development of the thymus gland, thus resulting in a faulty immune system. Second, testosterone is purported to slow down the growth of the left hemisphere, thereby rendering it more likely to undergo insult resulting in the cortical anomalies. Third, this slowing down of the left hemisphere leads to a compensatory growth of the right hemisphere resulting in symmetry -- the end result being a smaller than normal left hemisphere and a larger than normal right hemisphere.

MECHANISMS OF ASYMMETRY

Before we can ask questions about the mechanism of asymmetries, we must establish some facts. First, in the animal kingdom, cerebral anatomical asymmetry is the rule, not the exception. In most species examined, clearcut anatomical asymmetries are present (Sherman et al., 1982). Second, asymmetries (both functional and anatomical) can be categorized at the individual and population levels. In the human speech areas, for example, there are population biases in favor of the left hemisphere. Other animals, while maintaining asymmetry at the individual level, have no overall directional population bias (Collins, 1977).

Having established these facts, we can ask the following questions: When there is symmetry of an anatomical area, is it because the side that would normally be smaller (e.g., the right planum), is now larger and comparable in size to the left? or is the usually larger side (left planum) now smaller and therefore equivalent in size to the usually smaller side? or is there an increase in the small side and a concomitant decrease in the large side? These three alternatives are schematically represented in Figure 1.

To distinguish among these possibilities, we have reexamined the brains originally studied by Geschwind and Levitsky (1968). However, in contrast to Geschwind and Levitsky, who measured the length of the lateral border, we measured the area of the planum temporale on both the right and left sides as well as the areas of the superior temporal lobe. In spite of our differences in measurement, the results were essentially the same. Sixty-four percent of the cases had larger left planum, 15% were symmetrical, and the remaining 21% had larger right plana.

In order to help delineate possible mechanisms of asymmetry, we computed the correlation of the total planum area vs. the degree of asymmetry (irrespective of direction). Our reasoning was as follows: If symmetry is due to an increase in size of the normally smaller side, then symmetrical brains will be larger than asymmetrical brains and the correlation between total area and absolute asymmetry will be negative. Alternatively, if the normally larger side becomes smaller, then symmetrical brains will be smaller and the correlation between the area and the degree of asymmetry will be positive. Finally, if, as the Geschwind and Galaburda hypothesis would predict, symmetry results from a decrease in size of the larger side and an increase in the smaller side, then the total volume of the planum temporale should remain relatively constant and its correlation with asymmetry should be zero (see Figure 2).

We found that the asymmetry of the planum temporale and the total area were negatively correlated (r = -0.507, n = 100, $p <$ 0.001). This indicates that symmetrical brains are larger than asymmetrical brains in the region of interest, and that symmetry must be due to an increase in size of the normally smaller side. As further evidence showed, there was no correlation between the left planum area and the absolute asymmetry coefficient while there was a significant negative correlation between the right planum area and asymmetry (r = -0.651, n = 100, $p <$ 0.001). Thus, it is the smaller side (in the case or the planum temporale, the right for the majority of the

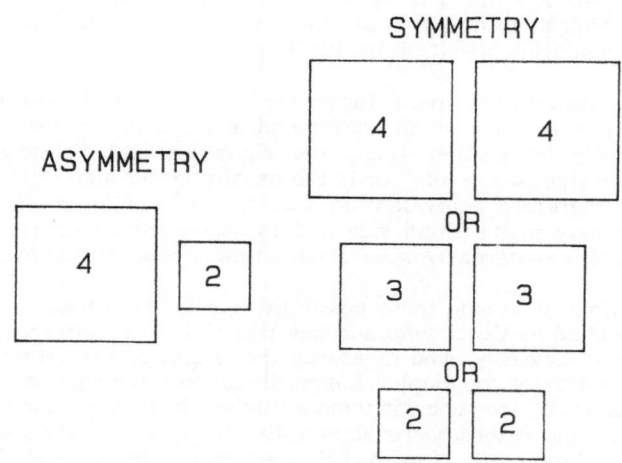

FIGURE 1

Schematic representation of possible differences between symmetrical and asymmetrical brain areas. Symmetry can result from an increase in the normally smaller side (top), a decrease in the normally larger side (bottom), or a combination of a decrease in the larger side with an increase in the smaller side (middle).

population) which varies to cause asymmetry. In fact, if one ignores direction and correlates the smaller side (either right or left) with absolute asymmetry, one finds a highly significant correlation (r = -0.852, n = 100, $p < 0.001$) while there is no correlation with the larger side.

HISTOLOGICAL BASIS OF ASYMMETRY

Having established that symmetry is the result of two large areas, and that asymmetry is therefore the result of a unilateral decrease in area on one side, we next asked what is the histological basis of this asymmetry. Specifically, when one side is smaller than its homologue, is it because neurons are more tightly packed (increased cell-packing density) on that side? or is there a decrease in numbers of neurons without a change in cell-packing

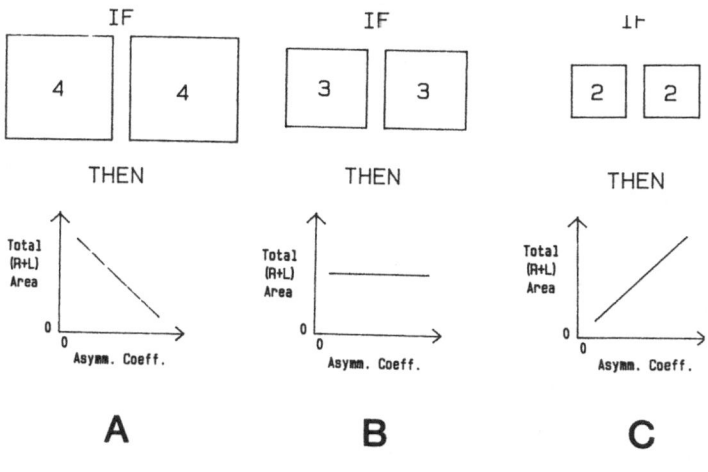

FIGURE 2

Schematic representation of the possible relationships between total volume and anatomical asymmetry of an area. **A.** If a symmetrical brain area is the result of an increase in the normally larger side, then asymmetrical brains would be smaller in total volume than symmetrical brains and the correlation between total volume and asymmetry would be negative. **B.** If symmetry is the result of a decrease in the normally larger side and an increase in the normally smaller side, then the total brain area should remain constant with respect to planum asymmetry. **C.** If symmetry results from a decrease in the normally larger side, then symmetrical brains should be smaller than asymmetrical brains and the correlation between total area and asymmetry should be positive.

density? or is there a change in both? We chose the rat as our experimental animal because of the increased ability to control for histological quality and because of the obvious anatomical asymmetries in this species (Sherman and Galaburda, 1984).

The brains of 19 Purdue-Wistar rats were serially sectioned coronally and stained with cresyl violet for Nissl substance. Parcellation of the entire neocortex and of area 17 (the primary visual cortex) of both hemispheres was performed under light microscopy. In addition, neurons within area 17 of both hemispheres were counted under 500X magnification (Galaburda et al., 1986).

In order to confirm that the rat and the human brain are comparable, we determined that there was a similar relationship to that seen in the human planum temporale between asymmetry of an architectonic area and its volume. Thus we correlated the absolute asymmetry of area 17 with the total volume of area 17 and found, as in the human, a significant negative correlation (r = -0.457, n = 19, p < 0.05). In addition, there was a significant negative correlation between absolute asymmetry and the smaller area 17 hemisphere (r = -0.592, n = 19, p < 0.05) while there was no correlation with the larger side. We can conclude, therefore, that asymmetrical brain areas have a smaller total volume of visual cortex than symmetrical brains, and that it is the smaller side that varies - a finding identical to that seen in the human planum temporale.

Having established that the rat neocortical asymmetry behaves analogously to that of the human planum temporale, we then sought to determine the histological basis of asymmetry in this animal. Neocortical volume is determined by two factors: cell-packing density (CPD) and cell numbers. Thus, when an area in one hemisphere is smaller than its homologue on the opposite side, it may be because there are 1) fewer numbers of cells in the smaller hemisphere with the CPD remaining constant, or 2) the same number of cells with an increase in cell-packing density, or 3) changes in both CPD and cell-numbers. In order to distinguish among these possibilities, we computed the correlation between CPD asymmetry and area 17 volumetric asymmetry. If there was a significant positive correlation between these two variables then asymmetry would be the result of changes in CPD. We found no correlation between these two variables indicating that asymmetry results from changes in cell numbers and not cell-packing density (Galaburda et al., 1986).

There are a number of reasons why one might expect asymmetry to reflect changes in cell numbers and not cell-packing density. If cell numbers were to remain constant, then any changes in volume must be due to changes in neuropil and/or in various intercellular elements, e.g. glia, myelin, blood vessels, all of which affect CPD. Yet, left-right side differences large enough to produce the large architectonic volumetric asymmetries seen in the human would significantly distort the cytoarchitecture and thus render impossible the task of assigning the same architectonic label to an area in each hemisphere. It is therefore logical that changes in cell numbers, as long as they affect all cell layers nearly equally, would not affect the cytoarchitectonic appearance of an area, and could underly anatomical asymmetries.

CONCLUSIONS

It is clear from our results that symmetrical architectonic areas are larger than asymmetrical areas. It is equally clear that asymmetries result from changes in cell numbers, not cell packing density. These facts lead us to a number of suppositions and questions concerning the development of asymmetry. Geschwind and co-workers hypothesized that symmetry was due to the effect of interactive factors altering the development of a normally asymmetrical brain. Thus, testosterone was hypothesized to act by slowing down the growth of the left hemisphere leading to a compensatory growth of the right hemisphere. We have found instead that the larger (i.e., left) side does not vary significantly with respect to asymmetry, and that changes seem to occur mainly on the smaller (right) side. Asymmetrical brains, therefore, have a relatively small right side, which presumably means that it is missing a full complement of neurons. It is not possible to tell at this time whether insufficient numbers of neurons are produced on the right, or if they are produced and then destroyed during development. Ongoing work in our laboratory is aimed at addressing these questions.

REFERENCES

Collins, R.L. (1977). Toward an admissible genetic model for the inheritance of the degree and direction of asymmetry. In Lateralization in the Nervous System. (eds. S. Harnad, R.W. Doty, L. Goldstein, J. Jaynes, and G. Krauthamer). Academic Press, New York.

Galaburda, A.M. (1980). La région de Broca: Observations anatomiques faites un siècle après la mort do son découvreur. Rev. Neurol., 136, 609-616.

Galaburda, A.M., Sanides, F. and Geschwind, N. (1978a). Human brain: Cytoarchitectonic left-right asymmetries in the temporal speech region. Arch. Neurol., 35, 812-817.

Galaburda, A.M., Aboitiz, F., Rosen, G.D. and Sherman, G.F. (1986). Histological asymmetry in the primary visual cortex of the rat: Implications for mechanisms of cerebral asymmetry. Cortex, 22, 151-160.

Galaburda, A.M., LeMay, M., Kemper, T.L. and Geschwind, N. (1978b). Right-left asymmetries in the brain. Science, 199, 852-856.

Galaburda, A.M., Sherman, G.F., Rosen, G.D., Aboitiz, F. and Geschwind, N. (1985). Developmental dyslexia: Four consecutive cases with cortical anomalies. Ann. Neurol., 18, 222-233.

Geschwind, N. and Behan, P.O. (1982). Left-handedness: Association with immune disease, migraine, and developmental learning disorder. PNAS, 79, 5097-5100.

Geschwind, N. and Behan, P.O. (1984). Laterality, hormones and immunity. In Biological Foundations of Cerebral Dominance. (eds. N. Geschwind and A.M. Galaburda). Harvard University Press, Cambridge, USA.

Geschwind, N. and Galaburda, A.M. (1985). Cerebral lateralization. Biological mechanisms, associations, and pathology: I. A hypothesis and a program for research. Arch. Neurol. 42, 428-459.

Geschwind, N. and Levitsky, W. (1968). Human brain: Left-right asymmetries in temporal speech region. Science, 161, 186-187.

Herron, J. (Editor) (1980). Neuropsychology of Left-Handedness. Academic Press, New York.

Orton, S.T. (1928). Specific reading disability - strephosymbolia. J. Am. Med. Assoc., 90, 1095-1099.

Sherman, G.F. and Galaburda, A.M. (1984). Neocortical asymmetry and open-field behavior in the rat. Exp. Neurol., 86, 473-482.

Sherman, G.F., Galaburda, A.M. and Geschwind, N. (1982). Neuroanatomical asymmetries in non-human species. Trends Neurosci., 2, 429-431.

Sperry, R.W. (1974). Lateral specialization in the surgically separated hemispheres. In Hemispheric Specialization and Interation. (ed. B. Milner). MIT Press, Cambridge, USA.

Springer, S.P. and Deutsch, G. (1985). Left Brain, Right Brain. W.H. Freeman and Co., New York.

Wada, J.A., Glarke, R. and Hamm, A. (1975). Cerebral hemispheric asymmetry in humans. Arch. Neurol., 32, 239-246.

Witelson, S.F. (1977). Developmental dyslexia: Two right hemispheres and none left. Science 195, 309-311.

4
Asymmetry in the Cerebral Cortex: Development, Estrogen Receptors, Neuron/Glial Ratios, Immune Deficiency and Enrichment/Overcrowding

Marian Cleeves Diamond

Unravelling the mysteries of the structural asymmetry in the cerebral cortex is proving to be an exciting challenge. So many factors need to be taken into consideration in order to determine both the consistency and the complexity of the subject. In previous review chapters written on rodent cerebral cortical asymmetry from the data gathered in our laboratory, the following subjects were discussed: 1. the developing and aging pattern in Long-Evans male and female rats. 2. a comparison between the intact and gonadectomized males. 3. a comparison between intact and ovariectomized females. 4. enriched and impoverished environmental influences on four different age groups (Diamond 1984; Diamond 1985).

This chapter will differ from the above by presenting newer and more detailed data on: 1. the thickness of the developing female rat right and left cerebral cortex. 2. the concentration of estrogen receptors in the developing male and female rat right and left postnatal cortex. 3. neuron/glial ratios in area 39 of the adult male and female rat right and left cortices. 4. cerebral cortical deficits in the immune-deficient nude female mouse. 5. overcrowding in enriched conditions in male rats.

1. THE THICKNESS OF THE DEVELOPING FEMALE RAT RIGHT AND LEFT CEREBRAL CORTEX

Before presenting the new data separately on the developing right and left cerebral cortex of the female, it might be of interest to offer the results from the developing male and female cortex with both hemispheres combined because they show such different sex patterns. It is evident from Figure 1 that all parts of the male Long-Evans rat cerebral cortex (including the combined right and left samples from the frontal, somatosensory, and occipital cortex) grow very rapidly after birth until somewhere between 26 and 41 days of age and then begin to decline. For all

37

38

Fig. 1

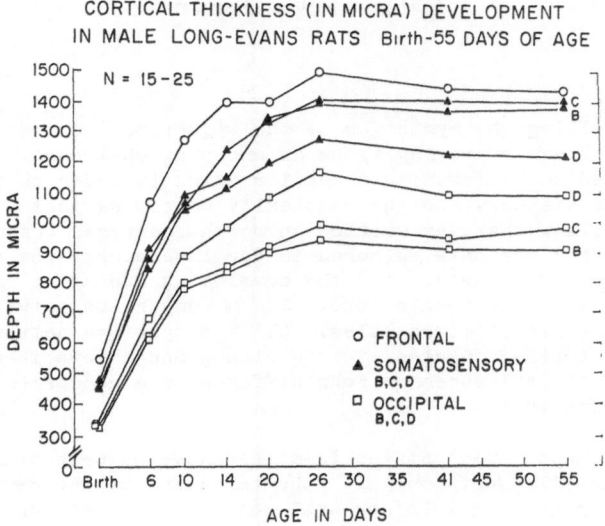

CORTICAL THICKNESS (IN MICRA) DEVELOPMENT
IN MALE LONG-EVANS RATS Birth-55 DAYS OF AGE

cortical regions the increase in thickness amounts to about 45%
until the peak is reached. In contrast to the male, as shown in
Figure 2, the female, Long-Evans rats' cortical areas have very
different patterns of growth when the data from the two hemispheres
are combined. If one examines the medial frontal region only, it
grows by 2% (NS) from 7 days of age until 14 days of age and
reaches its peak by 18 days of age. At the same time, another area
39 (according to the designation of Krieg 1946) is growing by 40%
(p<0.005) between 7 and 14 days of age and reaches its peak by
about 33 days of age. From these few examples it is clearly
evident that there are sex differences in regional rates of
cortical development.

The right and left data on the growth of only the female
cortex have been completely analyzed at this date (Lewis et al.
1986 in preparation). In order to study the differences in rate of
postnatal growth between the two hemispheres many age groups with

Fig. 2

Long-Evans Female Cortical Thickness Growth Curve

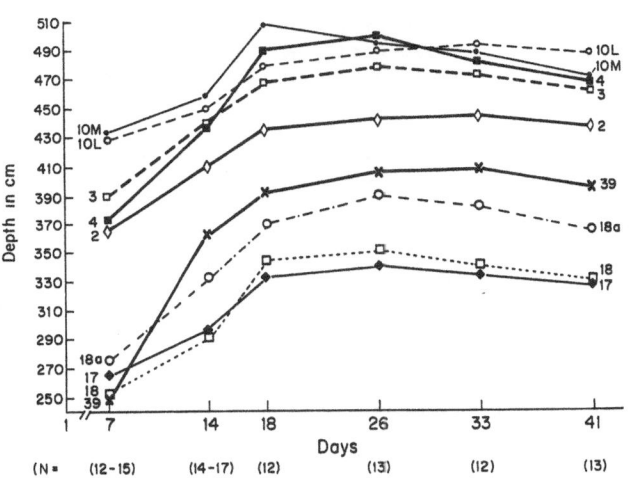

large samples of animals were collected - 7 days of age (N=12-15), 14(N=14-17), 18(N=12), 21(N=14), 24(N=11), 26(N=13), 28(N=13), 33(N=12) and 41(N=13) (Diamond et al., 1984. Lewis et al. 1986).

At the present time it appears that by 18 days of age, the left cortex has reached its peak of development. When comparing the right cortex at 18 days with the right cortex at 24, 33 and 41 days, a significant difference is seen. The same is not true for the left cortex. In other words the left reaches its peak before the right in the female. The cortical areas which are more laterally placed (10L, 3, 2, 18a and 39) show these significantly different growth rates between the hemispheres.

2.   CONCENTRATION OF ESTROGEN RECEPTORS IN THE DEVELOPING MALE AND FEMALE RIGHT AND LEFT POSTNATAL CEREBRAL CORTEX.

In 1975 we first reported that in the male Long-Evans rat the thickness of the right cerebral cortex was in general greater than the left (Diamond et al. 1975a). Further studies supported this earlier finding and also showed that in the female rat the left cortex was on the average thicker than the right but the values were not statistically significantly different (Diamond et al., 1981). Structures which account for cortical thickness have been demonstrated to include: neuronal soma and nuclear area (Diamond et al., 1967); number and length of dendritic branching (Holloway, 1966; Connor et al., 1980, 1981); dendritic spines (Globus et al., 1973); axo-dendritic postsynaptic density length (Mollgaard et al., 1971;Diamond et al., 1975b), glial cell number (Altman and Das, 1964; Diamond et al., 1966) and blood vessel diameter (Diamond et al., 1966).

Removal of the gonads at birth alters cortical development (Pappas and Diamond 1978) as well as cortical asymmetry (Diamond et al., 1981). Also ovariectomy at 300 days of age affects cortical thickness when measured 90 days later (Diamond et al., 1971). In addition, exogenous estrogen and progesterone administered to ovariectomized rats affect cerebral morphology (Pappas et al., 1979). Further evidence demonstrating the correlation between cerebral cortical cells and the sex hormones has been shown by the presence of cortical estrogen receptors during the first few weeks of post-natal life in both males and females (Barley et al., 1974; MacLuskey et al., 1979).

With all of this information indicating a role for estrogen in determining morphogenesis in the cerebral cortex, we wished to learn about the concentration of estrogen receptors in the right and left cerebral cortices. The complete results of this experiment have been published by Sandhu et al., 1986. In brief, the experiment proceeded as follows. On postnatal days 2-3, 7-8, 14-15 and 25-26, male and female Long-Evans rats were gonadectomized 24 hours before their brains were removed for study. (Eight rats of each sex were used at each age group). The dorsal right and left cerebral cortices were dissected free from the underlying white matter and placed on dry ice. Sephadex LH-20 gel filtration chromatography was used to dissociate the majority of alpha-fetoprotein-bound (3_H) estradiol while leaving the receptor (3_H) estradiol complex intact. The correction for the non-receptor binding, including alpha-fetoprotein, was made using parallel incubation containing unlabeled diesthystilbesterol.

The results are presented in Figures 3 and 4. It is clearly seen that in both sexes the estrogen receptor concentration is highest in days 2-3 and has disappeared by day 25-26, thus supporting the results of others. In the male rat (Fig. 3), the estrogen receptor concentration is significantly greater in the left cortex, with the opposite being true in the female (Fig 4). We postulate that the presence of estrogen receptors during the development of the cerebral cortex may play a role in establishing

Fig. 3

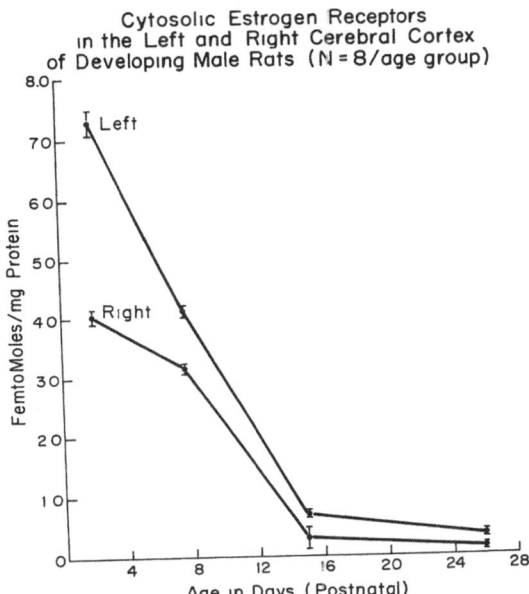

Cytosolic Estrogen Receptors
in the Left and Right Cerebral Cortex
of Developing Male Rats (N = 8/age group)

the hemispheric asymmetrical patterns.

We previously demonstrated that exogenous estrogen
administered to adult female rats decreased cortical thickness
(Pappas et al., 1979). In the male, testosterone has been shown to
change to estrogen to influence brain cells. If estrogen acts in a
similar fashion on early postnatal cortical cells as it does on
adult cortical cells, then one might hypothesize that the
hemisphere with the greatest number of estrogen receptors would
have the thinner cortex. This is precisely what we have found.
For the present, we can say for certain that estrogen receptors are
found in different concentrations in the right and left cortices in
males and females. Our projected plans will be directed toward
localizing these receptors more specifically. First, are they
found in a greater concentration in the frontal, parietal, temporal
or occipital cortex? Is their distribution different in the
various cortical layers? Upon what types of cells are the

Fig. 4

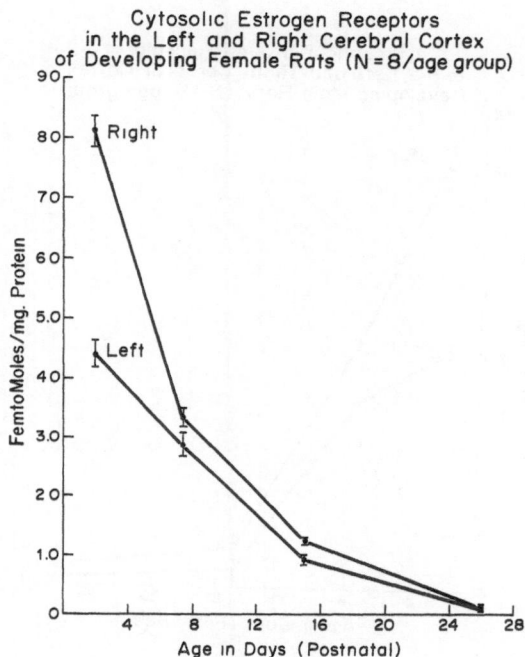

Cytosolic Estrogen Receptors
in the Left and Right Cerebral Cortex
of Developing Female Rats (N = 8/age group)

receptors found? These are some of the questions for which we hope
to find answers in the months ahead.

3.   NEURON/GLIAL RATIOS IN AREA 39 OF THE ADULT MALE AND FEMALE
RAT AND HUMAN RIGHT AND LEFT CORTICES.

Several years ago we were interested to learn whether the
human superior frontal cortex has more glial cells/neuron than does
the inferior parietal cortex. The reasoning which initiated this
study was as follows. Experiments with male rats provided
information showing that if the animals lived in stimulating,
active environments, samples from their cerebral cortex showed more
glial cells per neuron in comparison with cortical samples from
animals living in impoverished conditions (Diamond et al. 1966).
Using these findings as guidelines, we then asked whether
neuron/glial ratios would indicate areas of greater activity in the

it might have more glial cells/neuron than the frontal cortex which is more highly evolved in man than in lower primates?

We then undertook a study of neuron/glial ratios in non-neurologically diseased human male brains. From counts in eleven brains, we learned that the right and left frontal cortices did have more glial cells per neuron than did the right and left parietal cortices. However, upon requesting further information about the specific ages of the human brain samples, we were surprised to learn that two of the brains from which we had taken samples were female. As an example of our findings, Table 1 shows the comparison of neuron/glial ratios in the human female inferior parietal cortex (area 39) with that of males. Since the samples from the females are so small, only the suggested trends are demonstrated and no statistics are offered. (We are presently enlarging our human female sample.)

Table 1   Neuron/glial ratios in area 39 from human brains

N=11 male; N=2 female

	Left	Right
Male	2.25	1.99
Female	2.02	1.51
Einstein[*]	1.12	0.92

[*]Ref. (Diamond et al., 1985)

With these apparent sex differences in neuron/glial ratios in area 39 in human beings, our next step was to attempt to substantiate the results by counting nerve and glial cells in area 39 (Krieg, 1946) in male and female rats where more stringent controls can be utilized. McShane et al. (1986 in preparation) have recently completed these counts in 90-day-old male and female Long-Evans rats. The methods for differentiating nerve cells from glial cells in our laboratory have been previously published and will not be duplicated here. (Diamond et al., 1966 and 1985). Figure 5 and tables 3 and 4 indicate the findings. Figure 5 clearly demonstrates that the direction of differences between male and female rat neuron/glial ratios in area 39 agrees with that found in human males and females. The neuron/glial ratio in the female rat is significantly smaller than in the male in both the right and left hemispheres.

44

Fig. 5

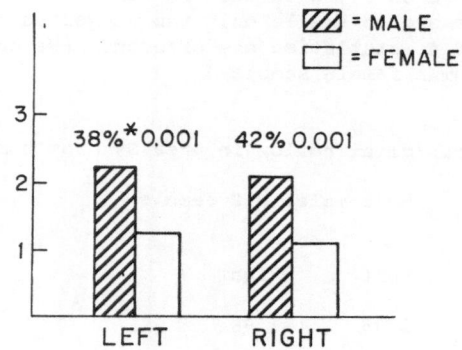

NEURON/GLIAL RATIOS IN AREA 39
FROM 90 DAY OLD LONG-EVANS RATS

✳ % difference between male and female

female rat is significantly smaller than in the male in both the
right and left hemispheres.

The neuron counts in males and females shown in Table 2
support the asymmetrical patterns seen in previous cortical
thickness measurements (Diamond et al., 1981). The left area 39 of
the female has significantly more neurons than the right, with the
opposite being true for the male. Table 3 indicates that glia are
more numerous in left area 39 of the female than the right. In the
male, the opposite is true with more glia in the right area 39 than
in the left. (The cell counts between males and female rats can
only be considered preliminary at present because celloidin sections were studied in males and frozen in females, Figure 5. Yet,
the right-left differences in Table 2 are considered final because
both hemispheres were treated similarly.)

Table 2. Neuron Counts in Area 39 from 90-day-old Long-Evans Rats

| | Left | Right | | |
	X̄ SEM	X̄ SEM	% diff L > R	p
female	1007±84	933±104	13	0.05
male	826±21	923±47	-12	0.05

Table 3. Glial Counts in Area 39 from 90-day-old Long-Evans Rats

| | Left | Right | | |
	X̄ SEM	X̄ SEM	% diff L > R	p
female	812±70	749±48	8	NS
male	388±17	450±36	-16	0.05

The sex differences in rat cortical neuron/glial ratios support our earlier findings in the human cerebral cortex. Whether these differences will be found in other cortical regions remains to be seen. Since we had the data from Albert Einstein's area 39, they were included in Table 2. The true significance of these findings will become apparent as we obtain more information about the function of glial cells. Their suggested role with the immune system (Kerns and Frank, 1980; Merrill et al., 1984; and Diamond et al., 1986), the relationships between the hemispheres and the immune system (Renoux et al., 1982), and the differences between male and females and the immune system make all of these findings of greater significance. The eventual integration of these factors will undoubtedly lead to a better understanding of the subtleties of interplay among the brain cells with our internal environment.

4. CEREBRAL CORTICAL DEFICITS IN THE IMMUNE-DEFICIENT NUDE MOUSE

In 1980 Renoux et al., reported that lesions in the left cerebral cortex affected the activity of natural killer cells. In 1983 the original report was modified and stated that "natural killer reactivity of mouse spleen T cells is controlled primarily by the left neocortex with an inductive influence from the right neocortex." In their original work they produced lesions in the frontal, parietal, and occipital cortex, but their later work included only the frontal and parietal cortices.

After having mapped the cerebral cortical thickness for so many years with our enriched environmental studies, we were curious to learn if we could localize more specifically which cortical areas were playing a role in this reported cortical immune response. Initially we attempted to correlate various cortical

lesions with T-cell activity and tumor growth. Having little
success with this approach, we examined the problem from another
direction. Namely, we looked for cortical deficiencies in an
immune deficient mouse.

We were interested in comparing cerebral cortical morphology
between the BALB/c (nu/nu) mouse and in the BALB/c (+/+) mouse. In
addition to studying the cortical thickness of the frontal,
parietal and somesthetic cortex in these experiments, we were also
curious to know about the neuronal and glial cell populations in
the cortex of these two strains of mice. From the studies of Kerns
and Frank (1980) we read that the (nu/nu) mice showed a 29%
decrease in lumbar gray oligodendrocytes and a 52% increase in
lumbar gray astrocytes compared with (nu/+) mice. Would we also
find such changes in the cerebral cortex in immune deficient
animals?

Merrill et al., (1984) have reported that glial growth-
promoting lymphokines have been shown to be products of activated T
lymphocytes. Concanavalin A binding sites have been identified on
the surface of oligodendroglia; whereas, other lectin receptors
have been expressed on astrocytes.

In our experiment, female nude (nu/nu, N=18) and BALB/c (+/+,
N=11) mice, 120 days of age, were prepared for light microscopic-
histological study. From transverse, 20 micra, frozen sections
stained with thionine, cortical thickness measurements and
neuron/glial counts were determined in our usual fashion (Diamond
et al.,1985). The full report on the findings from this experiment
is found in Diamond et al., in 1986. Briefly, in the BALB/c mouse,
the right cortex was thicker than the left in four of nine areas
measured, showing no significant trend in asymmetry. In the nude
mouse, however, the right hemisphere was thicker than the left in
seven of the nine cortical areas measured, with one area, 18,
becoming highly statistically significantly different.

We were surprised to learn that only two areas of the nine
measured for cortical thickness were significantly smaller in the
nude compared to the BALB/c mouse. (Figure 6.) These were the
frontal (8%, p>0.005) and lateral parietal (11%, p>0.005) areas.
Both of these regions had been lesioned by Renoux et al. and
created an immune deficient response. However, it is possible that
the lateral parietal area, which includes area 2, is less well
developed in the nude mouse because of reduced sensory input from a
hairless skin.

Since the female nude mouse, area 18 reached a statistically
significant right-greater-than-left difference (p > 0.005), this
area was selected for the differential cell counts. Both nerve
cell number and glial cell number were less per microscopic field
in the left occipital cortex of the nude mouse, BALBc (nu/nu),

Fig. 6

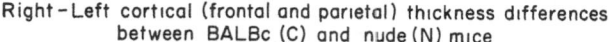

Right-Left cortical (frontal and parietal) thickness differences
between BALBc (C) and nude (N) mice

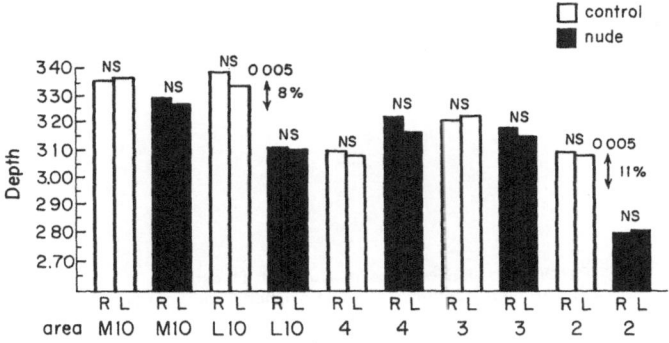

Cortical area and condition

5. OVERCROWDING IN ENRICHED CONDITIONS IN MALE RATS

Two questions were being addressed in this particular
experiment. Was one hemisphere affected differently from the other
when male rats lived in enriched environments? Would crowded-
enriched conditions alter the brain differently from noncrowded
enriched or nonenriched conditions?

Three environments were established to answer the above
questions. 1. the enriched condition, consisting of 12 rats
living in a large cage (70 x 70 x 46 cm) with access to "toys"; 2.
a nonenriched or standard colony condition with three rats living
in a small cage (32 x 20 x 20 cm) with no toys; 3. a crowded-
enriched condition with 36 rats living in a large cage (70 x 70 x
46 cm) with "toys". The animals lived in their respective
conditions from 60 to 90 days of age when the brains were collected
for cortical thickness measurements according to our usual

histological techniques.　The results are presented in Figures 7
and 8.

　　Even though the enriched rats were living in a multisensory
environment, it is the posterior section of the cortex which
demonstrates the most marked effects, and these are evident in both
the right and left cortices to a similar degree (Fig. 7).　With the
crowded-enriched condition, the changes in the cortex agree closely
with those found in rats from the non-crowded, enriched condition.
(Fig. 8.)　There is a tendency for the left area 39 to show a
slightly greater change in the crowded-enriched animal's brain
than in the enriched animal's.　Marked brain changes due to
overcrowding were not encountered in these experiments (Diamond et
al., submitted).　However, it will be of interest to learn if
overcrowding without the presence of "toys" will bring about
similar results.

Fig. 7

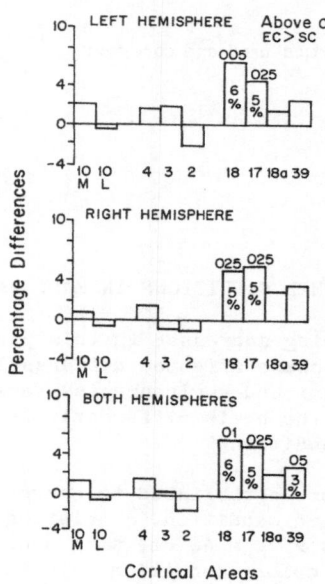

Percentage Differences in Cortical Thickness
Between Standard Colony (12) and Enriched Conditions (11)
Male, Long-Evans Rats

Fig. 8

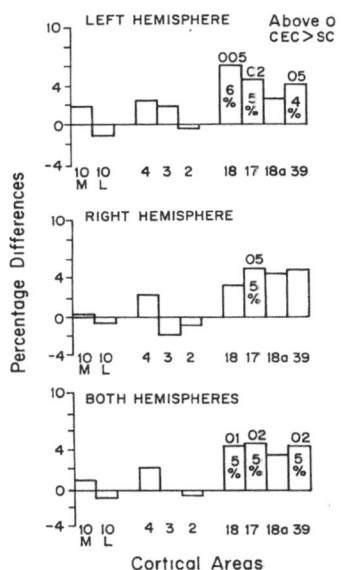

Percentage Differences in Cortical Thickness
Between Crowded-Enriched (16) and Standard Colony (12)
Male, Long-Evans Rats

CONCLUSIONS

Some of the conclusions to be drawn from the five experimental conditions presented in this paper are the following:

1. In the female Long-Evans rat, the right and left cerebral cortices do have different postnatal development patterns as measured by cortical thickness, with the left cortex reaching its peak before the right.

2. In the early postnatal, male, Long-Evans rat, estrogen receptors are in greater concentration in the left cerebral cortex compared to the right. The opposite is true in the female. In both sexes these receptors have disappeared before the end of the first postnatal month.

3.   The neuron/glial ratio is smaller in the 90 day-old, female cortical area 39 than in the male. Both neurons and glial cells are in greater numbers in the female left area 39 than in the right. The opposite is true in the male at 90 days of age.

4.   In the immune deficient female mouse the cerebral cortex is thinner on the left side than on the right, with area 18 being statistically significantly thinner. Oligodendrocytes are significantly fewer in left cortical area 18 in the female nude mouse compared to area 18 in the control BALBc mouse.

5.   Cerebral cortical thickness in the left and right hemispheres increases similarly in the male rat living in the enriched environment. Overcrowding plus enrichment affects cerebral cortical thickness in a pattern similar to that of the enrichment alone, but area 39 in the left hemisphere appears to respond to the crowded enriched condition more than to the enriched condition alone.

References

Altman, J. and Das, G.D. (1964). Autoradiographic examination of the effects of enriched environment on the rate of glial multiplication in the adult rat brain. Nature (London), 204,1161-1163.

Barley, J.M., Ginsberg, B., Greestein, B., MacLusky, N.J., Thomas, P.T., (1974). A receptor mediating sexual differentiation? Nature, 258, 259-260.

Connor, J.R., Diamond, M.C. and Johnson, R.E. (1980). Occipital cortical morphology of the rat: alterations with age and environment. Exp. Neurol., 68,158-170.

Connor, J.R., Melone, J., Yuen, A. and Diamond, M.C. (1981). Terminal segments' length of dendrites in aged rats: an environmentally induced response. Exp. Neurol., 73, 827-830.

Diamond, M.C., Law F., Rhodes, H., Lindner, B., Rosenzweig, M.R., Krech, D. and Bennett, E.L., (1966). Increases in cortical depth and glial numbers in rats subjected to enriched environment. J. Comp. Neurol., 128, 117-126.

Diamond, M.C. (1967). Extensive cortical depth measurements and neuron size increases in the cortex of environmentally enriched rats. J. Comp. Neurol., 131, 357-364.

Diamond, M.C., Johnson, R.E. and Ingham C.A. (1971). Brain plasticity induced by environment and pregnancy. Intern. J. Neurosci., 2, 171-178.

Diamond, M.C., Johnson, R.E. and Ingham, C.A. (1975a). Morphological changes in young, adult and aging rat cerebral cortex, hippocampus, and diencephalon, Behav. Biol., 14,163-174.

Diamond, M.C. Lindner, B., Johnson R., Bennett, E.L. and Rosenzweig, M.R., (1975b). Differences in occipital cortical synapses from environmentally enriched, impoverished and standard colony rats. J. of Neurosci. Res.,1, #2, 109-119.

Diamond, M.C., Dowling, G.A., Johnson, R.E. (1981). Morphologic cerebral cortical asymmetry in male and female rats. Exp. Neurol., 71, 261-268.

Diamond, M.C. (1984). Age, Sex and Environmental Influences. In Cerebral Dominance - The Biological Foundations. (eds. N. Geschwind and A.M. Galaburda.) Harvard Univ. Press, Cambridge, Mass.

Diamond, M.C. (1985). Rat Forebrain Morphology: Right-Left; Male-Female; Young-Old; Enriched-Impoverished. In, Cerebral Lateralization in Nonhuman Species. (eds. S. Glick) Academic Press, New York.

Diamond, M.C., Scheibel, A.B., Murphy, G., and Harvey, T. (1985). The Brain of a Scientist: Albert Einstein. Exp. Neurol., 88, 198-204.

Diamond, M.C., Rainbolt, R. D., Guzman, R., Greer, E.R., and Teitelbaum, S. (1986). Regional Cerebral Cortical deficits in the immune-deficient nude mouse: A preliminary study. Exp. Neurol., 92, 311-322.

Globus, A., Rosenzweig, M.R., Bennett, E.L., and Diamond, M.C. (1973). Effects of differential experience on dendritic spine counts in rat cerebral cortex. J. Comp. Physiol. Psych., 82, 175-181.

Holloway, R.L. (1966). Dendritic Branching: some preliminary results of training and complexity in rat visual cortex. Brain Research, 2, 393-396.

Krieg, W.J.S. (1946). Connections of the cerebral cortex. I. The albino rat. A. Topography of the cortical areas. J. Comp. Neurol., 84, 221-275.

Kerns, J.M. and Frank, M. (1980). A quantitative study of lymphocytes and neuroglia in the nude mouse spinal cord. Anat. Rec., 196, 96A

Maclusky, N.J., Chaptal, C. and McEwen, B.S. (1979). The development of estrogen receptor systems in rat brain and pituitary: post-natal development. Brain Res., 178, 143-160.

Merrill, J.E., Kutsumal, S., Mohlstron, C., Hofman, F., Groopman, J., and Golde, D.W. (1984). Human T lumphocytes promoted proliferation and maturation of oligodendroglia and astroglial cells. Science, 224, 1428-1430.

Mollgaard, K., Diamond M.C., Bennett, E.L., Rosenzweig, M.R. and, Lindner, B. (1971). Quantitative synaptic changes with differential experience in rat brains. Int. J. Neurosci., 2, 113-128.

Pappas, C.T.E., Diamond, M.C. and Johnson, R.E. (1978). Effects of ovariectomy and differential experience on rat cerebral cortical morphology. Brain Res., 154, 53-60.

Pappas, C.T.E., Diamond, M.C. and Johnson, R.E. (1979). Morphological changes in the cerebral cortex of rats with altered levels of ovarian hormones. Beh. Biol., 26, 298-310.
Renoux, G. (1980). The cerebral cortex regulates immune responses in the mouse. C. R. Acad. Sci. D., (Paris), 290, 719-722.

Renoux, G. (1983). The mode of action of imuthiol (sodium diethyldithiocarbamate): a new role for the brain neocortex and the endocrine liver in the regulation of the T-cell lineage. In Mechanisms of Immune Modulation. (ed. M. A. Chirigos). Marcell Dekker, New York.

Sandhu, S., Cooke, P., and Diamond, M.C. (1986). Rat cerebral cortical estrogen receptors: male-female, right-left. Exp. Neurol. 92:186-196.

Session II
Handedness and Hemispheric Specialization

Chairman: D. Schalling

## 5

Handedness and Cerebral Organization: Data from Clinical and Normal Populations

M.P. Bryden

Handedness is generally considered to be an important variable in neuropsychological studies of cerebral organization. Although the precise relationship between handedness and brain organization has yet to be specified, the assertion that left-handers are somehow different is rarely questioned. The present paper has two main goals. The first is to review the evidence that handedness and other lateral preference variables are, in fact, related to cerebral organization. The second is to examine the way in which handedness is measured, in order to see which measures of lateral preference are most likely to be related to functional brain lateralization.

One of the most obvious ways of investigating the relation between handedness and the lateralization of language functions is to study the incidence of aphasia following unilateral brain damage in left- and right-handers. In recent years, there have been at least two major attempts to summarize the literature on this topic, one by Carter, Hohenegger, and Satz (1980), later modified by Carter, Satz, and Hohenegger (1984), and one by Segalowitz and Bryden (1983). In both studies, data from a number of large-scale studies of apahsia were combined to generate estimates of the incidence of left hemisphere, right hemisphere, and bilateral language representation. Working with similar assumptions but with a somewhat different selection of data, Carter et al. (1984) estimate the incidence of right-hemisphere language to be only 1% in right handers, while Segalowitz and Bryden (1983) obtain a figure of 5%. Both assume that the incidence of bilateral language representation in right-handers is zero, although this assumption is evidently incorrect. Mateer and Dodrill (1983), for instance, report that 4 of their 6 cases of bilateral language representation were right-handed. The alternative proposals are shown in Table 1, along with data from Rasmussen and Milner's (1977) study with sodium amytal.

Among left-handers, the two studies differ even more in their estimates of the incidence of bilateral language representation (Table 1). Carter et al. (1984) obtain a very high figure for bilateral representation, 66%, with 23% having left-hemisphere

*Preparation of this paper was aided by a grant from the Natural Sciences and Engineering Research Council of Canada to the author.

Table 1

Language Lateralization and Handedness

	Lateralization		
	Left	Bilateral	Right
Right-handers			
Carter et al. (1984)	99%	0%	1%
Segalowitz-Bryden (1983)	95%	0%	5%
Rasmussen-Milner (1977)	96%	0%	4%
Left-handers			
Carter et al. (1984)	23%	66%	11%
Segalowitz-Bryden (1983)	61%	20%	19%
Rasmussen-Milner (1977)	70%	15%	15%

representation, and 11% having right-hemisphere representation. Segalowitz and Bryden's (1983) comparable figures are 20% bilateral, 61% left, and 19% right.Both studies do agree that a significant proportion of left-handers have bilateral speech representation. Furthermore, the Segalowitz and Bryden (1983) results are in close agreement with those reported from sodium amytal studies by Rasmussen and Milner (1977).

There are a variety of problems that arise from such surveys. The first is that of knowing which data to include. Large-scale surveys of the incidence of aphasia differ in many ways: some involve patients with penetrating head injury, others with focal epilepsy, still others with stroke. The age and sex distribution of the subjects may differ dramatically, and the data are rarely presented in a form in which one can isolate individual cases. While there are usually at least reasonable grounds for classifying a particular patient as aphasic, the grounds for including nonaphasic cases are often obscure, and the estimated incidence of aphasia will be profoundly affected by which nonaphasic cases are included. For instance, the inclusion of data from studies in which there were very few left-handed nonaphasics clearly inflates the Carter et al. (1984) estimates of bilaterality. In the earlier studies, verification of damage was often poor, and the classification of individuals as having unilateral damage may not have been accurate. Furthermore, in many of the early studies, determination of handedness was quite casual, and made reference only to writing hand or to a few simple motor activities. Worse yet, the "left-handed" group often included a mixed group of people, some of whom were genuinely left-handed, while others were simply people with familial sinistrality or who performed a few activities with the left hand.

Even now this problem of classifying handedness arises. Many investigators consider only those who use the right hand for virtually all activities to be right-handed, while all others are classified as "left-handers". I am sympathetic to such a classification: it is virtually the only way one is ever going to get a sufficient number of entries in the "left-hander" group to carry out meaningful statistical analyses. At the same time, one should be fully aware of just exactly what one is doing. The subdivision into what should properly be called "right-handers" and "non-right-handed group consists of people who use the right hand consistently. The "left-handed" group consists of a mixture of stongly left-handers, weak left-handers, weak right-handers, and truly ambidextrous individuals. On measures of hand preference or hand proficiency, this group will not only be more "left-handed" than the right-handed group, but they will also be less differentiated with respect to handedness. Thus, if we find differences between the so-called left and right handers, we cannot be certain as to whether it is determined by direction of lateral preference, degree of lateral preference, or both.

Because the data on handedness in aphasics are so weak, there is little to be gained by reanalyzing the existing literature once again. In order to obtain more accurate estimates of language lateralization, we need several thousand verified cases of unilateral brain damage, carefully assessed for aphasia, with each individual given a good test of lateral asymmetry that would permit a distinction between degree and direction.

DICHOTIC LISTENING AND LANGUAGE LATERALIZATION

If the clinical literature is not going to provide a clear statement about how handedness is related to language lateralization, then to what source of data can we turn? In the early 1960's, Doreen Kimura (1961) demonstrated that a relatively simple behavioural test, termed dichotic listening, provided a reasonable assessment of language lateralization in a clinical population, and that it could realistically be used with normal individuals with intact brains. In the dichotic listening procedure, subjects listen to two different speech messages simultaneously, one arriving at each ear, and are asked to report what they hear. In general, normal right-handed subjects are more accurate at reporting the items arriving at the right ear (Kimura, 1967; Bryden, 1982).

Kimura's study showed that neurological patients with known right-hemisphere speech lateralization, as determined by the sodium amytal test (Wada & Rasmussen, 1960), were more accurate on the left ear in dichotic testing, while those with known or presumed left-hemispheric language were more accurate on the right ear, regardless of hand preference. Subsequent studies have verified this relation between dichotic performance and measures of cerebral speech lateralization (Geffen, Traub, & Stierman, 1978; Strauss, Lapointe, Wada, Gaddes, & Kosaka, 1985; Witelson, 1983). By implication, one can therefore use the dichotic listening procedure to assess language lateralization in normal individuals.

It is rather difficult to do a meta-analysis of the dichotic listening data, because different people have reported the data in different ways. However, in a survey of the literature I was able to find 15 studies in which both left and right-handers had been studied, and in which the number of subjects showing right-ear advantages (REAs) and the number showing left-ear advantages (LEAs) had been reported. In Table 2, I have summarized the results of these studies for left and right-handers separately. Overall, 82.2% of the right-handers show an REA, while only 69.3% of the left-handers do. The procedures used in these studies fall into four major categories. In the early studies (e.g., Curry, 1967; Zurif & Bryden, 1969), subjects heard lists of numbers or words and reported as many items as they could in any order they chose. In this group of studies, 84% of the right-handers and 64% of the left-handers

Table 2

Relation of Handedness to Dichotic Performance

	Right-handers		Left-handers		
	N	%REA	N	%REA	χ^2.
Lists	416	83.7	224	63.7	35.21
Pairs	538	81.0	263	72.2	8.01
Monitoring	55	90.9	71	70.4	7.94
TOTAL	1009	82.7	558	68.5	41.92

All χ^2 values p < .01.

den, 1973; Lake & Bryden, 1976) or with focussed attention to one ear (e.g., Bryden, 1986) leads to much the same results as obtained with lists: 81% of right-handers and 69% of left-handers showed REAs. The one procedure that leads to somewhat stronger effects is the dichotic monitoring procedure developed by Geffen (Geffen, et al., 1978). Although the sample is considerably smaller for this test, 91% of right-handers and only 70% of left-handers showed REAs.

If the dichotic listening data were to fully corroborate the clinical data, and if one assumes that people with bilateral language representation would divide equally into REA and LEA categories, then one would expect to find an REA in about 95% of right-handers and in about 70% of left-handers, figures remarkably close to those arising from the dichotic monitoring task, but rather higher than those obtained with the other procedures. While there are several possible reasons for this discrepancy, it does imply that we cannot infer that an observed LEA in a subject necessarily means that language is right hemispheric for that individual.

In the past, I have argued (cf. Bryden, 1978) that procedures in which the strategies of attention and recall are left to the subject are more likely to lead to extraneous influences than those in which the experimenter has control of such variables, and have therefore recommended that procedures involving lists of word or numbers not be used. However, the data of Table 2 indicates that one obtains much the same difference between left and right handers with lists as one does with CV pairs. Thus, both procedures seem to be equally appropriate. One difficulty may lie in the error of measurement, in that test-retest reliability of the dichotic tests runs in the range of .7 to .85 (Bryden, 1986). A second problem may be that clinical tests such as the sodium amytal test assess expressive language functions, while the dichotic tests assess more receptive functions, and the two may not be perfectly correlated. For example, Strauss et al. (1985) have shown that dichotic performance is related to temporal planum size, as assessed by cerebral angiograms, but not to receptive language lateralization determined by sodium amytal testing. The best conclusion, therefore, would seem to be that the observation of an REA in a right-hander provides strong confirming evidence that the individual is left-hemispheric for speech, but that the observation of an LEA provides only a hint of right-hemispheric or bilateral speech and would need to be supported by other evidence.

Many other variables are correlated with handedness, and have been suggested as providing possible information about cerebral organization. Perhaps the most popular of these is familial sinistrality (or FS), defined as the incidence of one or more left-handers among first-order (and sometimes second-order) relatives (cf., Hécaen & Sauguet, 1971). Other potentially relevant variables include sex, eye dominance, footedness, arm crossing position, degree of hand preference or proficiency, and handwriting posture.

OTHER FACTORS RELATED TO HANDEDNESS

Over the past several years, my students and I have carried out a number of experiments using a controlled-attention dichotic listening procedure (cf., Bryden, Munhall, & Allard, 1983). In these studies, we have often acquired information about various potential correlates of left-handedness: familial sinistrality, eye dominance, arm folding position, degree of hand preference and proficiency, and handwriting posture. By now, we have data on 302 young adults, all tested with the same dichotic listening task: 120 trials of CV pairs, with attention directed to the left ear on half the trials and to the right ear on half. The data from this study are summarized in Tables 3 and 4.

First of all, it should be noted that there is a significant difference between left and right-handers in the data. A larger proportion of right-handers show the dichotic right-ear effect, as expected ($\chi^2 = 4.45$, p = .03), and the mean laterality score or lambda value is higher for right-handers (.557) than for left-handers (.166). Lambda is a log odds-ratio measure of the relative superiority of performance on the right ear, and is a statistic intended to correct for individual differences in overall performance level (Bryden & Sprott, 1981).

Table 3 shows the performance of right-handers in this sample. It is clear that degree of hand preference or proficiency, arm crossing, handwriting position, and familial sinistrality do not correlate significantly with language lateralization as measured by this dichotic listening procedure. For eye dominance, there is a significant effect for left- and right-handers combined, with right-eyed people yielding larger lambda values than left-eyed people (p = .04), although this is not statistically significant in either left-handed or right-handed groups alone. To make the picture more complex, there is a strong interaction (p = .015) between eye dominance and familial sinistrality in the right-handers: FS- individuals show the eyedness effect quite strongly, but in FS+ individuals, left-eyed people are actually more lateralized than right-eyed people. Furthermore, there is also a borderline (p = .052) interaction between sex and familial sinistrality: FS- women are less lateralized than FS- men, but the reverse is true with FS+ subjects.

Table 4 shows the comparable data for left-handers. Here, none of the factors investigated is a significant predictor of language lateralization. However, the eye dominance effect is in the same direction as that for right-handers, and at least approaches significance (p = .11). Clearly, the smaller sample size contributes to the lack of statistically significant effects, but in some cases one should not expect to find the same effects in left-handers as in right-handers. For instance, if Levy and Reid (1978) were correct in their assertions about the relevance of handwriting posture,

Table 3

Dichotic Listening Performance and Secondary Measures of Lateral Preference

Right-handers

	N	Mean Lambda	%REA	p
Eye Dominance				
Left	50	0.437	74.0	
Right	172	0.592	76.5	n.s*
Arm Crossing				
Left	58	0.583	75.9	
Right	46	0.393	75.0	n.s.
Handwriting Position				
Upright	212	0.562	75.8	
Inverted	9	0.540	87.5	n.s.
Familial Sinistrality				
FS+	76	0.619	77.0	
FS-	143	0.520	75.5	n.s.

* Eye dominance X Familial Sinistrality: p < .02
  Eye Dominance for both hands combined: p < .04

Table 4

Dichotic Listening Performance and Secondary Measures of Lateral Preference

Left-handers

	N	Mean Lambda	%REA	p
Eye Dominance				
Left	47	0.044	62.8	n.s
Right	33	0.340	65.6	
Arm Crossing				
Left	38	0.117	57.1	n.s
Right	34	0.177	68.8	
Handwriting Position				
Upright	49	0.135	60.9	n.s
Inverted	28	0.261	69.2	
Familial Sinistrality				
FS+	39	0.260	70.3	n.s
FS-	39	0.007	55.6	

the effects would be opposite in left- and right-handers. Similarly, most genetic models of handedness (e.g., Annett, 1985) would lead one to expect larger differences between FS+ and FS- in right-handers than in left-handers.

How do these data relate to other experimental and clinical data? We are certainly not the first to report data on a large sample of normal subjects given a verbal dichotic listening test, although others have not looked at the same variables we have. Probably the most complete study is that of Searleman (1980), who studied 117 left-handers and 256 right-handers with a task very similar to ours. Searleman reported that footedness differentiated subjects better than did handedness, with a fairly high percentage of left-footed left-handers showing an LEA. In addition, Searleman found no clear effect of FS nor of eyedness. He also reported an interaction of handedness and writing position, in the direction postulated by Levy and Reid (1978), and an interaction between handedness and degree of hand preference. In general, this interaction is consistent with the hypothesis that strongly right-handed people are the most lateralized to the left hemisphere and strongly left-handed people the least lateralized.

Other studies have provided variable results. Lake and Bryden (1976) reported that right-handed males were more likely to show an REA than right-handed females (not, as is often claimed, that men showed a larger absolute EA than women). Piazza's (1980) data seem to show the same effect. However, most other large-scale studies have not reported significant sex differences (cf., Briggs & Nebes, 1975; McKeever, Nolan, Diehl, & Seitz, 1984; Hiscock & MacKay, 1985). At the present time, it is probably safe to say that sex differences in dichotic listening performance have not been clearly established. Within the clinical literature, there is also no strong evidence for a simple sex difference in lateralization. There may, however, be within-hemisphere differences between men and women in speech organization. Kimura (1983) has reported that language disorders are much more common in women with anterior damage than with posterior damage, while such differentiation is not true for men. Similarly, Harasymiw, Halper, & Sutherland (1981) have reported that the relative incidence of Broca's aphasia is higher in women than in men. The issue of sex and language lateralization is a complex one (cf., McGlone, 1980), and a proper treatment of the topic would require far more time than I have here.

Familial sinistrality is also a popular variable. We found FS to be significant only in interaction with other variables, such as sex and eye dominance, and such seems to be the case in other dichotic listening studies. Certainly Briggs and Nebes (1976), McKeever et al. (1984), and Orsini, Satz, Soper, and Light (1985) failed to find major effects of FS. However, Piazza (1980) does show a small effect of FS in right-handers, as do Lake and Bryden (1976).

In the clinical literature, Hécaen et al. (1981) suggest that bilateral representation of language functions is more common in FS+ individuals, especially left-handers. My own analysis of the same data (Bryden et al., 1983) would lead to the same conclusion. Thus, the evidence with respect to FS is somewhat equivocal. Much may depend on the specific definition of FS+ employed: some people define familial sinistrality only in terms of parents, others include both parents and siblings, and still others include second-order relatives.

The factor of arm crossing was included largely because of the report by Sakano (1981) that arm crossing and hand folding patterns were predictive of hemispheric asymmetries. We were able to replicate Sakano's data on the distribution of these behaviours in left- and right-handers, but found no association with dichotic lateralization.

Handwriting posture is a variable popularized by Levy and Reid (1978), and much research has been devoted to this topic. The argument is that people who write in an inverted fashion have ipsilateral control of hand skills and of language. Thus right-handed inverters would be right-hemispheric for language, while left-handed inverters would be left-hemispheric for language. One of the attractive points about this proposal is that the distributions come reasonably close to matching those for left- and right-hemispheric language. Note that the hypothesis predicts that inversion will have statistically opposite effects in left- and right-handers. Precisely this interaction was reported by Searleman (1980) In general, however, handwriting posture has been found to be related to visual asymmetries more than to auditory ones (cf., Smith & Moscovitch, 1979; Weber & Bradshaw, 1982), although Tapley and Bryden (1983) found that right-handed inverters frequently exhibited an LEA. Furthermore, clinical studies (Ajersch & Milner, 1983; Strauss, Wada, & Kosaka, 1984) have failed to find any relation between handwriting posture and amytal-assessed speech lateralization.

Finally, a comment is due on Searleman's (1980) finding that footedness was a better predictor of the dichotic REA than handedness. Unfortunately, this is not a factor we have pursued, and so I cannot present any new data. However, most measures of footedness involve relatively gross movements of the whole leg, as in stepping or kicking. Thus "footedness" measures may actually be measures of "leggedness", and may correlate more closely with whole arm movements, as in throwing or hammering. As we shall see, this distinction between fine movements of the distal musculature and gross movements of the more proximal limb muscles may prove to be important (cf. Healey, Leiderman, & Geschwind, 1986).

HANDEDNESS AND RIGHT HEMISPHERE FUNCTIONS

It is easy to assume that the functions of the right hemisphere are related to handedness in much the same way that language functions are, but in reverse. That is, if the incidence of right-hemispheric or bilateral representation of language is higher in sinistrals, so is the incidence of left-hemisphere or bilateral control of visuospatial or musical functions. This may not, in fact, prove to be the case. In the Hécaen and Sauguet (1971) study, for example, comparable visuospatial deficits were observed following unilateral brain damage in left- and right-handers.

There are no large scale surveys of visuospatial or musical deficits following unilateral brain damage that compare to the aphasia studies. However, I have analyzed data provided to me by the late Henri Hécaen to estimate the incidence of visuospatial representation in the two hemispheres. These data are shown in Table 5. As can be seen, the pattern is by no means the mirror image of that observed for language functions.

There is at least reasonable support for this general pattern in other literature. Masure and Benton (1983) found unilaterally damaged left-handers performed much like right-handers on tests of visuospatial ability, and DeRenzi (1982) has also concluded that handedness is not critical to visuospatial lateralization.

Table 5

Handedness and Lateralization of Visuospatial Ability

	Lateralization		
	Left	Bilateral	Right
Right-handers	32%	0%	68%
Left-handers	30%	32%	38%

Data from Bryden et al. (1983).

There are also relatively few large-scale studies of handedness effects on right-hemisphere tasks in normal subjects. Piazza (1980) did find a handedness by visual field interaction in a lateralized face recognition task, but neither she nor Bryden (1986) found such an interaction in a musical dichotic listening task. The data of Table 5 would suggest that there are handedness effects for visuospatial functioning, but that they follow a quite different pattern than do those for language functions. In other words, language and visuospatial functions are not immutably assigned to opposite hemispheres. However, the particular way in which handedness is related to the lateralization of nonverbal functions remains a fruitful area for research.

THE MEASUREMENT OF HANDEDNESS

Some of the confusion in the literature may result from the varying measures of handedness that have been employed, particularly in the clinical literature. It is easy to observe or ask which hand is used for writing. It is more difficult to carry out a systematic assessment of hand preference or hand proficiency. For that matter, it is difficult to agree on just what should be assessed by a measure of handedness, since there is no absolute criterion against which to evaluate a measure of handedness.

In recent years, there have been a number of attempts to develop hand preference inventories that are reliable and easily administered. These have recently been reviewed by Fennell (1985). Perhaps the best known of these is the Edinburgh Handedness Inventory (Oldfield, 1971), although there are a variety of other materials (e.g., Annett, 1985; Bryden, 1977). In general, these questionnaires have two important properties: they are reliable, in the sense that people give the same score upon repeated testing; and they are valid, in that observational data tells us that people actually behave in the way that they claim to on the questionnaire. Some of the questionnaires (e.g., Bryden, 1977) are also unifactorial, in that the item intercorrelations are relatively high, and all questions seem to relate to a single underlying dimension. This may or may not be desirable: Geschwind and Galaburda (1985) have suggested that different systems may control gross arm movements, such as those used in hammering, and fine finger movements, as used in writing or picking up small objects, and Healey et al. (1986) have reported that there are separate proximal and distal factors to a hand questionnaire having a large number of items.

The problem is that questionnaires and factor analyses give you what you put into them. If you don't ask about hammering, you won't find out about hammering; if all of your questions are concerned with fine finger movements, you are unlikely to detect a gross movement factor. Without a good theory of handedness, we don't know quite what to look for, nor how to weight different types of responses.

An interesting possibility arises out of a conflict between work that I have been doing and the work of Marion Annett. In recent years, I have been interested in developing a group test of hand proficiency. Because it was cheap and easy to administer, I chose to use a task in which subjects were asked to place dots in small circles as rapidly as possible. When we looked at the data from a large number of undergraduates, we found the skewed distribution so typical of handedness tests (Tapley & Bryden, 1985). However, we also discovered that this distribution could be decomposed into two normal distributions, one with the mean shifted to the right and one with the mean shifted, although not quite so far, to the left. In other words, we seemed to have evidence for separate distributions for left- and right-handers.

Annett, however, has chosen to use a peg-moving task as a measure of proficiency (Annett, 1985). As Tapley and I were writing our paper, Annett and Kilshaw (1983) carried out a virtually identical analysis of peg-moving scores in a large number of subjects. They also reported two distributions, one shifted to the right and one with a mean of zero, thus supporting Annett's right-shift theory of handedness.

The two sets of data are quite different: there is no simple transformation that will turn one into the other. How then are we to account for the differences? One possibility is that peg-moving is essentially a gross arm-movement, with relatively little finger dexterity involved, although one does have to grasp and release the peg. Annett (1985) reports that this task correlates most highly with the preference questionnaire item dealing with hammering. In contrast, dot-filling is primarily a fine motor task, involving visually-guided movements of the hand and wrist. It is most closely related to the writing hand question on a preference inventory.

Thus, the two proficiency tests may measure somewhat different factors. It is not clear, at the moment, how these two separate factors relate to either clinically-determined language lateralization or to dichotic listening performance. I am encouraged by the fact that Kertesz, Black, Polk, and Howell (1986) have shown a very strong correlation between the Tapley-Bryden proficiency test and anatomical asymmetries as measured by magnetic resonance imaging. It may be, following along with the argument of Geschwind and Galaburda (1985), that distal and proximal proficiency relate to different aspects of brain lateralization. An answer to this question must await further research.

It may not be necessary to postulate separate factors to account for the differences between the Annett and Kilshaw (1983) and the Tapley and Bryden (1985) data. McManus (1985) has shown that the Annett and Kilshaw (1983) data can be quite adequately fit by postulating separate distributions for right- and left-handers. He argues that Annett's (1985) right-shift model is not an appropriate description of the data on human handedness, and that there are distinct subgroups of left- and right-handed people.

THE INHERITANCE OF HANDEDNESS

Any suggestion that familial sinistrality is a moderating variable in determining brain organization implies some genetic basis for handedness. Furthermore, the most generally accepted model of handedness, that of Annett (1985), is a genetic one, postulating separate genes for a right-shift and the absence of a right-shift.

In support of a hereditary basis for left-handedness, the incidence of left-handed offspring when one parent is left-handed is higher than when both parents are right-handed, and the incidence of left-handedness is still higher when both parents are left-handed (cf., Annett, 1985). In opposition to this is the observation that handedness in monozygotic twins follows a binominal distrubution, with many twins being discordant for handedness. Annett (1985) has tried to incorporate this into her right-shift model, but it remains a problem for any simple genetic theory.

Over the past several years, we have been collecting data on hand preference and hand proficiency in families in which we can test three generations. We have trained undergraduate students as our testers, and then asked them to administer our tests to their families. At present, we have collected data on some 110 families in which at least one of the grandparents was testable. By doing this, we hoped to escape from some of the environmental determination of handedness. In most cases, the grandparents did not raise the second generation, and therefore did not have any direct influence on the hand preference of the child.

Table 6

Associations Across Generations for Direction of Hand Preference

	Offspring of Left-handed Elders		Offspring of Right-Handed Elders	
	N	%LH	N	%LH
Grandparents-Children				
Males	14	7.1%	207	12.6%
Females	17	11.8%	305	16.7%
Total	31	9.7%	517	15.0%
Parents-Children				
Fathers-Sons	15	13.3%	155	18.7%
Fathers-Daughters	23	17.4%	236	17.8%
Mothers-Sons	15	20.0%	169	18.3%
Mothers-Daughters	18	16.7%	249	18.5%
Grandparents-Parents				
Fathers-Sons	0	------	19	5.3%
Fathers-Daughters	2	0.0%	33	9.1%
Mothers-Sons	3	0.0%	49	6.1%
Mothers-Daughters	3	0.0%	63	3.2%

Table 7

Associations Across Generations for Degree of Hand Preference

	Offspring of Weakly-handed Elders		Offspring of Strongly-Handed Elders		
	N	%Weak	N	%Weak	χ^2
Grandparents-Children					
Males	82	73.1%	139	53.2%	8.59**
Females	128	68.0%	194	55.7%	4.88*
Total	210	70.0%	333	54.7%	12.70***
Parents-Children					
Fathers-Sons	93	77.4%	77	50.6%	13.32***
Fathers-Daughters	126	66.7%	133	55.6%	3.31
Mothers-Sons	79	79.8%	105	55.2%	12.03***
Mothers-Daughters	109	70.6%	158	54.4%	7.13**
Grandparents-Parents					
Fathers-Sons	8	50.0%	11	36.4%	0.35
Fathers-Daughters	15	33.3%	20	30.0%	0.04
Mothers-Sons	23	69.6%	29	27.6%	9.10**
Mothers-Daughters	20	55.0%	46	19.6%	8.29**

* $p < .05$;   ** $P < .01$;   *** $P < .001$

The results of this study are most interesting. There seems to be relatively little evidence for a genetic determination of direction of handedness (see Table 6). The offspring of left-handed parents or left-handed grandparents are not significantly more likely to be left-handed than the offspring of right-handed parents or grandparents. In these data, the effect is even in the wrong direction, although I believe that to be a result of the relatively small number of left-handed grandparents we have tested to date.

Quite a different picture emerges if one classifies individuals as "strongly" or "weakly" handed according to either the preference inventory or the proficiency test. Here, strongly handed parents or grandparents are far more likely to produce strongly handed offspring than are weakly handed parents of grandparents (Table 7).

These data are in close accord with those reported by Collins (1985) for the mouse. Most mice exhibit a consistent paw preference; they will reliably use the same paw to reach for food. Some mice are very persistent in their paw preferences, and will use the dominant paw even when it it very awkward to do so; others are less persistent, and will shift to the nondominant paw if food is closer to that side. Collins has unsuccessfully tried to breed for paw preference in the mouse, but has found it quite easy to select for degree of pawedness. His data suggest that degree is genetically determined in the mouse, but that direction is not.

Our data are in full agreement with those of Collins. They clearly indicate that degree and direction of handedness are separate factors, and that there is a significant heritability to degree of handedness. Such data suggest that one should be assessing both degree and direction, and that the two factors might relate to quite distinct aspects of cerebral lateralization.

CONCLUSIONS

In this paper, I have tried to provide a general review of the relation between handedness and cerebral organization. It is clear that language lateralization differs in left- and right-handers, although even among left-handers the majority are left-hemispheric for language. Attempts to find other predictive variables, such as handwriting posture, eye dominance, or familial sinistrality have not met with overwhelming success. The data I have presented here suggest that eye dominance and footedness may be worth further examination.

Handedness also relates to the lateralization of right hemisphere functions, though in a different way than to the lateralization of language functions, and not so strongly. However, left-handers, especially those with a history of familial sinistrality, do seem more likely to have bilateral representaion of spatial function.

Many of the discrepencies between studies concerning the relation of handedness and cerebral organization may be resolved by a more sophisticated assessment of handedness and other lateral preference variables. I have presented some evidence to suggest that one should separate fine hand and finger movements from more gross arm movements in defining handedness, and that degree and direction of handedness should be considered as distinct variables.

REFERENCES

Ajersch, A.J., & Milner, B. (1983). Handwriting posture as related to cerebral speech lateralization, sex and writing hand. Human Neurobiol., 2, 143-146.

Annett, M. (1985). Left, right, hand and brain: The right shift theory. Lawrence Erlbaum, Hillsdale, NJ.

Annett, M., & Kilshaw, D. (1983). Right- and left-handed skill II: Estimating the parameters of the distribution of L-R differences in males and females. Br. J. Psychol., 74, 269-283.

Briggs, G.C., & Nebes, R.D. (1976). The effects of handedness, family history and sex on the performance of a dichotic listening task. Neuropsychol., 14, 129-134.

Bryden, M.P. (1973). Perceptual asymmetry in vision: Relation to handedness, eyedness, and speech lateralization. Cortex, 9, 418-435.

Bryden, M.P. (1977). Measuring handedness with questionnaires. Neuropsychol., 15, 617-624.

Bryden, M.P. (1978). Strategy effects in the assessment of hemispheric asymmetry. In Strategies of information processing. (ed. G. Underwood). Academic Press, London.

Bryden, M.P. (1982). Laterality. Academic Press, New York.

Bryden, M.P. (1986). Dichotic listening performance, cognitive ability, and cerebral organization. Canad. J. Psychol., in press.

Bryden, M.P., Hécaen, H., & DeAgostini, M. (1983). Patterns of cerebral organization. Brain and Lang., 20, 249-262.

Bryden, M.P., Munhall, K., & Allard, F. (1983). Attentional biases and the right-ear effect in dichotic listening. Brain and Lang., 18, 236-248.

Bryden, M.P., & Sprott, D.A. (1981). Statistical determination of degree of laterality. Neuropsychol., 19, 571-581.

Carter, R.L., Hohenegger, M., & Satz, P. (1980). Handedness and aphasia: An inferential method for determining the mode of cerebral speech specialization. Neuropsychol., 18, 569-574.

Carter, R.L., Satz, P., & Hohenegger, M. (1984). On the statistical estimation of speech-organization distributions from aphasia data. Biometrics, 40, 937-946.

Collins, R.L. (1985). On the inheritance of direction and degree of asymmetry. In Cerebral lateralization in nonhuman species. (ed. S.D. Glick). Academic Press, Orlando, FL.

Curry, F.K.W. (1967). A comparison of left-handed and right-handed subjects on verbal and non-verbal dichotic listening tasks. Cortex, 3, 343-352.

DeRenzi, E. (1982). Disorders of space exploration and cognition. Wiley, Chichester, UK.

Fennell, E.B. (1985). Handedness in neuropsychological research. In Experimental techniques in human neuropsychology. (ed. H.J. Hannay). Oxford University Press, New York.

Geffen, G., Traub, E., & Stierman, I. (1978). Language laterality assessed by unilateral ECT and dichotic monitoring. J. Neurol., Neurosurg., Psychiat., 41, 354-360.

Geschwind, N., & Galaburda, A. M. (1985). Cerebral lateralization: Biological mechanisms, associations, and pathology: I. A hypothesis and a program for research. Arch. Neurol., 42, 428-459.

Harasymiw, S.J., Halper, A., & Sutherland, B. (1981). Sex, age, and aphasia type. Brain and Lang., 12, 190-198.

Healey, J.M., Liederman, J., & Geschwind, N. (1986). Handedness is not a unidimensional trait. Cortex, 22, 33-54.

Hécaen, H., DeAgostini, M., & Monzon-Montes, A. (1981). Cerebral organization in left-handers. Brain and Lang., 12, 261-284.

Hécaen, H., & Sauguet, J. (1971). Cerebral dominance in left-handed subjects. Cortex, 7, 19-48.

Hiscock, M., & MacKay, M. (1985). The sex difference in dichotic listening: Multiple negative findings. Neuropsychol., 23, 437-440.

Kertesz, A., Black, S.E., Polk, M., & Howell, J. (1986). Cerebral asymmetries on magnetic resonance imaging. Cortex, 22, 107-128.

Kimura, D. (1961). Cerebral dominance and the perception of verbal stimuli. Canad. J. Psychol., 15, 166-171.

Kimura, D. (1967). Functional asymmetry of the brain in dichotic listening. Cortex, 3, 163-178.

Kimura, D. (1983). Speech representation in an unbiased sample of left-handers. Human Neurobiol., 2, 147-154.

Kimura, D. (1983b). Sex differences in cerebral organization for speech and praxic functions. Canad. J. Psychol., 37, 19-35.

Lake, D.A., & Bryden, M.P. (1976). Handedness and sex differences in cerebral asymmetry. Brain and Lang., 3, 266-282.

Levy, J., & Reid, M. (1978). Variations in cerebral organization as a function of handedness, hand posture in writing, and sex. J. exp. Psychol.: Gen., 107, 119-144.

Mateer, C.A., & Dodrill, C.B. (1983). Neuropsychological and linguistic correlates of atypical language lateralization: Evidence from sodium amytal studies. Human Neurobiol., 2, 135-142.

Masure, M.C., & Benton, A.L. (1983). Visuospatial performance in left-handed patients with unilateral brain lesions. Neuropsychol., 21, 179-181.

McGlone, J. (1980). Sex differences in human brain organization: A critical survey. Behav. Brain Sciences, 3, 215-227.

McKeever, W.F., Nolan, D.R., Diehl, J.A., & Seitz, K.S. (1984). Handedness and language laterality: Discrimination of handedness groups on the dichotic consonant-vowel task. Cortex, 20, 509-523.

McManus, I.C. (1985). Right- and left-hand skill: Failure of the right shift model. Br. J. Psychol., 76, 1-16.

Oldfield, R.C. (1971). The assessment and analysis of handedness: The Edinburgh Inventory. Neuropsychol., 9, 97-113.

Orsini, D.L., Satz, P., Soper, H.V., & Light, R.K. (1985). The role of familial sinistrality in cerebral organization. Neuropsychol., 23, 223-232.

Piazza, D.M. (1980). The influence of sex and handedness in the hemispheric specialization of verbal and nonverbal tasks. Neuropsychol., 18, 163-176.

Rasmussen, T., & Milner, B. (1977). The role of early left-brain injury in determining lateralization of cerebral speech functions. Ann. N. Y. Acad. Sci., 299, 355-369.

Sakano, N. (1982). Latent left-handedness: Its relation to hemispheric and psychological functions. VEB Gustav Fischer Verlag, Jena DDR.

Searleman, A. (1980). Subject variables and cerebral organization for language. Cortex, 16, 239-254.

Segalowitz, S.J., & Bryden, M.P. (1983). Individual differences in hemispheric representation of language. In Language functions and brain organization. (ed. S.J. Segalowitz). Academic Press, New York.

Smith, L.C., & Moscovitch, M. (1979). Writing posture, hemispheric control of movement, and cerebral dominance in individuals with inverted and noninverted hand postures during writing. Neuropsychol., 17, 637-644.

Strauss, E., Lapointe, J.S., Wada, J.A., Gaddes, W., & Kosaka, B. (1985). Language dominance: Correlation of radiological and functional data. Neuropsychol., 23, 415-420.

Strauss, E., & Wada, J. (1983). Lateral preferences and cerebral speech dominance. Cortex, 19, 165-177.

Strauss, E., Wada, J., & Kosaka, B. (1985). Visual laterality effects and cerebral speech dominance determined by the carotid amytal test. Neuropsychol. 23, 567-570.

Tapley, S.M., & Bryden, M.P. (1985). A group test for the assessment of performance between the hands. Neuropsychol., 23, 215-222.

Wada, J., & Rasmussen, T. (1960). Intracarotid injection of sodium amytal for the lateralization of cerebral speech dominance. J. Neurosurg., 17, 266-282.

Weber, A.M., & Bradshaw, J.L. (1981). Levy and Reid's neurological model in relation to writing hand/posture: An evaluation. Psychol. Bull., 90, 74-88.

Witelson, S. F. (1983). Bumps on the brain: Right-left asymmetry as a key to functional lateralization. In Language functions and brain organization. (ed. S. J. Segalowitz). Academic Press, New York.

Zurif, E.B., & Bryden, M.P. (1969). Familial handedness and left-right differences in auditory and visual perception. Neuropsychol., 7, 179-187.

6
Diversities in Right-handers in Left-hemisphere Processing

Jerre Levy and Wendy Heller

INTRODUCTION

Clinical studies of neurological patients with unilateral brain damage (Benson and Geschwind, 1969; Leischner, 1969; Lhermitte and Gautier, 1969) and examinations of epileptic patients using the Wada test (Rasmussen and Milner, 1977) have shown that in almost all right-handers, the left hemisphere is specialized for language comprehension and expression, both oral and written. However, when normal right-handers are examined on behavioral tasks that are designed to measure asymmetric cerebral functioning (Kimura, 1961, 1963, 1964, 1966), they manifest considerable variability in lateralized verbal performance, a variability that Tzeng and Hung (1985) perceive as a "discrepancy" (p. 124) in relation to the clinical data.

In order to resolve the discrepancy, Tzeng and Hung (1985) discuss a number of possible sources of the added variance that are unrelated to hemispheric specialization itself (e.g., a predominance of ipsilateral over contralateral auditory pathways, individual differences in characteristic asymmetries of hemispheric arousal, strategy or attentional variations that derive from experience, personality factors, or cognitive styles, etc.). They evidently believe that the clinical data rule out the possibility that right-handers might be variable in left-hemisphere verbal specialization per se. Yet, the clinical studies merely demonstrate that almost all right-handers have left-hemisphere specialization for language, but they do not imply that the level of specialized linguistic abilities in the left hemisphere is invariant across the dextral population.

Indeed, if the left hemisphere is the language specialist in essentially all right-handers, then given the fact that right-handers are variable in verbal abilities (e.g., vocabulary, fluency, verbal reasoning, etc.), it follows directly that there is a variability in the degree of linguistic specialization and

71

competency of the left side of the brain. Chase. Fedio, Foster.
Brooks, Di Chiro, and Mansi (1984) have shown, in fact, that
verbal IQ has a rather strong correlation with cerebral metabolism
in the left temporal and temporo-parietal areas. In brief, the
fact of reliable individual differences in verbal ability and the
fact that it is the left hemisphere that is the language
specialist necessarily mean that the left hemisphere's degree of
specialization and competency in verbal processing is variable
across people. The sole question is whether this variability has
effects on lateralized verbal performance.

Since many factors that are unrelated to linguistic
processing itself can affect lateralized performance on verbal
tasks (see Tzeng and Hung, 1985), it is necessary to define
performance characteristics that are specifically linguistic in
nature if this question is to be properly examined. When a word
or syllable is tachistoscopically flashed, the stimulus
information can either be encoded letter-by-letter in a serial
attentional scan or as a phonetic image in which attention is
almost globally distributed over the word (see Johnston, 1981).
When nonpronounceable letter strings are presented, only the
former encoding operation is available, and in consequence, people
are superior at identifying a target letter that is embedded in a
word as compared to string (the word-superiority effect) (Smith
and Spoehr, 1974). Also, since the right hemisphere has no
phonetic capacity (Levy and Trevarthen, 1977; Zaidel, 1978), it is
compelled to encode words in the same manner as it encodes
nonpronounceable strings, and as Krueger (1975) showed, there is
no word-superiority effect at all in the left visual field (LVF),
but a quite strong word-superiority effect in the right visual
field (RVF).

The differing encoding operations of the two hemispheres can
be seen from an examination of Krueger's (1975) Figure 3 for
vertically oriented stimuli. In the RVF, reaction times (RTs) for
identifying targets in words are no longer for targets in the
third position down than for targets in the top position, but for
targets in nonpronounceable strings, RTs are almost 140 msec
longer for target letters in the third position down as compared
to targets in the top position. In the LVF, RTs are longer for
targets in the third position down as compared to the top
position, regardless of whether stimuli are words or strings.
Thus, for words in the LVF, RTs to targets in the third position
are about 180 msec longer than to targets in the top position. In
brief, strings in the RVF and both words and strings in the LVF
are encoded in a letter-by-letter fashion, in which each letter is
encoded in a serial attentional scan, but words in the RVF are
encoded phonetically, and attention is distributed almost globally
over the stimulus.

The encoding differences of the two hemispheres can also be
observed in the very different error patterns for the two visual

fields when vertically oriented syllables are flashed for
identification. Levy, Heller, Banich, and Burton (1983)
classified erroneous responses as top-letter errors when the top
letter was missed and the bottom was correct, as bottom-letter
errors when the bottom letter was missed and the top was correct,
and as other-type errors. Their central findings were that
top-letter errors were equally common in the two visual fields,
bottom-letter errors were much more prevalent in the LVF than RVF,
and the difference in bottom- and top-letter errors (bottom minus
top) was much larger in the LVF than RVF. The difference in
frequency of bottom- and top-letter errors therefore provides a
direct index of the Linguistic Encoding Strategy Score (LESS): as
LESS becomes smaller, linguistic encoding increases.

Our question was whether variations among right-handers in
left-hemisphere specialization and competency for verbal
processing, as measured on standard psychometric instruments of
verbal ability, would covary negatively with the value of LESS,
especially and perhaps only for the RVF. If so, then part of the
variance in lateralized verbal performance must be attributed to
individual differences in left-hemisphere specialization for
language, and not only to factors that are independent of
specialization itself. In sections to follow, we describe a study
of right-handed women that examined this issue.

METHOD

Subjects

The subjects were 32 normal right-handed women and girls,
ranging in age from 15 through 30. Handedness was assessed by an
8-item questionnaire, and only those subjects were selected who
showed a right-hand preference on all 8 items. All subjects had
normal or corrected-to-normal vision, as determined by the Titmus
Vision Tester. These subjects were selected from a larger sample
of 45 on the basis of their overall visual-field asymmetry on a
tachistoscopic syllable-identification task (to be described), and
were restricted to those who showed better overall RVF than LVF
performance. The restriction of our subjects to those with a RVF
advantage on the syllable test was a conservative measure that was
undertaken to assure that all our subjects had the typical
direction of lateral specialization for language, since the
central question was whether variations in the verbal
specialization of the left hemisphere, as indexed by measures of
verbal ability, had affects on linguistic processing on verbal
laterality tasks.

Apparatus, Stimuli, and Materials

A Gerbrands 3-channel tachistoscope was used for stimulus

presentation. A fixation card, consisting of a black card with white dot, was placed in one channel. A second channel contained the stimulus cards. Stimulus cards were black with a white center digit and a laterally displaced and vertically oriented consonant-vowel-consonant (CVC) syllable. The syllable was 1 degree of visual angle lateral to fixation. There were 24 practice cards, 12 with a syllable to the right and 12 with the syllable to the left, in addition to 120 experimental cards with 60 different syllables. Each syllable occurred once on the right and once on the left.

Other materials consisted of the Vocabulary and Similarities subtests of the Wechsler Adult Intelligence Scale (WAIS) or the Wechsler Intelligence Scale for Children - Revised (WISC-R).

Procedure

All subjects were first tested for vision and handedness, and then were given a test of verbal fluency in which they were asked to name as many words as they could in 1 minute that began with the letter "d". The numbers of words generated by subjects were then Z-transformed over the group to mean zero and standard deviation unity. Subjects were then given the Vocabulary subtest from the WAIS or WISC-R, followed by the Similarities subtest. Raw scores on these two subtests were first converted to age-appropriate scaled scores, and each set of scaled scores was then Z-transformed to mean zero and standard deviation unity. We therefore had three verbal measures, one of word knowledge (vocabulary), one of verbal reasoning (similarities), and one of word production (fluency) that were all standardized to the same scale. It might be noted that if a prorated verbal IQ is derived from the Similarities and Vocabulary subtests, the standard deviation across subjects was 14.8, which is very close to the standardization value. The variation of our subjects in verbal ability was, therefore, the same as that for the population at large.

At a second session, subjects were brought in for tachistoscopic testing. The fixation field was present continuously except during stimulus presentation. Subjects were instructed to depress a pedal once they were on fixation, which presented the stimulus field after a 1 sec delay. Subjects first named the center digit, then pronounced the lateral syllable. If the center digit was missed on any of the experimental trials, the trial was discarded and replaced. After a practice run, experimental trials were presented in two blocks of 60 trials/block. Based on practice performance, exposure durations were set individually for each subject, but never over 200 msec, in an attempt to control overall performance at around 50% accuracy.

For each visual field, trials on which errors were made were classified into those in which the bottom letter was wrong, but the top was correct (bottom-letter errors), those in which the top letter was wrong, but the bottom was correct (top-letter errors), and all other error types (other-type errors).

RESULTS

For each subject, the difference in frequency of bottom- and top-letter errors was calculated for each visual field as an index of the Linguistic Encoding Strategy Score (LESS). Recall that linguistic encoding increases as the value of LESS becomes smaller. For the LVF, the mean value of LESS was 17.2, with a standard deviation of 7.8, and for the RVF, the mean value of LESS was 11.5, with a standard deviation of 5.2. The difference between LESS for the two fields (5.7) had a standard deviation of 7.4 and a standard error of 1.3, which shows that LESS was quite significantly smaller in the RVF than LVF [$t(31) = 4.252$, $p <$ .0001, 1-tailed], although the magnitude of the asymmetry was quite variable across subjects.

Our major question concerned whether the value of LESS for the two fields was correlated with our measures of verbal function (vocabulary, similarities, fluency) or with some average of the Z-transformed values of these measures. If variations among subjects in LESS, especially for the RVF, derive from real individual differences in the verbal functioning of the left hemisphere, then LESS for the RVF, but not LESS for the LVF, should be significantly negatively correlated with our measures of verbal ability.

First, neither the zero-order correlation of Similarities with LESS-LVF [$r = -0.144$, $t(30) = -0.795$, ns] nor that with LESS-RVF [$r = -0.280$, $t(30) = -1.595$, ns] was significant. However, both Fluency and Vocabulary were significantly correlated with LESS in both fields. For Fluency, the correlation with LESS-LVF was -0.326 [$t(30) = -1.889$, $p <$ .036, 1-tailed] and with LESS-RVF was -0.447 [$t(30) = -2.737$, $p =$ .0052, 1-tailed]. For Vocabulary, the correlation with LESS-LVF was -0.429 [$t(30) =$ -2.601, $p <$ .0075, t-tailed] and and with LESS-RVF was -0.539 [$t(30) = -3.505$, $p =$ .0012, 1-tailed]. These zero-order correlations tell us very little, however, with respect to the contributions of each cerebral hemisphere to the variance in verbal ability since there was a low, but significant, between-field correlation in LESS scores [$r = 0.366$, $t(30) =$ 2.157, $p = 0.042$, 2-tailed].

The between-field correlation implies that, at least in part, the covariation of LESS with verbal ability for one or both fields derives from the effects of the opposite visual field. In particular, since the final verbal output on the tachistoscopic task depends on the left hemisphere, the left-hemisphere capacity to reconstruct callosally received information from the LVF may mediate the correlation between LESS-LVF and verbal ability. Further, if this capacity of the left hemisphere covaries with its capacity to reconstruct directly received information from the RVF, then a significant portion of the variance in verbal ability that is predicted by LESS-LVF may, in fact, derive from the covariance with LESS-RVF. In other words, there may be little independent contribution from the right hemisphere. In order to understand the underlying relations, multivariate procedures must be used.

We performed three analyses, in all of which LESS-LVF and LESS-RVF were multiple predictors of verbal ability. Note that in such analyses, the predictor variables are effectively combined into a single score that is the weighted sum of the two predictors. The weights are so assigned that the correlation between the predictors and the dependent variable is maximized, and the weights determine the relative contributions of each predictor in accounting for variance in the dependent variable. The weighted sum of LESS for the two fields, with each in standard units, provides the Hemispheric score. In one analysis, the dependent variable was the Fluency measure, and in a second analysis, the dependent variable was Vocabulary. In the third analysis, a canonical correlation, the dependent variable was a general Word Score composed of the weighted sum of the Vocabulary and Fluency measures, in which the weights maximized the correlation of the Word Score with the predictors. Table 1 shows the Beta coefficients for each of the two predictors for each dependent variable, the multiple correlations, and the proportions of variance in the dependent variables that are accounted for by the independent effects of LESS-LVF, LESS-RVF, and by the covariance between the two fields.

As Table 1 shows, the multiple correlation was significant for all dependent variables, and in each case, the independent contribution of LESS-RVF was significant, the independent contribution of LESS-LVF was not, and the covariance was greater than the independent variance of LESS-LVF. The variance explained by the independent effects of LESS-RVF is almost certainly an underestimate of the true effects of left-hemisphere functioning. On theoretical grounds, it is reasonable to attribute the covariance to left-hemisphere processes on the supposition that it reflects a positive covariation between the left-hemisphere's ability to reconstruct callosally received information from the LVF and directly received information from the RVF. Thus, the sum of the variance predicted by the independent effect of LESS-RVF and the variance predicted by the covariance between fields provides a measure of the true left-hemisphere effect. For Fluency, Vocabulary, and the Word score, the sums are 19.53%, 27.63%, and 30.95% of the variance, respectively.

TABLE 1. Multiple correlation analysis in which LESS-LVF
and LESS-RVF are predictors of verbal ability.

	BETA COEFFICIENTS		
	FLUENCY	VOCABULARY	WORD SCORE
LVF	-0.1871	-0.2686	-0.2703
RVF	-0.3787	-0.4364	-0.4660
	VARIANCE PREDICTED		
LVF	3.50%	7.21%	7.30%
RVF	14.34%*	19.04%**	21.69%***
LVF x RVF	5.19%	8.59%	9.22%
TOTAL	23.03%*	34.84%***	38.21%***
	CORRELATION		
	-0.4799*	-0.5903**	-0.6181***

* $p < .05$, ** $p < .02$, *** $p < .01$

Figure 1 shows the scatter plot between the Hemispheric Score
and the Word Score, with each standardized to mean zero and
variance unity. This standardization means that with the two LESS
measures each expressed in Z-score units, the weight assigned to
each predictor in deriving the Hemispheric Score is equal to its
Beta coefficient divided by the correlation between the predictors
and dependent variables. Similarly, with Vocabulary and Fluency
each expressed in Z-score units, the weight of each dependent
variable in deriving the Word Score is its Beta coefficient
(-0.4722 for Vocabulary, -0.2192 for Fluency) divided by the
correlation between the predictors and dependent variables. A
frequency analysis of the number of subjects in each of the four
categories, defined by whether they are above or below the mean in
each measure, shows that the distribution is highly significant in
the predicted direction (p = .0021, 1-tailed).

DISCUSSION

Our analyses show that the difference in frequency of bottom-
and top-letter errors in the RVF, which Levy et al. (1983) and we
interpret as an index of the level of linguistic encoding, is
quite significantly correlated with certain standard measures of
verbal ability, and in particular, Vocabulary, Fluency, and a

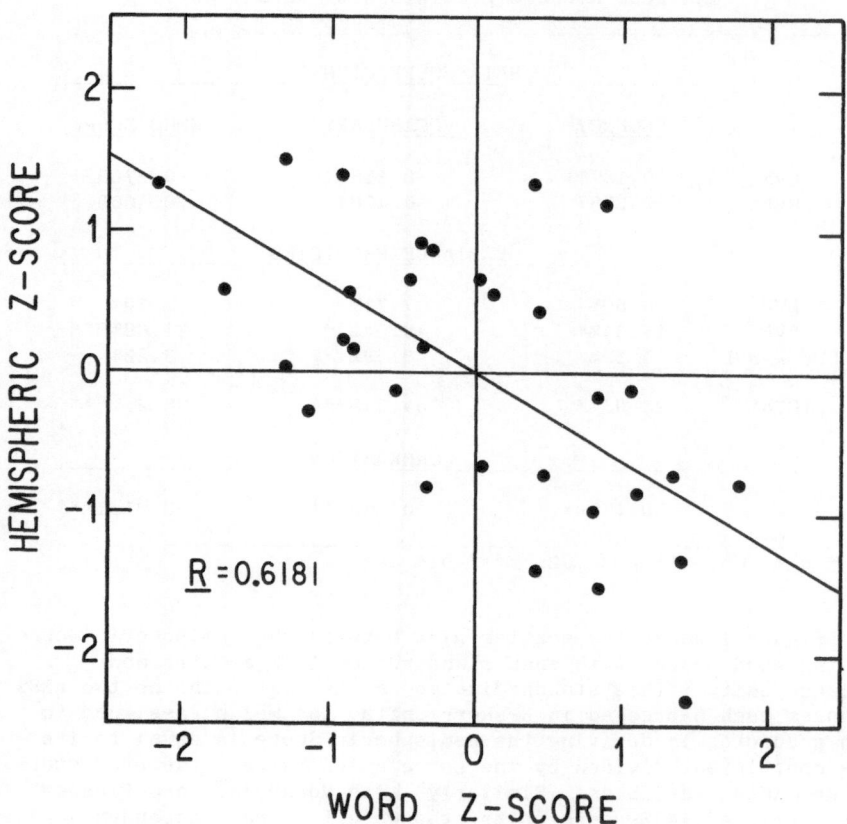

FIGURE 1. Scatter plot between the Hemispheric Score, as
the optimally weighted sum of LESS-LVF and
LESS-RVF, and the Word score, as the optimally
weighted sum of Vocabulary and Fluency.

composite measure of the two. These measures, as well as the linguistic encoding index itself, all reflect, to one degree or another, the capacity to generate, store, and access words as phonetic representations. It is not irrelevant that the Similarities measure was unrelated to LESS.

Unlike the Vocabulary and Fluency measures, which both reflect word representation itself. Similarities predominantly indexes the ability to extract relations among meanings, and in principle, at least, the test could equally well be administered by presenting pictures of the pairs of items whose similarity is to be determined. Although the Similarities test certainly indexes certain aspects of verbal functioning, it does not index those aspects that affect how well a briefly flashed syllable is linguistically encoded and represented phonetically. Evidently, however, the Vocabulary and Fluency tests do index these aspects of verbal processes.

The complete absence of any relation between verbal ability and the independent variance of LESS-LVF and the presence of a strong relation with LESS-RVF shows that in our sample, subjects were homogeneous in having left-hemisphere specialization for language, precisely as the clinical data would predict. These data, however, do not address the issue of individual differences in the level of linguistic specialization and competency of the left side of the brain, as revealed in standard measures of verbal ability, nor the issue of whether such individual differences affect lateralized performance. Our findings demonstrate that beyond the many other variance-producing factors on lateralized tasks that Tzeng and Hung (1985) discuss, and which are independent of hemispheric specialization itself, diversities in the linguistic organizational properties of the left hemisphere also produce considerable variation among right-handers in lateralized tachistoscopic performance. This conclusion, rather than being in conflict with clinical findings, simply elaborates upon them. We suggest, in fact, that since variations in verbal ability indicate variations in left-hemisphere organization, the premorbid ability profile in neurological patients would, to some extent, predict the nature of symptoms following left-hemisphere damage.

It is relevant that differences in the patterns of verbal ability in men and women, in which males surpass females in mathematical reasoning and females surpass males in verbal fluency and speed of reading (see Frieze, Parsons, Johnson, Ruble, and Zellman, 1978) co-occurs with a gender difference in left-hemisphere organization. Kimura (1983) found that the aphasia frequency was equal for men and women with anterior left-hemisphere damage, and was equal to the aphasia frequency for men with posterior left-hemisphere damage. However, in women with posterior damage, aphasia was rare, and much rarer than in either women with anterior or males with posterior damage. If

between-sex differences in verbal ability are associated with
between-sex differences in left-hemisphere organization and in
symptoms following left-hemisphere damage, then similar relations
might hold for individual differences within genders.

One caution concerning our observations derives from the fact
that our sample was restricted to female subjects, and given the
gender differences in the organization of language, no
generalizations can be made with respect to males. It is
possible, for example, that even if the mechanisms used by men and
women to generate phonetic representations of RVF syllables are
identical, their verbal abilities, as assessed on standard
instruments, reflect different underlying processes. This
possibility does not seem unlikely since the early speech
development of girls, but not that of boys, is correlated with
later verbal and intellectual ability (Cameron, Livson, and
Bayley, 1967; Moore, 1967), which indicates that the genders
differ in the mechanisms of language function. The nature of the
relations between verbal ability and linguistic encoding on
lateralized tachistoscopic tasks in male subjects is a matter for
future investigation.

REFERENCES

Benson, D. F. and Geschwind, N. (1969). The alexias. In
Handbood of Clinical Neurology, Vol. 4, Disorders of Speech,
Perception, and Symbolic Behaviour. (eds. P. J. Vinken and
G. W. Bruyn). North-Holland, Amsterdam.

Cameron, J., Livson, N., and Bayley, N. (1967). Infant
vocalizations and their relationship to mature
intelligence. Science, 157, 331-332.

Chase, T. N., Fedio, P., Foster, N. L., Brooks, R., Di
Chiro, G., and Mansi, L. (1984). Wechsler adult intelligence
performance: Cortical localization by fluordeoxyglucose F 18
- positron emission tomography. Arch. Neurol., 41,
1244-1247.

Frieze, I. H., Parsons, J. E., Johnson, P. B., Ruble, D. N.,
and Zellman, G. L. (1978). Women and Sex Roles. Norton, New
York.

Johnston, J. C. (1981). Effects of advance precuing of
alternatives on the perception of letters alone and in
words. J. Exp. Psychol.: Hum. Percept. Perform., 7,
560-572.

Kimura, D. (1961). Cerebral dominance and the perception of
verbal stimuli. Can. J. Psychol., 15, 156-165.

Kimura, D. (1963). Speech lateralization in young children as determined by an auditory task. J. Comp. Physiol. Psychol., 56. 899-902.

Kimura, D. (1964). Left-right differences in the perception of melodies. Quar. J. Exp. Psychol., 16, 355-358.

Kimura. D. (1966). Dual functional asymmetry of the brain in visual perception. Neuropsychologia, 4, 275-285.

Kimura. D. (1983). Sex differences in cerebral organization for speech and praxic functions. Can. J. Psychol., 37, 19-35.

Krueger, L. E. (1975). The word-superiority effect: Is its locus visual-spatial or verbal? Bull. Psychonom. Soc., 6, 465-468.

Leischner. A. (1969). The agraphias. In Handbood of Clinical Neurology, Vol. 4, Disorders of Speech. Perception, and Symbolic Behaviour. (eds. P. J. Vinken and G. W. Bruyn) North-Holland. Amsterdam.

Levy. J., Heller. W., Banich, M. T., and Burton, L. A. (1983). Are variations among right-handed individuals in perceptual asymmetries caused by characteristic arousal differences between hemispheres? J. Exp. Psychol.: Hum. Percept. Perform . 9, 329-359.

Levy, J. and Trevarthen, C. (1977). Perceptual, semantic, and phonetic aspects of elementary language processes in split-brain patients. Brain, 100, 105 118

Lhermitte, F. and Gautier, J.-C. (1969). Aphasia. In Handbook of Clinical Neurology, Vol. 4, Disorders of Speech, Perception, and Symbolic Behaviour. (eds. P. J. Vinken and G. W. Bruyn). North-Holland, Amsterdam.

Moore. T. (1967). Language and intelligence: A longitudinal study of the first 8 years. Hum. Devel.. 10, 88-106.

Rasmussen. T. and Milner. B. (1977). The role of left-hemisphere injury in determining lateralization of cerebral speech functions. In Annals of the New York Academy of Sciences, Vol. 299. Evolution and Lateralization of the Brain. (eds. S. J. Dimond and D. A. Blizard). New York Academy of Sciences, New York.

Smith, E. E. and Spoehr, K. T. (1974). The perception of printed English: A theoretical perspective In Human Information Processing: Tutorials in Performance and Cognition. (ed. B. H. Kantowitz). Erlbaum. Potomac, Md.

Tzeng, O. J. and Hung, D. L. (1985). From hemispheric to perceptual asymmetry: From where has all the variance come? In Individual Differences in Cognition, Vol. 2. (ed. R. Dillon). Academic Press, New York.

Zaidel, E. (1978). Concepts of cerebral dominance in the split brain. In Cerebral correlates of conscious experience. (eds. P. A. Buser and A. Rouguel-Buser). Elsevier/North Holland, Amsterdam.

ACKNOWLEDGEMENTS

We thank the Spencer Foundation for support of this research.

7
Normal Variation in Human Brain Organization: Relation to Handedness, Sex and Cognitive Abilities

Richard A. Harshman and Elizabeth Hampson

If the brains of some left handers are organized differently from the brains of right handers, might not their patterns of cognitive abilities also be different? While this seems plausible, it has been surprisingly difficult to demonstrate; previous findings have been inconsistent and often contradictory. Nonetheless, we will present evidence that substantial handedness-related cognitive differences do exist, and that previous inconsistencies can be resolved by taking other characteristics of the individual into account. The most interesting implications of this evidence go beyond the question of mere handedness effects. It turns out that we may have been operating under overly restrictive assumptions about the functional structure of the "normal" brain.

In this short paper we can only provide an overview of the evidence for these suggestions. For details, the reader is referred to the original articles by Harshman, Hampson, & Berenbaum (1983) and Harshman, Hampson, & Lundy (in preparation).

BACKGROUND

It is well-established that left handed persons are more likely than right handed persons to have some language functions vested in the right cerebral hemisphere (e.g. Milner, 1974). However, those left handers with unusual localization of language are not simply "mirror images" of the typical right handed person, in that the left hemisphere does not necessarily subserve those functions usually subserved by the right hemisphere in right handers (Bryden, Hecaen, & DeAgostini, 1983). Nevertheless, there is some evidence that lateralization of nonverbal functions may also differ in some left handers (Hecaen, DeAgostini, & Monzon-Montes, 1981), and even callosal interconnections may show differences according to handedness (Witelson, 1985). In all likelihood, multiple different brain patterns are associated with left handedness.

83

84

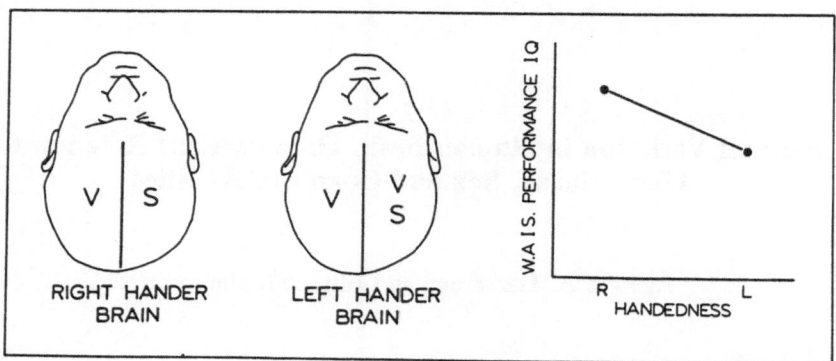

Figure 1. Brain organization and spatial ability in right and left handers, according to Levy (1969).

In 1969, Jerre Levy suggested that left handedness might be associated with lowered spatial ability (Levy, 1969). She postulated that the evolutionary basis for the separation of verbal and spatial abilities into different hemispheres is that they require different kinds of neural architecture, which do not coexist efficiently in the same hemisphere. Differential specialization of the two hemispheres allows each type of ability to develop optimally. In some left handers, however, this specialization is incomplete. As a result, the intrusion of the structural organization optimal for language into the right hemisphere might reduce the neural space that otherwise could be devoted to spatial processing, resulting in a decrement in spatial skills. This has come to be known as the "crowding" hypothesis (see Figure 1).

To test her prediction, Levy (1969) compared scores on the Verbal and Performance subscales of the Wechsler Adult Intelligence Scale in a group of left handed graduate students and a comparable group of right handed graduate students. As predicted, the left handers' Performance scores were depressed relative to their Verbal scores, and relative to the Performance IQ of the right handers. Moreover, in a reanalysis of her data, Levy (1974) also reported that the left handers were marginally superior to the right handers in their Verbal scores.

Subsequent attempts to replicate and extend Levy's (1969) findings produced frustrating inconsistencies. Because it had been suggested that male and female brains might not have the same typical pattern of lateralization (Bryden, 1979; Harshman & Remington, 1976), some investigators began to look at handedness effects in males and females separately. Levy and Gur (1980) revised and extended Levy's earlier reasoning to predict different effects for males and females.

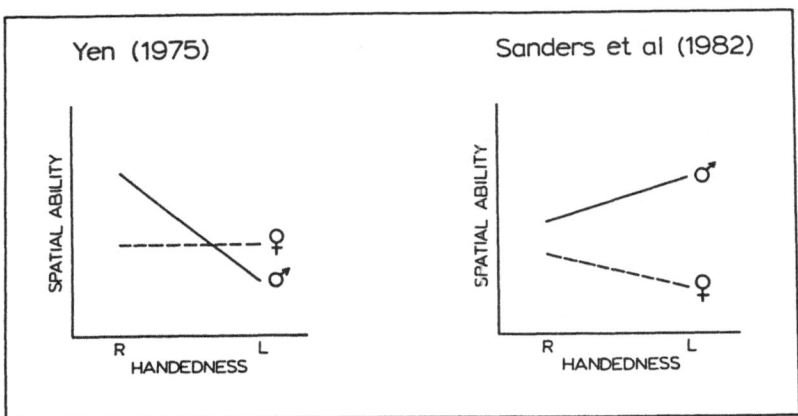

Figure 2. Conflicting results of two large sample studies.

However, some very large-scale studies showed no effect of handedness on cognitive functioning (Hardyck, Petrinovich, & Goldman, 1976), and other studies continued to show inconsistent patterns. This prompted some observers to suggest that the inconsistent handedness effects were merely Type I errors. However, some of these contradictory results were obtained with such large samples that an explanation in terms of Type I errors seems most unlikely. For example, Yen (1975) tested almost 2500 high school students, and Sanders, Wilson, & Vandenberg (1982) analyzed data on almost 900 individuals, yet these two studies obtained opposite patterns of sex and handedness effects for performance on spatial tasks (see Figure 2). More recently, Inglis and Lawson (1984) found minor sex differences and no effect of handedness on Performance IQ scores in the 1880 men and women who made up the WAIS-R standardization sample.

A MULTI-SAMPLE STUDY

In an attempt to shed light on the inconsistencies in the literature, Harshman, Hampson, and Berenbaum (1983) undertook a large multi-sample analysis of individual differences in cognitive abilities and brain organization (as inferred from dichotic listening scores). We examined data from three fairly large samples, collected for three different purposes. One study was carried out at the University of Western Ontario (UWO sample, N=195), one at the State University of New York at Stony Brook (SUNY sample, N=174), and one at the University of California, Berkeley (UC sample, N=430). Details of methodology and a description of the cognitive tests employed can be found in Harshman et al. (1983), Springer & Searleman (1978), and Watkins & Meredith (1981), for the three samples.

In all, we examined 41 different cognitive variables, including 12 different tests of spatial abilities, as well as multiple measures of various verbal, perceptual, and other abilities. Because our original hypotheses were formulated in terms of differences in spatial abilities, our focus in this paper will be largely on the spatial tests. Dichotic listening scores were available for two of the three samples (UWO and SUNY), and for a very small subset of the UC sample (as tested by Berenbaum, 1978). The dichotic listening data allowed us to examine the hypothesis that variations in brain organization might underlie some of the sex and handedness differences in abilities that we observed. Finally, subjects were classified according to their sex, handedness, familial sinistrality, and other grouping variables.

Initial Results

In our preliminary analyses, we tested for independent effects of sex and handedness on each cognitive variable, as well as for sex by handedness interactions, using analysis of variance techniques. The disappointing result was that we in effect "replicated the inconsistencies" in the literature. There were no main effects of handedness. The UWO sample yielded some potentially promising sex by handedness interaction patterns (e.g. DAT Space Relations), but the UC sample, which was twice as large, did not replicate those patterns. The SUNY sample was intermediate, with some borderline effects.

This raised the question of whether there were systematic differences between the samples that might somehow account for these disparities. One likely difference between the UWO and UC samples was in the level of general intelligence. Presumably, the intellectual requirements involved in admission and progress in the university would bias the UWO sample in favor of higher reasoning ability, whereas the UC sample, which was comprised of newly married couples obtained from marriage license records in two California counties, would be more representative of the full range of intellectual variations found in the general population.

To explore the effects of differences in reasoning ability, we divided each of the three samples into high- and low-reasoning subsamples by splitting them at the median on a test of reasoning ability. Fortunately, one test of reasoning ability had been administered to each sample: the ETS Inference Test to UWO, ETS Nonsense Syllogisms to SUNY, and modified Raven Progressive Matrices to UC. Because we had 3 different tests, we were forced to split each sample at its own median, rather than adopting a common level for all three groups. This undoubtedly resulted in a higher cutpoint in the UWO sample than in the UC sample, therefore the lower half of the UWO sample might better be viewed as a mixed group.

We then repeated the same analyses of variance as before, but performed them separately in each subsample, looking for effects of handedness, sex, and possible interactions. In other analyses, we added reasoning ability as a third factor and looked for three-way interactions in the sample as a whole.

Results after Splitting on Reasoning Ability

Analyzing the above- and below-median reasoning groups separately produced striking results. Similar patterns began to emerge in all three samples. The general pattern for spatial ability is exemplified by the UC principal component scores on the spatial tests (see Figure 3a). The effects of handedness interact with sex, but the interaction has the opposite form in the higher reasoning half of the sample (top diagram) as compared to the lower reasoning half (bottom diagram). When all the subjects are considered together, these effects "cancel out"; yet in the separate half-samples, the effects are sometimes as large as .8 standard deviations.

Similar results were found with all three samples across different spatial tests, e.g. ETS Card Rotations, Shepard-Metzler Mental Rotations, DAT Space Relations, PMA Spatial Relations, etc. The pattern for the high reasoning half of each sample was most consistent: left handed males performed worse than right handed males in all 15 tests across all three samples; left handed females performed better than right handed females 12 out of the 15 times (Harshman et al., 1983). This pattern tended to reverse in the lower-reasoning subsamples. As might be expected, the clearest reversal occurred in the low reasoning half of the UC sample, (newly married couples), with slightly less clear-cut results in the low-reasoning half of the partly-student SUNY sample. The UWO sample, being more homogeneous in reasoning level, showed relatively little moderating effect of reasoning ability on spatial test performance. The low-reasoning half of the sample tended to resemble the high-reasoning half of the sample in some of the two-way interaction patterns found. Again, this homogeneity probably explains why the UWO sample was the only one to show a significant sex by handedness interaction in our initial analyses.

RECONCILING PAST CONFLICTS IN THE LITERATURE

The interaction with reasoning ability helps to explain many of the past inconsistencies in the literature. Let us take, as an example, the four studies mentioned earlier. Levy's (1969) original study was conducted with male graduate students at the California Institute of Technology. The reduction in spatial performance that she found among left handers is typical of the pattern for high reasoning males in our data. Yen's (1975) interaction was obtained in high school students. This would appear to be an unselected group, but, as Yen notes, the PMA scores for her sample put them three-fourths of a standard deviation above the national average for subjects of comparable age. These students apparently came from a

88

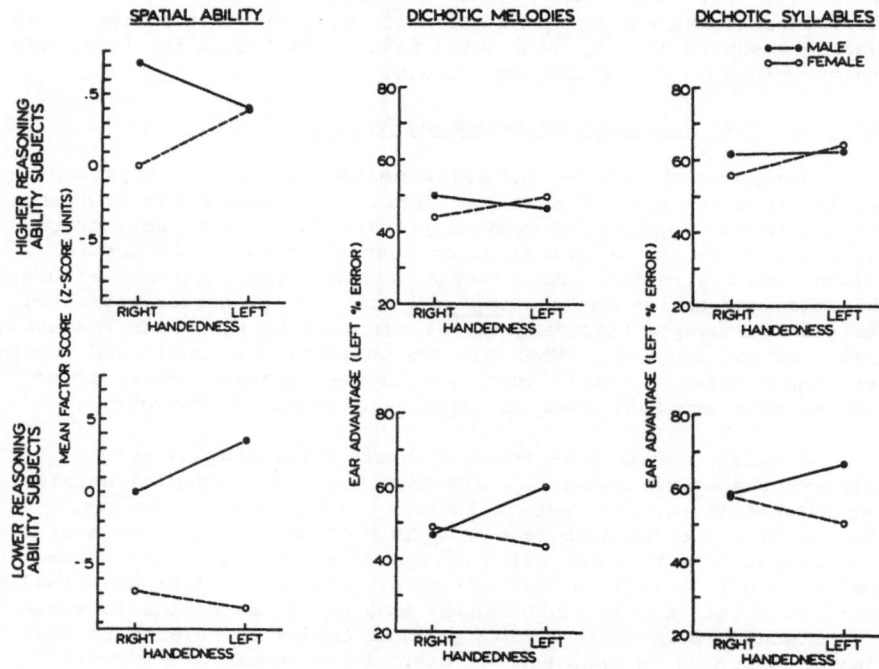

Figure 3. Effects of handedness, sex, and reasoning level on
spatial ability (a) and dichotic listening performance (b,c).

neighborhood that was, for some reason, above average in ability.
Thus it is consistent that their pattern most closely resembles that
of our higher-reasoning-ability adults. On the other hand, the data
of Sanders et al (1982) were taken from the Hawaii family study,
which showed Raven test performance closer to the lower-reasoning
half of the UC sample (Wilson & Vandenberg, 1978). Consistent with
this, the Sanders et al (1982) sample showed the reversed sex by
handedness interaction on spatial performance which we found in the
lower-reasoning half of the UC sample. Finally, the Inglis and
Lawson (1984) sample showed no handedness effects whatsoever, which
is plausibly explained by the relatively broad and unselected nature
of their sample. This might also explain the null results obtained
by Hardyck et al (1976), although in this case the young age of the
sample is another complicating factor.

Many of the inconsistencies in the literature can now be
interpreted as part of a larger, consistent pattern. And since our
initial results were published, parts of this pattern have already
been replicated (Lewis & Harris, 1986). The broad picture thus
changes from one of confusion into one which gives substantial

support for handedness effects on cognition. We must now ask to what extent these systematic variations in spatial ability can be attributed to group differences in brain organization.

EVIDENCE FOR A NEUROLOGICAL BASIS

Several lines of evidence suggest that these group differences in spatial (and other) abilities may be due, at least in part, to group differences in brain organization. What is arguably the most important reason for favoring a neurological interpretation is that only an implausibly elaborate sociocultural explanation could account for the complexities of the interaction patterns we have discovered. The effects of sex, reasoning ability, and familial sinistrality interact with, rather than simply add to, the effects of handedness. The complex form of these interactions, and the variations of the form both across cognitive tasks (see below), and when other moderator variables are substituted for reasoning ability, makes it very difficult to imagine a plausible social-cultural account of how they arise (this point is argued in detail in Harshman et al., 1983, p. 184-6). Although environmental factors play an undeniable role in cognitive performance, biological predispositions might provide a simpler explanation for some of the patterns we have observed.

Differences between right and left handers are by now well documented, and similarly, the evidence for sex differences in brain organization has been accumulating for some time (for reviews see Bryden, 1979; Harshman & Remington, 1976; McGlone, 1980; and for more recent developments see Kimura, 1983; Kimura & Harshman, 1984). Other studies suggest that differences in brain organization may also be correlated with reasoning level. For example, Mateer, Polen, & Ojemann (1982) reported a difference in the distribution of naming errors evoked by electrical stimulation of left dominant cortex in patients with a high versus a low Verbal IQ. And in the studies of Benbow and colleagues, students selected on the basis of extreme mathematical and/or verbal giftedness showed unusual field advantages on tachistoscopic testing (Benbow, this volume). It is also possible, if not probable, that these various sources of biological variance might interact within the individual, and that there may be cognitive consequences of the resulting differences in functional organization. Indeed, Benbow has previously shown an unusually high incidence of left handedness among gifted students (Benbow, in press), and also, at the highest levels of mathematical aptitude, a preponderance of males (Benbow & Stanley, 1983). Kertesz and his colleagues have been studying anatomical asymmetries as revealed by magnetic resonance imaging (Kertesz, Black, Polk & Howell, 1986), and have recently found a sex by handedness interaction in the left-right asymmetry of the parietal lobe (Kertesz, personal communication, June 1986). The reduced asymmetry found in left handed males most closely resembles that of right handed females.

A final, and important source of evidence for group differences in brain organization in our particular samples comes from dichotic listening. The pattern of dichotic results is particularly noteworthy, and is discussed in more detail below.

Dichotic Listening

Because each ear is more effectively connected to the contralateral hemisphere, we can gain some index of the relative specialization of the two hemispheres for a given type of perceptual processing by examining the relative performance of the two ears when they are simultaneously presented with competing stimuli (see Bryden, 1982; Kimura, 1967). Various formulae have been proposed for converting the accuracy scores for the two ears into a single measure of brain asymmetry (see e.g., Bryden & Sprott, 1981; Harshman & Krashen, 1972) but this controversy becomes moot when results are "robust" with respect to measure (Harshman, 1974). Fortunately, in our multi-sample study a similar picture is obtained with all of the common methods of computing asymmetry. The results reported below are in terms of the left ear's percentage of the total errors (POE scores). With this measure, 50% indicates no asymmetry, scores above 50% indicate a right ear advantage (typical for language stimuli) whereas scores below 50% indicate a left ear advantage (typical for some nonverbal tasks such as dichotic melodies).

Because dichotic data were not collected in the UC newlywed sample, except for supplementary testing of a very small subset by Berenbaum (1978), we had to concentrate on the UWO sample composed of university students and the SUNY twin sample, some of whom were students. These samples were not as balanced in terms of the distribution of reasoning ability. Therefore, to clarify possible reasoning effects, we changed from a median to a tertile split on reasoning ability, and concentrated on comparing the upper and lower thirds of each sample.

As an example, Figure 3b shows the results obtained with Kimura's dichotic melodies test (1964) in the UWO sample. Let us consider first the results for the high-reasoning third of the sample (top diagram). We see that among right handed subjects, the expected perceptual asymmetry (i.e. POE below 50%) is only obtained for the females, whereas among left handed subjects this pattern is somewhat reversed. Thus, there is a sex by hand interaction in which the lateralization of left handed males is increased (i.e. is shifted further below 50%), but that of females is reduced (shifted closer to 50%), relative to their right handed counterparts.

On the other hand, roughly the opposite interaction pattern is seen in the low-reasoning third of the sample. Among left handed males the lateralization is not just reduced (compared to the right handers), it is actually reversed (shifted over 50%); at the same time, the left ear advantage of left handed females is increased (shifted further below 50%) compared to their right handed counter-parts. This reversal of the two-way interaction results in a three-way interaction of sex, handedness, and reasoning ability (p=.011).

Figure 3c shows the similar but complementary pattern that is seen in the POE scores for the UWO sample on a standard verbal dichotic test (the Kresge Hearing Labs consonant-vowel monosyl-lables). Because this is a verbal task, the POE scores now tend to be above 50%. In the high-reasoning third of the sample (top diag-ram), we see that the right handers show the frequently reported sex difference in degree of language lateralization, with the males more lateralized, but among left handers this difference disappears. The interaction takes the following form: left handed females are more lateralized (further above 50%) than their right handed counter-parts, while the left handed males show little difference from right handed males. In the lower-reasoning third of the sample (bottom diagram), there are no sex differences among right handers, and the opposite interaction with handedness is observed: the left handed females show a sharp drop in degree of lateralization, whereas the left handed males show an increase, relative to their right handed counterparts. In this case, the three-way interaction only has a p-value of .087. Nonetheless, the pattern appears to replicate very well, in that the same interaction pattern also appears on the dichotic syllables test in the SUNY sample, when corrected for the main effect of handedness.

We thus have shown that for both verbal and nonverbal dichotic tests, there is a three-way interaction of sex, handedness, and rea-soning level, which parallels the form of the three-way interaction for spatial ability to a surprising degree. To the extent that the dichotic listening test is a valid index of relative hemispheric specialization, these data provide evidence that consistent group differences in functional brain organization are associated with group differences in spatial ability. Moreover, the form of the parallel dichotic and spatial interactions is consistent with pre-dictions derived from Levy's theory (Levy, 1969; Levy & Gur, 1980).

According to Levy and Gur (1980), stronger lateralization of language accompanied by weaker lateralization of nonverbal functions should be advantageous to spatial ability, whereas the opposite pat-tern of lateralization should be advantageous to verbal ability. Our dichotic results show that for females in the high-reasoning group, left handedness is associated with patterns of brain organ-ization which should be advantageous to spatial ability, but for males in this group, left handedness is associated with the oppo-site, spatially disadvantageous (but verbally advantageous), kinds of patterns. And as Levy would predict, left handedness is assoc-iated with improved spatial performance in females, but impaired spatial performance in males, in this subsample. Although it had not been predicted in advance, our dichotic data indicate that the opposite lateralization patterns are associated with left handedness in the low reasoning subsample, and likewise, we find the opposite

patterns of effects on spatial performance. Thus our dichotic
results parallel our spatial results in a manner consistent with a
"crowding" hypothesis.

In several respects, however, we are painting an overly
simplistic picture. Our dichotic tasks tap quite specific verbal
and nonverbal perceptual functions and should not be taken as an
index of the global organization of a whole hemisphere. We are
using the terms "spatial ability" and "verbal ability" as if they
were single entities, when in fact these terms represent classes
of abilities with multiple exemplars which may differ in many
ways. Not all verbal or spatial tasks show the same pattern of
interactions in our data, although the degree of agreement among
different spatial tasks is surprisingly high. Finally, we must
recognize that although we favor a neurological interpretation,
other interpretations of the parallel between cognitive and
dichotic interactions cannot be ruled out completely at the
present time. For example, the cognitive differences between
groups might somehow promote the dichotic differences, through
strategy effects or other factors.

OTHER COGNITIVE TESTS AND OTHER MODERATORS

The three samples in our study were administered tests which
tapped a number of different cognitive abilities in addition to
spatial ability. These test results also were analyzed by sex,
handedness, and reasoning level, to see if differences in brain
organization among the subgroups would manifest themselves in these
aspects of cognitive performance as well. To illustrate the variety
of patterns we found, we present in Figure 4 the sex by handedness
by reasoning ability interactions for the verbal, perceptual speed,
and visual memory component scores extracted by principal component
analysis from the data of the UC sample.

The interaction patterns among the verbal tests are of
particular interest, because Levy's "crowding" metaphor would
predict a complementary relationship between verbal and spatial
skills. For example, the encroachment of verbal processing
structures into the nonverbal hemisphere should simultaneously be
beneficial to verbal ability and detrimental to spatial ability.
Applied to our data, we would therefore expect the interaction
pattern for the verbal tasks to be simply a mirror image of the
interaction pattern for the spatial tasks. If we focus on the
broadest and most stable verbal variable, the verbal principal
component score from the UC sample (see Fig. 4a), there is a sex by
handedness interaction in the high reasoning half of the sample,
which _is_ in fact complementary to the spatial interaction pattern
obtained for these same individuals. However, a complementary
pattern is not obtained for the lower reasoning half of the sample,
in which there is no interaction at all. Rather, left handed
subjects tend to do worse than right handed subjects, regardless of

sex, although this pattern is not statistically significant.
Moreover, when we looked at the different verbal tasks individually,
there was much more variation in the pattern of sex and handedness
effects than we had found among the spatial tasks. Different verbal
measures showed somewhat different patterns of results, undermining
the notion of an undifferentiated "verbal ability".

The second composite variable, illustrated in Fig. 4b, is based
on the perceptual speed principal component scores from the UC
sample. Perceptual speed shows an interaction pattern similar to
that for spatial ability in the high reasoning half of the sample,
but in contrast to spatial ability, shows a strong main effect in
favor of females, rather than males. Once again, the low reasoning
half of the sample shows no interaction, but instead just main
effects of sex and handedness. This time, however, the effect of
left handedness is to improve rather than to impair performance.

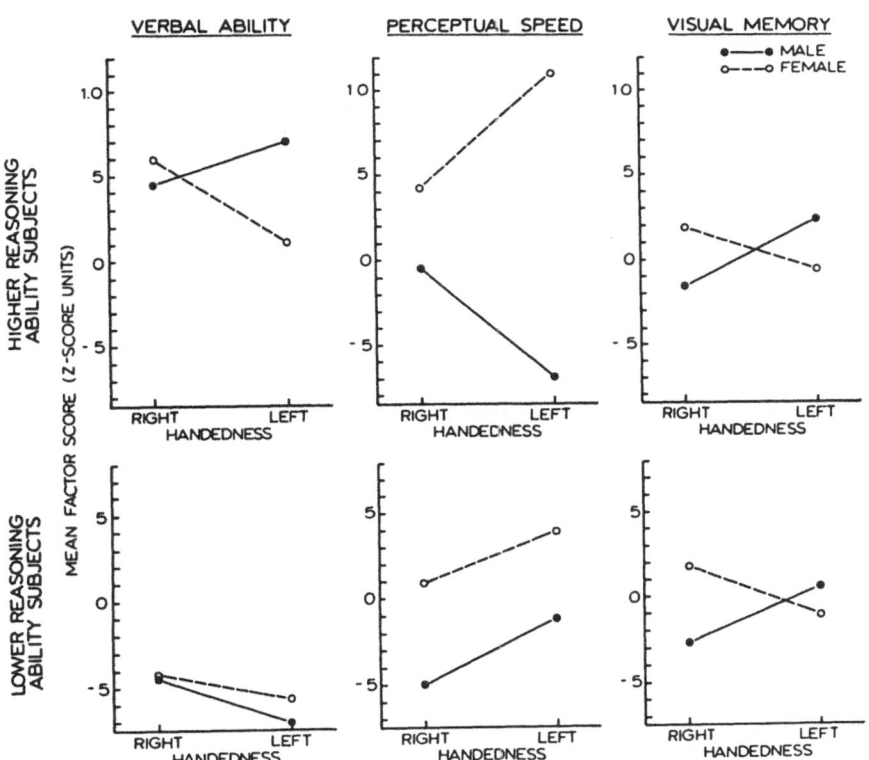

Figure 4. Effects of handedness, sex, and reasoning level on three
nonspatial cognitive variables.

Finally, in Fig. 4c, we see the interaction pattern for the visual memory component. Here we find a sex by handedness interaction in the absence of any moderating effect of reasoning ability.

If there are differences in brain organization across our subgroups, it appears that they have consequences not only for spatial and verbal skills, but for a variety of other cognitive abilities as well. There did not appear to be a simple reciprocal relationship between verbal and spatial skills, except to a very restricted extent. Moreover, among other skills, a surprising variety and complexity of effects was noted, including a tendency for subgroups that deviated in the same direction on one particular cognitive skill to dissociate on other cognitive skills, and align themselves with different subgroups. One possible interpretation of this latter result is that although some aspects of brain organization were shared in common among certain groups, and resulted in similar consequences for particular cognitive skills, these same groups were not the same in other aspects and therefore differed in other cognitive patterns. To the extent that these interactions are a function of differences in underlying brain organization, these data suggest that we may need to expand our conception of how the normal human brain may vary to include a multiplicity of possible variations.

Having found that the form of the sex by handedness interactions may differ according to the level of reasoning ability in a sample, we next asked whether reasoning is special in this regard. Might other cognitive abilities also act as moderator variables? For example, can we find different patterns of sex and handedness effects on cognition by comparing individuals that are high versus low in their level of spatial ability, perceptual speed, or vocabulary? We explored these possibilities by reclassifying members of our largest sample, the UC subjects. While none of these variables showed moderator effects as widespread as those of reasoning ability, each did produce some interesting results (see Harshman et al., 1983). It appears that reasoning is not special, and that by selecting brains with particular cognitive specializations, one may sometimes also be implicitly selecting brains with particular structural specializations.

VARIATION IN THE NORMAL HUMAN BRAIN

It appears that there is no single pattern of cognitive strengths and weaknesses associated with left handedness. The relationship between handedness and ability varies as a function of the subject's sex, reasoning level and/or other cognitive specializations. What, then, does this tell us about variations in brain organization in left handers?

We have argued above, and, in more detail, elsewhere (Harshman et al, 1983, p. 184-185) that these group differences in cognitive abilities are due, at least in part, to group differences in brain organization. A biological explanation of the cognitive group differences seems particularly plausible in light of the growing evidence for both sex and handedness differences in brain organization; it seems reasonable that these two sources of variation might interact as they exert their influence on the developing brain. Genetic factors which will ultimately give rise to one or another kind of left-handed brain are probably active early in development, laying down a somewhat distinctive left-handed organization in the embryo. As this proto-brain develops it is exposed in utero to sex hormones and possibly other factors which influence brain development. The neurological effects of these sex-related factors (e.g. hormones) may depend on which substrate they encounter, and so could easily result in a different modification of a developing "left-handed" brain than they would produce in a developing "right handed" brain.

What, then, are we to make of the interaction of these factors with reasoning level (or other indices of cognitive specialization)? We proposed in Harshman et al (1983) that individuals at different reasoning levels may have brains which differ in their structural organization. We did not postulate a unitary "high reasoning" brain pattern, nor a single factor that promoted the development of a brain with high reasoning ability. Instead we postulated that there could be a number of different brain organizations that promote high reasoning, and a number of other brain organizations that are inconsistent with it (but perhaps are advantageous to some other cognitive abilities). Our interaction effects would then arise because we are dividing a heterogeneous collection into contrasting but more homogeneous subsets.

A simple analogy helps to clarify how this can happen. Suppose we were studying the characteristics of motor vehicles, and we looked at a heterogeneous sample which included tractors, heavy trucks, jeeps and regular automobiles of varied manufacture such as Honda, Ford, and Corvette. This sample contains "several different kinds of vehicle organization". Suppose further that we wanted to see how an external trait such as top speed is related to an internal anatomical feature, such as engine horsepower (or cylinder volume). If we simply compared the average top speed of high power and low power vehicles, using the sample as a whole, we would find little or no difference. But if we split the sample into two distinct subsets based on a "moderator" variable such as "body streamlining", we could find an interaction. In the "High-Streamline" group (Honda, Ford, Corvette) there would be a positive relationship between power and top speed. In the "Low-Streamline" group (Jeep, Truck, Tractor) there would be a negative relationship.

The "crowding hypothesis" (Levy, 1969; Levy & Gur, 1980) has given us a simple but useful model for understanding how brain variation may be linked to cognitive variation, and in so doing has stimulated a good deal of research. As we saw earlier, the hypothesis gains some support from our dichotic results, where group differences in dichotic lateralization seem to parallel differences in spatial ability in just the way Levy would predict. In addition, the pattern of verbal and spatial means in the high-reasoning subsample corresponds closely to the predictions of Levy and Gur (1980). On the other hand, the pattern in the lower reasoning subsample was not predicted, nor is it clear how they could account for it. More generally, since their hypothesis is formulated in terms of "verbal" versus "spatial" functions, it does not account for the fact that different interaction patterns are found for other cognitive abilities such as perceptual speed. Clearly, more is going on than a shift in the relative amount of neural space devoted to linguistic vs spatial processing.

In order to retain a crowding account we must at minimum extend the hypothesis to include reduction or enlargement of neural areas mediating other cognitive functions besides "spatial" and "verbal" abilities. The more complex modification of available neural tissue could arise either as a direct consequence of the 'intrusion' of one function into the region of another, or as a secondary result of such intrusions, through rearrangement of processing areas and re-allocation of neural space.

At some point, however, we will need to consider additional mechanisms by which brain anatomical differences can result in performance differences. For example, consider some possible consequences of changes in the brain's structural "topology": (a) differences in the patterns of intercommunication between regions, (b) changes in ease of selective activation or inhibition of regions due to differences in cortical adjacency, (c) changes in competition or interference between regions (e.g. reciprocal inhibition), etc. Each of these changes could benefit some skills at the expense of others, particularly since many complex cognitive tasks are likely to involve cooperative interaction of several neural regions or systems.

To date, individual differences research in neuropsychology has focussed almost exclusively on variations in "degree of lateralization", perhaps because procedures for measuring left-right asymmetries in normal subjects were readily available. But left-right asymmetry is only one of many ways that brains might vary. Note, for example, that our dichotic lateralization data parallel some cognitive variations, such as spatial ability, but not others, such as perceptual speed. The variations in perceptual speed might be related to degree of lateralization on some other kind of task, but they might just as well be related to variations in intrahemispheric organization, or to variations in the connections between hemispheres. To study such variations we will need to develop new experimental techniques. There already has been some progress in this direction. For example, Lewis and Christiansen (1986) used an interference paradigm to test Kimura's (1983) claim of sex differences in the anterior-posterior organization of speech in the left hemisphere. They demonstrated greater interference between a sequential tapping task and verbal performance in females than in males, consistent with Kimura's data suggesting that speech and praxic functions are more focally organized in females. Wood and Harshman (1986) used a tachistoscopic task requiring comparison of two visual stimuli presented simultaneously to the left and right fields, in order to test for sex differences in interhemispheric transfer of visual information; they found evidence suggesting greater callosal effectiveness in females. Other fruitful techniques are just now becoming available. PET scans and special EEG techniques can help us study individual brain functioning in detail. Study of anatomical variations via autopsy and in normals via MRI scans can also contribute. Concrete data which these methods provide will permit the development of more explicit theories explaining how individual differences in cognitive performance result from differences in functional architecture.

To speculate for a moment, the data reviewed here could be taken to suggest a modified concept of normal brain variation. Instead of one "standard" pattern (e.g. the right handed male brain) which is modified a particular way in left handers and in females, we might envision a developmental process which has multiple branches resulting from the interaction of various genetic and environmental factors. Judging from recent evidence concerning sex and handedness, differences between some branches of the developmental "tree" could involve fairly substantial alterations in localization of functions, connections between and within hemispheres, etc. Indeed, variation in human brain organization could to some extent mirror the rich variation of human traits and abilities.

REFERENCES

Benbow, C.P. (in press). Physiological Correlates of Extreme Intellectual Precocity. Neuropsychologia.

Benbow, C.P. and Stanley, J.C. (1983). Sex Differences in Mathematical Reasoning Ability: More Facts. Science, 222, 1029-1031.

Berenbaum, S.A. (1977). A Model of Individual Differences in Hemispheric Specialization and Cognitive Abilities, with Particular Reference to Sex and Handedness. Unpublished Doctoral Dissertation, University of California at Berkeley. (1978). Diss. Abstr. Int., 39. (University Microfilms, Ann Arbor, No. 7812488).

Bryden, M.P. (1979). Evidence for Sex-Related Differences in Cerebral Organization. In Sex-Related Differences in Cognitive Functioning: Developmental Issues. (eds. M.A. Wittig and A.C. Petersen). Academic Press, New York.

Bryden, M.P. (1982). Laterality: Functional Asymmetry in the Intact Brain. Academic Press, New York.

Bryden, M.P. and Sprott, D.A. (1981). Statistical Determination of Degree of Laterality. Neuropsychologia, 19, 571-581.

Bryden, M.P., Hecaen, H. and DeAgostini, M. (1983). Patterns of Cerebral Organization. Brain & Lang., 20, 249-262.

Hardyck, C., Petrinovich, L.F. and Goldman, R.D. (1976). Left-

98

Handedness and Cognitive Deficit. Cortex, 12, 266-279.
Harshman, R.A. (April, 1974). Meaning and Measurement of "Degree of Lateralization". Paper presented at the 87th meeting of the Acoustical Society of America, New York. (1974). J. acoust. Soc. Am., 55, S10 (Abstract).
Harshman, R. and Krashen, S. (1972). An "Unbiased" Prodecure for Comparing the Degree of Lateralization of Dichotically Presented Stimuli. UCLA Working Papers in Phonetics, 23, 3-12; 17-21. (University Microfilms, Ann Arbor, No. 10,085).
Harshman, R.A. and Remington, R. (1976). Sex, Language and the Brain, Part I: A Review of the Literature on Adult Sex Differences in Lateralization. UCLA Working Papers in Phonetics, 31, 86-103. (University Microfilms, Ann Arbor, No. 10,085).
Harshman, R.A., Hampson, E. and Berenbaum, S.A. (1983). Individual Differences in Cognitive Abilities and Brain Organization, Part I: Sex and Handedness Differences in Ability. Can. J. Psychol., 37, 144-192.
Harshman, R.A., Hampson, E. and Lundy, M.E. Individual Differences in Cognitive Abilities and Brain Organization, Part II: Dichotic Listening. Manuscript in preparation.
Hecaen, H., DeAgostini, M. and Monzon-Montes, A. (1981). Cerebral Organization in Left-Handers. Brain & Lang., 12, 261-284.
Inglis, J. and Lawson, J.S. (1984). Handedness, Sex and Intelligence. Cortex, 20, 447-451.
Kertesz, A., Black, S.E., Polk, M. and Howell, J. (1986). Cerebral Asymmetries on Magnetic Resonance Imaging. Cortex, 22, 117-127.
Kimura, D. (1964). Left-Right Differences in the Perception of Melodies. Q. Jl. exp. Psychol., 16, 355-358.
Kimura, D. (1967). Functional Asymmetry of the Brain in Dichotic Listening. Cortex, 3, 163-178.
Kimura, D. (1983). Sex Differences in Cerebral Organization for Speech and Praxic Functions. Can. J. Psychol., 37, 19-35.
Kimura, D. and Harshman, R.A. (1984). Sex Differences in Brain Organization for Verbal and Non-Verbal Functions. In Progress in Brain Research, Vol. 61. (eds. G.J. DeVries, J.P.C. DeBruin, H.B.M. Uylings and M.A. Corner). Elsevier, Amsterdam.
Levy, J. (1969). Possible Basis for the Evolution of Lateral Specialization of the Human Brain. Nature, 224, 614-615.
Levy, J. (1974). Psychobiological Implications of Bilateral Asymmetry. In Hemisphere Function in the Human Brain. (eds. S.J. Dimond and J.G. Beaumont). Wiley, New York.
Levy, J. and Gur, R.C. (1980). Individual Differences in Psychoneurological Organization. In Neuropsychology of Left-Handedness. (ed. J. Herron). Academic Press, New York.
Lewis, R.S. and Christiansen, L. (1986). Sex Differences in Intrahemispheric Representation of Language in Normal Individuals. Society for Neuroscience Abstracts, 12, Part 1, 720.
Lewis, R.S. and Harris, L.J. (1986). The Relationship Between Sex, Handedness, and Spatial Ability in High Academic Achievers. Manuscript submitted for publication.
Mateer, C.A., Polen, S.B. and Ojemann, G.A. (1982). Sexual Variation in Cortical Localization of Naming as Determined by Stimu-

lation Mapping. Behav. & Brain Sciences, 5, 310-311.

McGlone, J. (1980). Sex Differences in Human Brain Asymmetry: A Critical Survey. Behav. & Brain Sciences, 3, 215-263.

Milner, B. (1974). Hemispheric Specialization: Scope and Limits. In The Neurosciences: Third Study Program. (eds. F.O. Schmitt and F.G. Worden). M.I.T. Press, Boston.

Sanders, B., Wilson, J.R. and Vandenberg, S.G. (1982). Handedness and Spatial Ability. Cortex, 18, 79-90.

Springer, S.P. and Searleman, A. (1978). The Ontogeny of Hemispheric Specialization: Evidence from Dichotic Listening in Twins. Neuropsychologia, 16, 269-281.

Watkins, M.P. and Meredith, W. (1981). Spouse Similarity in Newlyweds with respect to Specific Cognitive Abilities, Socioeconomic Status, and Education. Behav. Genet., 11, 1-21.

Wilson, J.R. and Vandenberg, S.G. (1978). Sex Differences in Cognition: Evidence from the Hawaii Family Study. In Sex and Behavior: Status and Prospectus. (eds. T.E. McGill, D.A. Dewsbury and B.D. Sachs). Plenum Press, New York.

Witelson, S.F. (1985). The Brain Connection: The Corpus Callosum is Larger in Left-Handers. Science, 229, 665-668.

Wood, G.J. and Harshman, R.A. (1986). Individual Differences in Interhemispheric Transfer of Simple Visual Information. Can. Psychol., 27(2a), 633.

Yen, W.M. (1975). Independence of Hand Preference and Sex-Linked Genetic Effects on Spatial Performance. Percept. Mot. Skills, 41, 311-318.

8
The Evolution of Handedness in Primates

Peter F. MacNeilage

INTRODUCTION

It is almost universally believed that non-human primates do
not have population-level hand preference asymmetries. The most
influential proponent of this view has been J. M. Warren who has
stated it in several recent reviews. (Warren, 1977 a, b; Warren
and Nonneman, 1976; Warren, 1980). However this conclusion is
probably wrong. There is, in fact, a good deal of evidence for
population-level hand preferences in non-human primates. At least
a dozen statistically significant asymmetries have been reported,
and some of these have been in the literature for many years. When
examined closely these results reveal a pattern. The earliest primates, the prosimians, show a <u>left</u> hand preference in reaching for
and prehending food. This is also observable in monkeys. But in
monkeys there is also evidence for development of a <u>right</u> hand
preference for practiced stereotyped acts involving more than simple
reaching, in laboratory situations; and there is strong evidence in
one study (Beck and Barton, 1972) of a right hand preference for
complex manipulative acts. In this study, the left hand reaching
preference and the right hand manipulative preference were observed
in the same animals. The left hand reaching preference may not be
present in apes. There is some evidence for a <u>right</u> hand reaching
preference in males. And there is strong evidence in 4 species of
apes for a right limb (hand and foot) preference in the initiation
of locomotion. In humans of course the left hand preference is
absent under ordinary circumstances, but right hand and foot preferences are dominant in the population.

This evidence is consistent with the following view of the evolution of handedness in primates. The first specializations developed in early prosimians in the context of the development of the
prehensile hand, and they developed in the service of unimanual predation. A right hemisphere specialization for visually guided
movement may have developed concurrently with a left hemisphere
specialization for postural control of the right side and axial

100

structures in the prosimians classified as vertical clingers and
leapers. With the development of freedom from the demands of
vertical clinging, in monkeys and apes, the right side of the body
may have become the operative side for foraging, including manipulation, and bimanual coordination, with the left side becoming
supportive for these functions. In hominids, in whom amount of
skilled hand control increased enormously, the left hemisphere
specialization for this control developed further. In summary, it
is claimed that handedness, and hemispheric specializations related
to it, evolved across the entire primate order, but primarily in
the context of use of the entire body.

The remainder of this paper will have three parts. First I
will briefly suggest some reasons why practically nobody has emphasized the presence of any hand preferences in other primates, even
though evidence for them has been available. Then I will review
the evidence itself by summarizing and updating a more comprehensive
review (MacNeilage, et al, 1987). Finally, I will develop in some
detail the hypothetical view of the evolution of the specializations
of both hemispheres in primates which I have just outlined.

REASONS FOR PREVIOUS NEGATIVE CONCLUSIONS

Probably the main reason that handedness in other primates has
not been emphasized is that their preferences are for the most part
neither as strong nor as homogeneous at the group level on any given
task as those of humans. Consequently they rarely show up in small-
scale studies. A typical result tends towards a bimodal distribution of left and right hand preferences with a sizeable minority
lacking preferences. Another possible reason is that the relatively
frequent finding of a greater number of _left_ hand preferences has
been ignored because it seems counter-intuitive from an anthro-
pocentric perspective. A review of _tasks_ used in previous studies
suggests that simple reaching to foodwells in front of a cage - a
paradigm that has been widely used - is not demanding enough to produce evidence of specialized manual function. Widespread use of
relatively young monkeys has probably produced lower preference
levels and lower population levels of asymmetry than would have been
obtained with older animals. Use of different hands for different
acts in single animals has been taken to indicate inconsistency
rather than a principled division of labor such as the one between
left and right hands described earlier, or the one that humans use
in bimanual coordination. Serial trends, such as the one towards
greater right hand use, mentioned earlier, have also been taken as
evidence for inconsistency. I suspect that if all these considerations are taken into account, and diet, habitat, locomotion pattern
and hand structure are also examined, at least one specialized manual function may be demonstrable in all species of primates.

THE EVIDENCE FOR HANDEDNESS

Prosimians

The more well documented of two studies of prosimians showing a left hand preference for food reaching and prehension is one on the lesser bushbaby (Galago senegalensis) by Sanford et al, (1984). These animals live on a diet of live insects and small animals which they typically catch with one-handed, prehensive movements while clinging to a vertical support with the other three limbs. They are highly specialized feeders, and are capable of ballistic "smash-and-grab" movements (Bishop, 1964) that can be faster than the eye can see. Sanford et al found that twelve of twenty-five animals made more than 80% of reaching movements with the left hand from their typical feeding position while the preferences of the other 13 animals were distributed over the other preference levels. The typical bimodal distribution was not present. This left hand effect was not observed when the animals fed one-handed from a more atypical quadrupedal stance. In the other study of prosimians, Subramoniam (1957) observed virtually exclusive use of the left hand for prehensive food-reaching in 8 slender lorises (Loris tardigradus). As bushabies in particular are considered to be close to the earliest primates in form and function (Charles-Dominique and Martin, 1970) we have suggested (MacNeilage et al 1987) that the left hand reaching preference originally arose with the prehensile hand as part of a specialized right hemisphere capacity for manual predation in the earliest primates; and as the non-reaching right hand concurrently served a postural role, the reaching specialization may have been accompanied by a postural specialization involving the left hemisphere. We predict that this postural asymmetry has the consequence that the right limbs will be typically placed above the corresponding left limbs when prosimians rest in an upright position on vertical supports.

We are according great importance to the evolution of manual predation by suggesting that it may have been responsible for the first hemispheric specializations to develop in primates. Specifically we would suggest that the combined demands of a live insect and small animal diet and the three-dimensional arboreal habitat may have evoked consistent asymmetry in brain-behavior relations. The emphasis on the importance of manual predation as a formative influence in primate evolution is shared by Cartmill (1972) who considers the evolution of frontal eyes (with their possibilities for depth perception) and the prehensile hand in the earliest primates to be adaptations for manual predation. As part of his evidence he notes that both these specializations are absent in many other arboreal but nonpredatory animals.

Monkeys

A left hand preference for reaching for and prehending food was observed in three field studies of Japanese macaques (Macaca fuscata) (see Fig. 1), and in one open field study of marmosets (Callithrix jacchus) (Box, 1977). In addition, a trend towards a left hand preference for catching food was noted in one study of Japanese macaques (Kawai, 1967).

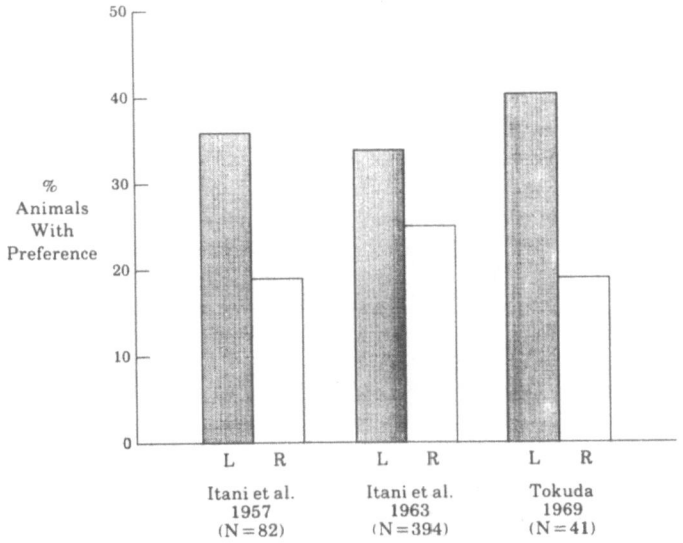

Fig. 1. Hand preferences in three field studies of Japanese Macaques picking up food.

Left hand food-reaching preference was also noted in earlier stages of four two-choice visual and tactile discrimination studies of rhesus monkeys (Macaca mulatta) by Ettlinger and his colleagues (see Fig. 2). The animal was required to decide which of two stimuli above food wells was the correct one, and then had to un-cover the correct well and reach for the food. The tactile tasks were typically preceded by visual pretraining trials.

In the study by Beck and Barton (1972) a left hand food-reaching preference was obtained in 10 stumptailed macaques (Macaca speciosa) performing 17 tasks, including a simple reach task, various complex (indirect or impeded) reach tasks, tasks involving simultaneous stabilization of the apparatus with the other hand, and tasks that involved manipulation and stabilization preceding the reach (see Fig. 3). Twelve of the fifteen tasks in which there was a reaching preference gave rise to a left hand preference, and there

was a strong tendency for the left hand preference level to increase
with the complexity of the reach, with the presence of an unstabil-
ized food locus, and with the number of other acts involved in the
task.

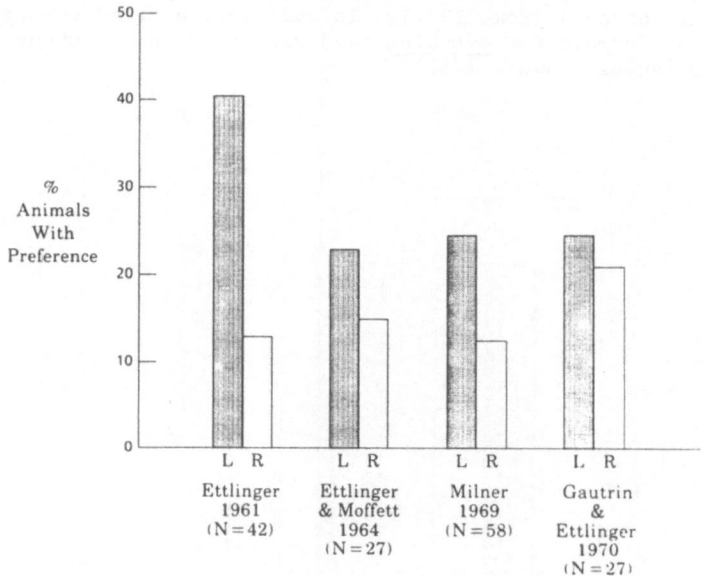

Fig. 2. Hand preferences in early trials of four
two-choice discrimination studies.

We have argued that all of the situations in which there is a
left hand reaching preference in monkeys have in common the need for
a substantial amount of visual guidance. This is produced by the
varied location of the food relative to the animal in field studies,
the demand for visual stimulus analysis preceding reaching in the
visual discrimination studies, and the need for visual guidance of
sets of acts in the study of Beck and Barton.

The left hand reaching preference was accompanied by a __right__
hand preference for manipulation in the Beck and Barton study (see
Fig. 3). All eight of the manipulative acts in this study showed
the right hand preference, and on one task requiring two single
finger hasp opening movements, the median right hand preference
levels for the monkeys were 96% and 97%. As in the study of bush-
babies, bimodal distributions of preferences were absent in this
study. No animal showed a strong right hand preference for reaching
or a strong left hand preference for manipulation. Because this
study is by far the most demanding in the entire primate handedness
literature in terms of the range of tasks used and their difficulty,
we think the demonstration of concurrent left hand reaching and

right hand manipulation preferences in such a small group of animals
has a very considerable significance for the study of monkey manual
function.

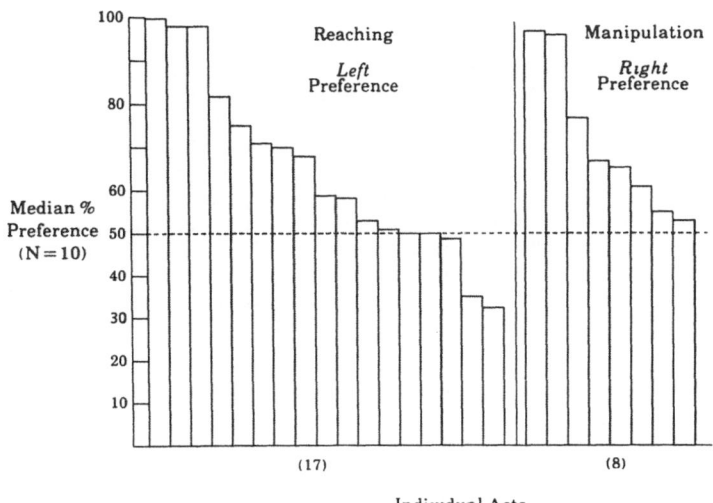

Fig. 3. Median preference levels of a group of ten
         stumptail Macaques for reaching and manipu-
         lation acts (Beck and Barton, 1972).

Significant serial trends towards right hand preference were
observed in three cases. In the tactile discrimination studies of
Ettlinger and his colleagues, there was a significant trend for
animals initially showing no preference to develop a right hand
preference. Significant trends towards right hand preferences with
practice were also present in studies of various tasks by Warren
(1977 b) using rhesus monkeys and Rothe (1973) using marmosets.

Apes

        Handedness of apes is not clearly documented. An oft-cited
study of 30 chimpanzees by Finch (1941) did not show a population-
level asymmetry of hand preferences. However Finch used simple
reaching tasks of the kind that do not evoke asymmetric preference
levels in monkeys. I have conducted an open field study of food
reaching and prehension in 51 chimpanzees. (Unpublished observations) The group as a whole showed no asymmetry of preferences in
a total of 591 reaches, but a T test comparing frequency of left

and right reaches showed a significant (.05 level) <u>right</u> hand preference in the 21 males in the study. In a very recent study by Heestand (1986) a significant tendency to use the right limbs (hand and foot) to initiate terrestrial locomotion was noted in 4 types of ape: siamangs, chimpanzees, orang-utans and gorillas.

Humans

In humans there is no left hand reaching preference; simply a typical right hand preference for practically everything unimanual, and a dominant right hand in bimanual coordination. However it is intriguing that Guiard, et al (1983) have shown that normal human right handers are more accurate with their left hand than with their right in ballistic reaching. This perhaps represents a vestige of an earlier left hand reaching capacity. Humans are also more often than not right footed (Peters and Durding 1979). This together with the evidence for a right foot preference in the initiation of loco-motion in apes makes it appear to be desirable to consider the evo-lution of <u>sidedness</u> in these taxa.

HANDEDNESS AND THE EVOLUTION OF HEMISPHERIC SPECIALIZATION

Right Hemisphere Specialization

We believe that these results have important implications for the evolution of specializations of both hemispheres in primates. The initial left hand reaching preference, presumed to be for manual predation, may be indicative of the evolution of a right hemisphere specialization for what could be called <u>off-line</u> perceptual-motor feeding-related functions involving interaction between intraperson-al and extrapersonal space. The perceptual-motor relationships are assumed to be primarily off-line for two reasons. First, once a stationary prey is sighted (and/or heard) movements of the predator into its vicinity are not dependant in a moment-to-moment manner on details of visual or auditory feedback from the prey. They are de-pendant on the structure of the arboreal environment between the predator and the prey. Nevertheless the fact that the reach will eventually be made with the left limb may give the right hemisphere the primary guiding role in this behavior. Second, and perhaps more important, once in the vicinity of the prey, the detailed "plan" for the forthcoming ballistic movement of the left arm is totally estab-lished off-line, that is, before the movement is made. Even compu-tations underlying slower reaches for inanimate objects, more promi-nent in monkeys, may be primarily made in advance. The evolution in the right hemisphere of this off-line relation between perception and action with the accompanying capacity for relative decoupling of the two may be partially responsible for the fact that many aspects of right hemispheric specialization in humans for both the extra-personal and the intrapersonal world <u>now</u> seem to be purely per-ceptual, and not related to action. The possibility that perceptual and motor functions could be decoupled may also have allowed the left limb reaching preference to drop out in hominids, perhaps under selection pressures favoring the manipulative capacity of the right limb, while retaining the right hemisphere perceptual specializations.

Left Hemisphere Specialization

Now consider the issue of why right hand preferences developed at all if the first primate hand preference for operating on the world was indeed a left hand preference. A background consideration here may be that parallel to the development of the left hand-right hemisphere system may have been a postural specialization of the right side and the left hemisphere involving what could be called on-line perceptual motor interaction between intrapersonal and extrapersonal space. In the maintenance of posture, every movement of the animal or its support presumably has direct immediate (on-line) perceptual implications for the control system. Perception and action are thus continuously relevant to each other on a moment-to-moment basis. When higher primates evolved quadrupedal locomotion, and consequently came to depend less on vertical supports, the earlier postural specialization may have found a new use, namely on-line perceptual-motor interaction of the right side with extrapersonal space primarily for foraging. With the development of the opposable thumb, which made manipulation possible, monkeys with more terrestrial habitats may have begun to engage in moment-to-moment interaction with the environment, in operations such as tearing off bark while searching for food objects, and exploring crevices with the fingers; and perhaps the more force or the more finesse required in these acts the more benefit might have accrued from having a system specialized for on-line interaction with the environment directly involved, via the right hand. The right hand preferences observed in monkeys for manipulation and stereotyped acts are consistent with this hypothesis.

Two further findings in the primate handedness literature support the conclusion of a right limb advantage for both force and finesse. Schaller (1963) reported that 59 out of 72 gorillas initiated chest-beating displays with the right hand, and Preilowski (1979) reported a significantly greater skill with the right hand in all members of a group of 8 rhesus monkeys producing and maintaining specific levels of pressure between the thumb and forefinger.

Anatomical evidence supports the claim that the right side may have become both the strong side and the operational side in monkeys. Dhall and Singh (1972) report that 18 of 21 forelimb muscles and bones are larger on the right side of rhesus monkeys than on the left. The possibility that the left upper limb may have come to serve a supporting function for the right when these animals are in suspension is suggested by the fact that the only significant left sidedness trend was found in the pectoralis major muscle. This muscle is considered by the authors to be the most important muscle in forelimb suspension.

The evidence for right sided locomotor initiation in apes suggests that they have come to favor the right side, at least for departures from stationary positions. Though we are not aware of any evidence of bodily anatomical asymmetries in apes, there is evidence of cerebral anatomical asymmetries in the great apes resembling those found in right handed humans. These include sylvian fissure length, (Yeni-Komshian and Benson, 1976) sylvian point height and right occipitopetalia (LeMay, 1976; LeMay et al, 1972).

There is a good deal of encouragement for considering human right handedness in the context of the more general concept of right sidedness. To begin with, continuity with the trends noted in other higher primates can be observed in a combination of functional and anatomical findings in human neonates. Pande and Singh (1971) have found greater muscle and bone weight on the right side in human fetuses, just as in rhesus monkeys. They also showed that the only exception to this trend, as in rhesus monkeys, was a greater weight of the pectoralis major muscle on the left side. Possible functional correlates of the left sided laterality trend for the pectoralis major muscle are given in the review of infant manual laterality by Young et al (1983). They observed that "...a left hand preference in suspension grasping seems evident..." (p. 26) in studies of infants in the age range of 1-4 months. They also reported greater involvement of the left hand in instances involving palmar (versus digital) contact and use of a closed hand. These instances of left hand preference in behaviors which can be said to involve the power grip, (Napier, 1962) which is necessary for suspension grasping, form the bulk of the cases for which left hand preferences are exhibited in human neonates according to the review by Young et al (1983).

The instances of neonatal left hand preferences coexist with a larger body of data on propensities to favor the right side (see) Young et al, 1983 for a summary). One frequent finding involves the tonic neck reflex, a tendency to orient the head towards the right when in supine position. This tendency is correlated with later right handedness (Michel, 1983). There is also a tendency towards a stronger right hand grasp and a longer duration of right hand grasps. Lower limb involvement is indicated in the favoring of the right limb in the initiation of the stepping reflex. Harris (1983) noted that in a number of earlier studies of child development a right hand preference was seen for extreme reaches in the first year but not for shorter reaches performed by the same children. Extreme reaches presumably involve more of the body than short reaches. In a summary of a comparison of the right upper limb with the left in the first few months Young et al conclude that "the right hand is preferred for more directed target-related activity whether the proximal hand-finger-wrist---or distil arm-shoulder-musculature---

is involved" (p. 27).

In human <u>adults</u> the greater size of muscles and bones of the
right side is for the most part maintained (Malina and Buschang,
1984). The greater weight of the left pectoralis major is no longer
observed. But the <u>left</u> <u>femur</u> becomes larger than the right. Von
Bonin (1962) cites evidence as to the time course of this developmental change. The evidence shows that while the right femur is
still larger than the left in the period between 6 and 12 years, the
<u>left</u> femur is larger in the period between 13 and 20 years. Thus it
would appear that as modern humans grow up - as terrestrial bipeds
- the left leg may take over the support function which the left arm
provided for in earlier primates with at least partially arboreal
habitats. Functional evidence to support <u>this</u> conclusion comes from
the work of Tan (1985). He found a higher amplitude and a longer
time course of decay of the Hoffman Reflex in the postural soleus
muscle of the <u>left</u> hindlimb of right handed human subjects than in
the right hindlimb, suggesting some sinistral postural specialization. The trend was not observed in left handers.

A number of years ago Geschwind (1975) suggested that there may
be lateralized control of axial function in humans. In a recent
factor-analytic study of handedness questionnaire data, following
Geschwind's suggestion, Healey et al (1986) found two factors with
loadings on tasks involving axial structures.

Some more direct evidence is available on the presence of left
hemisphere specialization for control of body motor functions in
humans. Glick and his collaborators (e.g. Glick and Shapiro, 1984)
have shown in rats that when there is a higher level of dopamine in
the basal ganglia of one side (the usual case) rats prefer to make
rotational movements to the contralateral side. Glick et al (1982)
have also shown higher dopamine levels in the left basal ganglia of
a group of 14 human subjects (post-mortem assessment) presumed to be
predominantly right handed. Most recently Bracha et al (1987)
showed that during a routine day: "Males that were consistently
right sided (left hemisphere dominant) for hand, foot and eye dominance rotated more to the right than to the left--." (Strangely
enough, they found that left hemisphere dominant <u>females</u> rotated
more to the left than to the right.) Proctor et al (1964) reported
that Parkinson's Disease patients with a primarily left hemisphere
dopamine deficiency exhibited bilateral deficits in orienting a line
to the vertical while sitting in a tilted chair. Patients with a
predominantly right hemisphere deficiency only showed a deficit when
tilted to the left.

In summary a diverse body of morphological physiological and
functional evidence derived from studies of human neonates and
impaired and normal adults points toward the continued existence of
a functional asymmetry of action involving the whole body for which
there is some evidence in monkeys and apes. The evidence points to
the right side of the body as the operative side and the left side

as the supportive side, with a transition from a supportive role of the left upper limb to the left lower limb accompanying the advent of bipedalism.

HEMISPHERIC MANUAL SPECIALIZATIONS: THE PRIMATE HERITAGE

It has been suggested here that part of our primate heritage consists of two originally complementary specializations, a left hand-right hemisphere specialization for off-line perceptual characterization of both extrapersonal and intrapersonal space in the service of subsequent predatory action, and a right side-left side hemisphere specialization for on-line perceptual-motor interaction with extrapersonal space. Except for the left hand reaching preference, this heritage is evident in modern man. It is possible that many of the present day capacities of the two hemispheres can either be conceptualized directly in these terms or can be viewed as developments for which the original capacities were preadaptive. If this is true, then human hemispheric specializations can be placed squarely in the context of the Darwinian theory of evolution by natural selection of adaptive functions. This possibility is explored in more detail elsewhere, particularly as it relates to communication. (Lindblom et al 1987; MacNeilage 1986).

SUMMARY

In summary, there is evidence for population-level laterality of manual function throughout the primate order. The initial development, in prosimians may have been for left hand prehension in the service of manual predation, with a complementary postural specialization for right limbs and axial structures. When the change to quadrupedal locomotion in monkeys eliminated the need for 3 limbs to provide support of vertical posture, the specialization for right side and axial control may have been adapted for the interaction with the environment required for foraging. There is evidence that the left hand preference for prehension remains in monkeys, but anatomical evidence suggests that the upper left limb also takes on a supportive role in suspension, presumably associated with below-branch feeding. There is some evidence that the right side is the operative side in apes but no evidence of a left hand preference for prehension. With the evolution of bipedality the right sided bias becomes much more predominant but the <u>lower</u> left limb now takes on a supporting function, at least in adults.

The developments in handedness were presumably associated with developments in hemispheric specialization which remain influential today; a) a right hemisphere specialization for off-line conceptualization of extrapersonal and intrapersonal space, and b) a left hemisphere specialization for on-line interaction with extrapersonal space.

The proposed evolutionary developments in handedness across the primate order are embedded in developments in whole-body motor organization. The common element is an across-the-body pattern in which one upper limb is the prime mover, and, depending on posture, one or more of the contralateral limbs is involved, with axial structures, in the support function. The conclusion that handedness developed in the context of whole-body action is not a surprising one when it is considered that the hands themselves evolved in that context.

REFERENCES

Beck, C.H.M. & Barton, R.L. (1972). Deviation and laterality of hand preference in monkeys. Cortex, 8, 339-363.

Bishop, A. (1964). Use of the hand in lower primates. In Evolutionary and Genetic Behavior of Primates. Vol. 2, (ed. J. Buettner-Janush). Academic Press, New York.

Bonin, G. von (1962). Anatomical asymmetries of the cerebral hemispheres. In Interhemispheric Relations and Cerebral Dominance. (ed. V. B. Mountcastle, Johns Hopkins University Press, Baltimore.

Box, H.O. (1977). Observations on spontaneous hand use in the common marmoset (Callithrix jacchus). Primates, 18, 395-400.

Bracha, H.S., Seitz, D.J., Otemaa, J. and Glick, S.D. (1987). Rotational movement (circling) in normal humans: sex difference and relationship to hand, foot and eye preference. Brain Res., (In Press).

Cartmill, M. (1972). Arboreal adaptations and the origin of the order Primates. In The Functional and Evolutionary Biology of Primates. (ed. R.H. Tuttle). Aldine-Atherton, Chicago.

Charles-Dominique, P. and Martin, R.D. (1970). Evolution of lorises and lemures. Nature, 227, 257-260.

Dhall, U. and Singh, I. (1977). Anatomical evidence of one-sided forelimb dominance in the rhesus monkey. Anat. Anz., 141, 420-425.

Ettlinger, G. (1961). Lateral preferences in monkeys. Behaviour, 17, 275-287.

Ettlinger, G. and Moffett, A. (1964). Lateral preferences in the monkey. Nature, 204, 606.

Finch, G. (1941). Chimpanzee handedness. Science, 94, 117-118.

Gautrin, D. and Ettlinger, G. (1970). Lateral preferences in the monkey. Cortex, 6, 287-292.

Geschwind, N. (1975). The apraxias. Am. Scient., 63, 188-195.

Glick, S.D., Ross, D.A. and Hough, L.B. (1982). Lateral asymmetry of neurotransmitters in human brain. Brain Res., 234, 53-63.

Glick, S.D. and Shapiro, R.M. (1985). Functional and neurochemical mechanisms of cerebral lateralization in rats. In Cerebral Lateralization in Nonhuman Species. (ed. S.D. Glick). Academic Press, Orlando.

Guiard, Y., Diaz, G. and Beaubaton, D. (1983). Left hand advantage in right-handers for spatial constant error: Preliminary evidence in a unimanual ballistic aimed movement. Neuropsych., 21, 111-115.

112

Harris, L.J. (1983). Laterality of function in the infant: histori-
cal and contemporary trends in theory and research. In Manual
Specialization and The Developing Brain. (eds. G. Young, S.J.
Segalowitz, C.M. Corter and S.E. Trehub). Academic Press,
Orlando.

Healey, J.M., Liederman, J.L. and Geschwind, N. (1986). Handedness
is not a unidimensional trait. Cortex, 22, 33-53.

Heestand, J.E. (1986). Behavioral lateralization in four species of
apes? Unpublished Ph.D. Dissertation, University of Washington.

Itani, J. (1957). Personality of Japanese monkeys. Iden., XI, 29-
33. (In Japanese).

Itani, J., Tokuda, K., Furuya, Y., Kano, K. and Shin, Y. (1963).
The social construction of natural troops of Japanese monkeys
in Takasakiyama. Primates, 4, 1-42.

Kawai, M. (1967). Catching behavior observed in the Koshima troop-
-a case of newly acquired behavior. Primates, 8, 181-186.

LeMay, M. (1976). Morphological cerebral asymmetries of modern man,
fossil man, and nonhuman primates. In Origins and Evolution of
Language and Speech. Ann. N.Y. Acad. Sci. (eds. S.R. Harnad,
H.D. Steklis and J. Lancaster). New York Academy of Sciences,
New York.

LeMay, M., Billig, M.S. and Geschwind, N. (1982). Asymmetries of
the brains and skulls of nonhuman primates. In Primate Brain
Evolution: Methods and Concepts. (eds. E. Armstrong and D.
Falk). Plenum, New York.

Lindblom, B., MacNeilage, P.F. and Studdert-Kennedy, M.G. (1987).
(In preparation) Evolution of Spoken Language. Academic Press,
Orlando.

MacNeilage, P.F. (1986). The evolution of hemispheric specialization for manual function and language. In Higher Brain Functions: Recent Explorations of the Brain's Emergent Properties.
(ed. S. Wise). Wiley, New York. (In Press).

MacNeilage, P.F., Studdert-Kennedy, M.G. and Lindblom, B. (1987).
Primate handedness reconsidered. Beh. and Br. Sci. (In Press).

Malina, R.M. and Buschang, P.H. (1984). Anthropometric asymmetry in
normal and mentally retarded males. Ann. of Hum. Biol., 11,
515-531.

Michel, G. (1983). Development of hand use preference during infancy. In Manual Specialization and the Developing Brain.
(eds. G. Young, S.J. Segalowitz, C.M. Corter and S.E. Trehub).
Academic Press, New York.

Milner, A.D. (1969). Distribution of hand preferences in monkeys.
Neuropsych., 7, 375-377.

Napier, J.R. (1962). The evolution of the hand. Scient. Am., 207,
56-62.

Pande, B.S. and Singh, I. (1971). One sided dominance in the upper
limbs of human fetuses as evidenced by asymmetry in muscle and
bone weight. J. Anat., 109, 457-459.

Peters, M. and Durding, B.M. (1979). Footedness of left and right
handers. Am. J. Psychol., 92, 133-142.

Preilowski, B. (1979). Performance differences between hands and
lack of transfer of finger posture and sensory-motor skill in

intact rhesus monkeys: possible model of the origin of cerebral asymmetry.Neuroscience L., Supp., 3, S89.

Proctor, F., Riklan, M., Cooper, I.S. and Teuber, H.L. (1964). Judgment of visual and postural vertical by parkinsonian patients. Neurology, 14, 287-293.

Rothe, H. (1973). Handedness in the common marmoset. (Callithrix jacchus). J. Phys. Anthro., 38, 561-566.

Sanford, C., Guin, K. and Ward, J.P. (1984). Posture and laterality in the bushbaby (Galago senegalensis). Br. Beh. and Evol., 25, 217-224.

Schaller, G.B. (1963). The Mountain Gorrila. Chicago University Press, Chicago.

Subramoniam, S. (1957). Some observations on the habits of the slender loris. (Loris tardigradus L) J. Bombay Nat. Hist. Soc., 54, 386-398.

Tan Ü.(1985). Relationships between hand skill and the excitability of motoneurons innervating the postural soleus muscle in human subjects. Int. J. Neurosci., 26, 289-300.

Tokuda, K. (1969). On the handedness of Japanese monkeys. Primates, 10, 41-46.

Warren, J.M. (1977a). Functional lateralization of the brain. In Evolution and Lateralization of the Brain. Ann. N.Y. Acad. Sci. (eds. S.J. Dimond and D.A. Blizard). New York Academy of Sciences, New York.

Warren, J.M. (1977b). Handedness and cerebral dominance in monkeys. In Lateralization in the Nervous System. (eds. S. Harnad, R.W. Doty, J. Jaynes, L. Goldstein and G. Krauthamer). Academic Press, New York.

Warren, J.M. (1980). Handedness and laterality in humans and other animals. Phys. Psychol., 8, 351-359.

Warren, J.M. and Nonneman, A.J. (1976). The search for cerebral dominance in monkeys. In Origins and Evolution of Language and Speech. Ann. N.Y. Acad. Sci. (eds. S.R. Harnad, H.D. Steklis and J. Lancaster). New York Academy of Sciences, New York.

Yeni-Komshian, G. and Benson, D.A. (1976). Anatomical study of cerebral asymmetry in the temporal lobe of humans, chimpanzees, and rhesus monkeys. Science, 192, 387-389.

Young, G., Segalowitz, S.J., Misek, P., Ercan Alp, I. and Boulet, R. (1983). Is early reaching left handed? Review of manual specialization research. In Manual Specialization and the Developing Brain. (eds. G. Young, S.J. Segalowitz, C.M. Corter And S.E. Trehub). Academic Press, New York.

ACKNOWLEDGMENT

This paper was prepared in part while the author was a Fellow at the Center for Advanced Study in the Behavioral Sciences. The support of the Alfred Sloan Foundation is gratefully acknowledged.

9
Spontaneous Drawing in an Unselected Sample of Patients with Unilateral Cerebral Damage

Doreen Kimura and Robert Faust

SUMMARY

Spontaneous drawing tasks were administered to a large series of patients selected only for unilateral damage to the left or right cerebral hemisphere. Aphasic patients were not excluded. Overall, patients with left-hemisphere lesions performed more poorly than those with right-hemisphere lesions, whether one considered crude ratings of the pictures, or detailed scoring measures. Patients who were manually apraxic and/or aphasic were by far the most impaired. Only a small number of patients with right-hemisphere lesions made very poor drawings, and as a group they tended to show neglect by other criteria, and to have a significant hemiparesis. Restricted damage within the left hemisphere, particularly of the anterior region, still yielded a high incidence of poor drawings, but this was much less true in the right hemisphere. It appears that drawing/constructional ability is more diffusely organized in the right hemisphere than in the left.

INTRODUCTION

Impairments in the ability to draw familiar objects have usually been described within the broader context of constructional disorders. Constructional ability has been defined by Benton (1979) as "-- any type of performance in which parts are put together or articulated to form a single entity or object, for example, assembling blocks to form a design, or drawing --." Its impairment is labelled as constructional apraxia, and such impairment is known to occur after damage to either the left or the right hemisphere. It has, however, often been claimed that difficulties in drawing are more fre-

114

quent and/or more severe after right-hemisphere damage
than after left-sided damage (Arrigoni and de Renzi,
1964; Piercy, Hécaen and de Ajuriaguerra, 1960). It has
also been reported that there are qualitative, if
subtle, differences between defects after lesions of
left and right sides (McFie, Piercy and Zangwill, 1950;
McFie and Zangwill, 1960; Piercy et al., 1960;
Warrington, James and Kinsbourne, 1966).

There are exceptions to the generally held view
that right-hemisphere lesions cause greater impairment
of drawing. Warrington et al. (1966) found no difference
in overall severity of drawing disturbance after left-
or right-hemisphere damage, although there was a tendency for different types of errors to predominate in one
or the other group. Arrigoni and de Renzi (1964) were of
the opinion that greater severity of lesions in their
right-hemisphere group may have accounted for the
greater drawing disability.

One possible factor contributing to lesser severity
of lesions after left-hemisphere damage in previous studies may have been the inadvertent or deliberate exclusion of some patients with severe aphasia. We have been
collecting samples of spontaneous drawings from a series
of research patients over several years. All patients
with unilateral lesions have been examined, and all
aphasics have been given the drawings to attempt, even
if the aphasia is severe. Subsequently, the second
author carried out a retrospective analysis of the drawings, using an objective scoring system, as well as
cruder overall rating measures. The findings indicate a
strong association of drawing disability with aphasia
and/or apraxia, and hence, with left-hemisphere pathology.

PATIENTS

The total number of right-handed patients in whom
drawings had been collected since 1974 was 331. These
were patients who had consented to take part in a larger
study on the neural mechanisms underlying aphasia,
apraxia, and constructional disorders. Only rarely did
a patient or patient's relative refuse consent to participate. The only criterion for inclusion in the study
was that the damage appeared to be restricted to one
hemisphere of the brain. Ultimate classification of uni-
laterality and of side of lesions, was based on all the
medical evidence available on the patient, and typically
included at least two of: neurological data (visual
fields, hemiparesis), angiography, CT scan, EEG, brain
scan, and surgery reports. Table 1 summarizes demograph-

TABLE 1. INFORMATION ON RIGHT-HANDED PATIENTS WITH UNILATERAL LESIONS

	Left-Hemisphere Lesions	Right-Hemisphere Lesions	Left-Hemisphere Lesions, Right-Handed Drawing	Right-Hemisphere Lesions, Left Hemiparetics Omitted
N =	191	140	149	90
Mean Age	48.4	47.6	47.3	44.1
WAIS Performance IQ	95.3*	93.0	96.6*	97.8
# Vascular Pathology	119 (62%)	93 (66%)	85 (57%)	55 (61%)
# Tumour Pathology	54 (28%)	36 (26%)	45 (30%)	26 (29%)
# Hemiparesis	51 (27%)	50 (36%)	---	---
# Visual Field Defects	40 (21%)	34 (24%)	26 (17%)	25 (28%)
# Aphasic	61 (32%)	2 (1%)	37 (25%)	1 (1%)
# Apraxic	41 (21%)	3 (2%)	27 (18%)	3 (3%)

* These scores are inflated because some aphasic patients early in our series, who would have obtained low scores, were not given this test.

ic and neurological data on the left and right-hemisphere groups.

Of the 331 right-handed patients, 140 had damage restricted to the right hemisphere, and 191 had left-hemisphere pathology. These two groups do not differ significantly in the frequency of any neurological characteristic except incidence of aphasia, but there is a trend for the right-hemisphere group to have a higher incidence of hemiparesis (X^2 = 3.0949, df = 1, P<.10).

Of the 191 patients with left-hemisphere lesions, 42 had hemiparesis severe enough to incapacitate them for drawing with their preferred hand; consequently they drew with the non-preferred left hand. This yielded 149 patients with left-hemisphere damage,who did draw with the preferred hand. Initial comparisons of drawing ability were made between these 149 patients with left-hemisphere damage and no or minimal hemiparesis, and the entire group of 140 patients with right-hemisphere damage, all of whom drew with their right hand. Perusal of Table 1, however, suggests, as might be expected, that the selected left-damaged group may be less impaired neurologically than the entire right-damaged group, even apart from the degree of hemiparesis. The incidence of vascular pathology shows a trend to be lower in this left-damaged group compared to the total right group (X^2 = 2.7964, df = 1, P<.10), and there is also a weak trend for visual field defects to be less frequent (X^2 = 2.471, df = 1, P<.15).

Therefore, comparisons were also made between the right-hand drawers in the left-damaged group and only those patients with right-hemisphere lesions who did not have a left hemiparesis, in an attempt to equate lesion size and/or extent. The two groups thus selected have an equivalent proportion of vascular pathology, but there is still a trend for them to differ significantly in incidence of visual field defects (X^2 = 3.5657, df = 1, P<.10).

Descriptions of subsamples of patients with a more restricted locus of damage, or with specific symptoms such as aphasia, apraxia, or unilateral neglect, will be given in the appropriate sections below.

METHODS

As part of a neuropsychological research assessment, all patients were required to draw, separately, a house and a person. The person-drawing was begun somewhat later in the series of patients, and therefore slightly fewer person-drawings were available for analysis. For each drawing the subject was provided with a pencil and a fresh sheet of 8 1/2 in. X 11 in. paper, placed horizontally on the table. No model was given for either drawing. A stopwatch was used to measure the time taken to draw each picture, but again the timing aspect was not recorded on the earliest patients in the series.

Additional information was available on each patient from the remainder of the research assessment, i.e., IQ measures and aphasia/apraxia scores.

Scoring of Drawings

Each patient's drawing was scored with respect to: recognizability, presence of neglect, amount of detail, distortion, number of lines (house only), size, rotation on page, and time per line and per detail.

Recognizability. This was a crude 3-point rating scale: 1 - if the drawing would not be identified as a house or person to an uninformed observer, 2 - for a satisfactory identifiable figure, 3 - if the drawing contained exceptional detail or was particularly well constructed, i.e., a presumptively "normal" drawing. Most drawings score "2" under this scale.

Neglect. Neglect was noted only as being present or absent, and was defined as significant omission or termination of lines appearing on the side of the drawing contralateral to the lesion.

Detail. A list was made of all possible appropriate items which might be included in either a house or person drawing. Most items were scored as either 1 (present) or 0 (absent), with the exception of doors, walls and windows of the house where a maximum score of 2 each was possible. Half-points might also be given for the person drawing if only one leg, arm or eye was drawn, etc. The maximum score for the house was 13 and for the person was 17.

Distortion. This was a ratio of spatially distorted or poorly drawn detail items over the total amount of detail as measured above.

Number of Lines. This was counted on the house drawings only, since it was difficult to judge on the person drawing. All lines were counted except chimney smoke and extraneous items.

Size. An 8 1/2 in. X 11 in. acetate sheet, imprinted with a grid of .05 inch squares, was aligned precisely over the drawing page. For the house the vertical length was measured from the highest point of the roof (exclusive of chimney) to the lowest part of the house (wall or baseline), and the horizontal length from the leftmost vertical wall line to the rightmost wall line. For the person, the vertical length was taken from the highest to the lowest point, and the horizontal length from the extreme left and right points. The total approximate area was calculated by multiplying the vertical and horizontal lengths.

Rotation on Page. (House only) Lines were drawn parallel to the two horizontal and vertical lines of the most prominent wall, and the angle between these and the true vertical and horizontal of the page were determined. Rotation was taken to occur only if both lines fell in the same direction (clockwise or counter-clockwise) to the page standards. The rotation score is a mean of the horizontal and vertical angles.

Time per Detail. The total time taken to complete the drawing, divided by the number of appropriate details as described above.

Time per Line. (House only) The total time taken to complete the drawing, divided by the number of lines.

RESULTS

Large Group Comparisons

The data for the four groups outlined in Table 1 were subjected to either a X^2 analysis, or two-tailed t tests (pooled variance). The four groups will be labelled Left Total (all patients with left-hemisphere lesions), Left Nonparetic (those without significant right hemiparesis), Right Total (all patients with right-hemisphere lesions), and Right Nonparetic (those without significant left hemiparesis). Comparisons were made in turn, between each of the mutually exclusive groups, but not within-left or within-right groups, since they are not mutually exclusive. Similarly, since subjects in the Total and Nonparetic groups overlap, no overall analysis across the four groups could be done.

TABLE 2. INCIDENCE OF UNRECOGNIZABLE DRAWINGS AND OF CONTRALATERAL
OMISSIONS (HOUSE AND PERSON COMBINED)

	Left Total	Right Total	Left Nonparetic	Right Nonparetic
# Drawings	362	265	279	170
# Unrecognizable[1]	40 (11%)	17 (6%)	20 (7%)	4 (2%)
# With Neglect on Drawings[2]	2 (1/2%)	16 (6%)	0	3 (2%)

1 x^2 (Total Groups) = 4.74, df = 1, P<.04
2 x^2 (Total Groups) = 16.51, df = 1, P<.01

Looking first at the cruder measures of drawing disability, Table 2 shows the occurrence of unrecognizable drawings and of drawings with omissions on the side contralateral to the lesion. More unidentifiable drawings are generated by patients with left-hemisphere lesions, but more instances of neglect in drawings occur after right-hemisphere damage. The latter is consistent with reports, both anecdotal and documented in the literature, although the incidence of gross neglect in drawings after right-hemisphere lesions in our patients is low (6%), in contrast to what might be expected from previous reports.

Table 3 shows the data on finer measures, for the four groups, and Table 4 shows the results of t test analyses for these data. Because of the multiple comparisons being made in this analysis (7 items on house drawings, and 4 on person drawings with 4 pairwise comparisons on each, or 44 t tests), it was decided that probability levels should be 2% or less, in order to reject the null hypothesis. Consequently, only those differences significant at less than .02 will be discussed.

Simple perusal of Tables 3 and 4 indicates that overall differences between left- and right-damaged groups are minimal, and generally favour the right-damaged groups. Thus, on all measures except house rotation, there is either no difference between groups, or the left-damaged group is the more impaired. Patients with left-hemisphere damage as a group tend to draw fewer details, to take longer per line and per detail, and to draw smaller drawings. In contrast, the right-

TABLE 3. MEAN SCORES ON DRAWINGS

	Left Total		Right Total		Left Nonparetic		Right Nonparetic	
Total N =	191		140		149		90	
	Mean	(SD)	Mean	(SD)	Mean	(SD)	Mean	(SD)
House Drawing:								
*Number of Details/17	8.2	(2.3)	8.7	(1.8)	8.4	(2.2)	8.9	(1.6)
*Distortion Ratio (max=1.00)	0.58	(0.27)	0.54	(0.23)	0.52	(0.24)	0.48	(0.21)
*Number of Lines	36.8	(23.8)	43.8	(30.1)	39.4	(25.2)	44.9	(29.9)
Size (in^2)	13.2	(11.2)	14.1	(11.5)	13.3	(10.9)	14.6	(11.2)
*Time per Line (sec)	3.2	(2.8)	2.8	(2.7)	3.1	(2.8)	2.3	(1.8)
Time per Detail (sec)	12.7	(9.9)	12.95	(13.0)	12.6	(10.2)	10.6	(8.3)
*Rotation (degrees)	4.1	(2.9)	5.1	(3.4)	3.7	(2.2)	4.25	(3.0)
Person Drawing:								
*Number of Details/13	9.3	(3.3)	10.1	(3.5)	9.9	(3.2)	10.8	(3.4)
Distortion Ratio (max=1.0)	0.57	(0.30)	0.50	(0.30)	0.51	(0.29)	0.44	(0.29)
*Size (in^2)	11.9	(12.0)	15.7	(12.0)	12.5	(12.4)	15.7	(12.0)
Time per Detail (sec)	12.9	(11.0)	14.1	(19.7)	13.7	(12.0)	11.7	(9.3)

*see Table 4 for significance levels.

TABLE 4. P VALUES FOR TWO-TAILED $\underline{t}$ TESTS (POOLED VARIANCE)

	L Total vs. R Total	L Nonparetic vs. R Nonparetic	L Total vs. R Nonparetic	L Nonparetic vs. R Total
House Drawing:				
Number of Details/17	.029	.059	.006**	.244
Distortion Ratio	.159	.223	.005**	.588
Number of Lines	.019*	.132	.016*	.175
Size	.483	.351	.331	.514
Time per Line	.218	.036	.016*	.368
Time per Detail	.906	.168	.133	.884
Rotation	.116	.017*	.888	.017*
Person Drawing:				
Number of Details/13	.053	.062	.002**	.636
Distortion Ratio	.057	.115	.002**	.888
Size	.009**	.070	.022	.041
Time per Detail	.570	.273	.487	.877

*p<.02
**p<.01

damaged group as a whole tends to rotate the drawing on
the page. However, the proportion of lines or details
which are distorted or poorly drawn does not differentiate the groups. Moreover, although many of the differences are statistically significant, none of them is
large, and some of the differences occur with only one
comparison, typically when the entire left-damaged group
is compared with the right nonparetic group.

Consequently, a further attempt to identify the
source of poor drawings was made, by perusing the salient neurological and neuropsychological symptoms associated with poor drawings. All patients who made unidentifiable drawings, as well as patients in each group who
were in the lowest 10th percentile on the Detail score
were examined for aphasia, apraxia, visual field defect
and hemiparesis. It became clear that in the left-damaged group, poor drawings were highly associated with
aphasia and apraxia, whereas in the right-damaged group
they were associated with left hemiparesis (but not with
visual field defects). Therefore, the patient groups
were further subdivided according to these symptoms, and
the data are presented in the following sections.

Role of Aphasia/Apraxia

Classification as aphasic or apraxic had been derived from cutoff scores developed in the past for clinical purposes, and was applied as such in this study.
Thus a patient had been classified as aphasic if the
score on an abbreviated aphasia screening test was less
than 87 out of 90, or on an abbreviated Token test
(de Renzi & Vignolo, 1962) was 75 or less out of 84
(Kimura, 1984a, p. 17, 22). Similarly, a patient had
been classified as manually apraxic if his/her score on
a manual Movement Copying test was below 12 out of 24
(Kimura, 1984a, p. 43). Many patients who were aphasic
were also apraxic, and it should be emphasized, particularly for the apraxia criterion, that it was quite ar-
bitrary, and that patients who scored 13 or 14 (above
the cutoff) would still be quite impaired, albeit classified as non-apraxic. Even so, there is a great deal of
overlap between aphasic and apraxic groups, with 29 of
the 61 aphasic patients also apraxic, and 29 of the 41
patients classified as apraxic also being aphasic.

Looking first at the crude recognizability measure,
Table 5 shows clearly that the aphasic or apraxic patients are significantly impaired relative to nonaphasic/
nonapraxic patients. This does not appear to be a function of which hand is used to do the drawing, since even

TABLE 5. INCIDENCE OF UNRECOGNIZABLE DRAWINGS IN APHASIC/APRAXIC PATIENTS (LEFT-HEMISPHERE LESIONS ONLY)

Total Group (N=191)

	Aphasic	Nonaphasic	Apraxic	Nonapraxic
House				
# Drawings	61	130	41	150
# Unrecognizable	18 (30%)	5 (4%)	15 (37%)	8 (5%)
Person				
# Drawings	54	117	36	135
# Unrecognizable	12 (22%)	5 (4%)	9 (25%)	8 (6%)

Right Hand Drawers (N=149)

	Aphasic	Nonaphasic	Apraxic	Nonapraxic
House				
# Drawings	37	112	27	122
# Unrecognizable	9 (24%)	3 (3%)	8 (30%)	4 (3%)
Person				
# Drawings	30	100	22	108
# Unrecognizable	5 (17%)	3 (3%)	4 (18%)	4 (4%)

All aphasic/nonaphasic and apraxic/nonapraxic differences are significant beyond the .01 level, using X^2 analyses.

if we consider only those patients who drew with the right hand, the contrast between aphasic/nonaphasic and apraxic/nonapraxic patients is quite striking.

One might argue that some of the unrecognizable drawings in aphasic patients were related to the fact that they simply did not understand what was required of them. A careful review of the drawings in the 18 aphasic patients who made unidentifiable drawings of the house indicates that only 4 bore no resemblance whatsoever to a house. (These 4 drawings were not included in the detailed scoring of drawing characteristics in Table 3, 6, 8 or 10.) In all 4 cases, patients wrote their names or attempts at their names, and appeared, despite urging, to be unable to produce anything else with a pencil. All four patients were also apraxic, and it is probable that their difficulties were related to motor control deficits as much as to speech comprehension difficulties. For the remaining 14 "unidentifiable" drawings, if one knows that the target object is supposed to be a house, one can see that in most drawings the production looks more like a house, or box, than like a person. (This would still have been classified as unidentifiable by the criterion originally stated, which implied no presumption about the target object.) It seems unlikely, therefore, that most of the gross impairments in drawing were due to comprehension problems. Moreover, the incidence of unrecognizable drawings is always slightly higher in the apraxic patients, arguing again that simple speech comprehension was not a major influence in these groups.

Detailed scoring of the drawings for aphasic and apraxic subgroups is shown in Table 6. Because with right-damaged patients included there are more than two mutually exclusive groups, an analysis of variance was employed, and the F ratios comparing the subgroups are shown in Table 7. There is a trend on almost all measures for aphasic or apraxic patients to perform worse than their comparison groups. The effect is particularly striking for apraxia, which affects every measure adversely, even angle of rotation, where left-hemisphere-damaged apraxics are as impaired as the total right-hemisphere-damaged group (see Table 3). The effect is also salient on size of drawings, with apraxics making by far the smallest drawings of any group, on both house and person.

The association of aphasia/apraxia with drawing difficulties is suggested also by the correlations, in the left-hemisphere group, of Movement Copying and Aphasia Screening scores with the number of details on

TABLE 6. MEAN SCORE ON DRAWINGS IN APHASIC/APRAXIC PATIENTS
(LEFT-HEMISPHERE DAMAGE ONLY)

	Aphasic		Nonaphasic		Apraxic		Nonapraxic	
Total N =	61		130		41		150	
	Mean	(SD)	Mean	(SD)	Mean	(SD)	Mean	(SD)
House Drawing:								
*Number of Details/17	6.9	(2.6)	8.7	(2.0)	6.5	(2.7)	8.6	(2.0)
*Distortion Ratio (max=1.00)	0.74	(0.30)	0.51	(0.23)	0.78	(0.29)	0.53	(0.24)
*Number of Lines	29.3	(21.7)	39.3	(24.0)	27.0	(18.4)	39.2	(24.4)
Size (in²)	12.5	(12.4)	13.6	(10.7)	8.8	(9.4)	14.3	(11.3)
Time per Line (sec)	4.0	(3.9)	2.9	(2.2)	3.9	(3.5)	3.0	(2.6)
Time per Detail (sec)	13.3	(11.2)	12.4	(9.4)	14.9	(11.8)	12.0	(9.3)
Rotation (degrees)	4.1	(2.5)	4.2	(3.2)	5.0	(3.4)	3.8	(2.7)
Person Drawing:								
*Number of Details/13	7.9	(3.1)	9.9	(3.3)	7.8	(2.6)	9.6	(3.4)
*Distortion Ratio (max=1.0)	0.72	(0.24)	0.50	(0.30)	0.71	(0.24)	0.54	(0.30)
*Size (in²)	11.2	(12.1)	12.2	(12.0)	8.1	(7.1)	12.8	(12.7)
Time per Detail (sec)	12.5	(13.0)	13.0	(10.6)	17.8	(18.3)	11.4	(7.2)

*See Table 7 for significance levels.

TABLE 7. F RATIOS FOR APHASIC OR APRAXIC GROUPS

	Left Aphasic/Left Nonaphasic Right Total	Left Apraxic/Left Nonapraxic Right Total
House Drawing:		
Number of Details/17	15.856***	17.837***
Distortion Ratio	17.949***	16.010***
Number of Lines	5.898**	5.985**
Size	.424	3.839
Time per Line	2.456	1.942
Time per Detail	.075	.665
Rotation	1.246	2.360
Person Drawing:		
Number of Details/13	8.082***	5.562**
Distortion Ratio	11.734***	6.207**
Size	3.576	5.503**
Time per Detail	.174	1.906

* P<.02	For each asterisked F ratio, apraxic or aphasic subgroups also
** P<.01	differ from each of the other two comparison groups, on a post-hoc
*** P<.001	Newman-Keuls test.

drawing. Aphasia Screening correlates .28 with number of details (N=180, P<.01), while Movement Copying correlates .42 (N=163, P<.01). (The Receptive Component of the Aphasia Screening test correlates .29 with number of details.) Within the aphasic or apraxic subgroup, the correlation of Aphasia Screening with number of details was .12 (N=63, NS), while Movement Copying and number of details correlated .26 (N=61, P<.05).

Hemiparesis and Neglect in the Right-Damaged Group

Table 8 shows the association of drawing characteristics with the presence of hemiparesis or visual field defects, in the right-hemisphere group. The presence or absence of these two symptoms was simply determined from the neurological history, sometimes supplemented by ophthalmological exam, done at about the same time as the research testing. Although the hemiparetic patients are slightly worse than those with visual field defects on 12 of the 13 measures, no one comparison actually reaches statistical significance. Probability levels are therefore not shown. Although presence of hemiparesis has sometimes been used as an index of location of damage in the anterior part of the hemisphere, a consideration of the extent of brain tissue which contributes to limb control (Brodal, 1981) suggests that the presence of hemiparesis probably indicates a significantly larger lesion, on average, than is present in its absence.

A symptom which is claimed to be more frequent after right-hemisphere damage, and to be associated with other visuospatial deficits is contralateral neglect, the tendency to ignore, overlook, or omit items in the half of space contralateral to the lesion. We have routinely assessed neglect on a visual search task (Kimura, 1984a; Kimura, Barnett & Burkhart, 1981). While moderate degrees of neglect on this task are equally common after right or left-hemisphere lesions, severe neglect is more common after right-hemisphere damage (Kimura, 1984a, b). By severe neglect is meant a difference between fields of 5 items found, out a total of 20, or a mean difference between fields of at least 5 seconds in locating objects.

Of the 140 patients with right-hemisphere lesions who were assessed on drawing, 18 showed severe neglect by either one of these criteria. These 18 patients, none of whom are manually apraxic, are compared with left-hemisphere-damaged apraxic patients, in Table 9. It is apparent that this right-hemisphere subgroup is more

TABLE 8. ASSOCIATION OF DRAWING DISABILITY WITH HEMIPARESIS OR VISUAL FIELD DEFECTS IN RIGHT-DAMAGED PATIENTS

	Paresis, No Field Defect (N=38)		No Paresis, Field Defect (N=22)	
# Drawings (House & Person)	73		43	
# Unrecognizable	11 (15%)		2 (5%)	
# Contralateral Neglect	7 (10%)		2 (5%)	
	Mean	(SD)	Mean	(SD)
House Drawing:				
Number of Details/17	8.2	(2.2)	8.3	(1.5)
Distortion Ratio	0.61	(0.25)	0.56	(0.21)
Number of Lines	43.2	(33.2)	42.5	(35.1)
Size (in^2)	14.0	(13.3)	17.2	(16.4)
Time per Line (sec)	4.0	(4.3)	2.4	(1.0)
Time per Detail (sec)	18.5	(19.5)	9.2	(3.6)
Rotation (degrees)	6.4	(3.9)	5.7	(4.8)
Person Drawing:				
Number of Details/13	9.1	(3.6)	9.6	(3.7)
Distortion Ratio	0.59	(0.29)	0.51	(0.34)
Size (in^2)	14.0	(9.7)	14.5	(13.3)
Time per Detail (sec)	20.5	(36.6)	12.1	(10.6)

TABLE 9. COMPARISON OF LEFT-DAMAGED APRAXIC PATIENTS AND RIGHT-DAMAGED PATIENTS WITH CONTRALATERAL NEGLECT

	Left-Hemisphere Lesions, Apraxic (N=41)	Right-Hemisphere Lesions, Visual Search Neglect (N=18)
Mean Age	57.3 (SD= 9.0)	57.5 (SD= 7.8)
***Mean Performance IQ	82.9 (SD=13.2)	73.6 (SD=11.6)
Restricted Anterior Damage	8 (20%)	0
Restricted Posterior Damage	8 (20%)	3 (17%)
Visual Field Defect	17 (41%)	9[1] (50%)
***Hemiparesis	14 (34%)	15 (83%)

Data on Drawings (House & Person)

# Drawings	77	33
*# Unrecognizable	24 (31%)	4 (12%)
***# Neglect on Drawings	0	9 (27%)

House	Mean (SD)	Mean (SD)
Number of Details/17	6.5 (2.7)	7.7 (2.2)
Distortion Ratio	0.78 (0.29)	0.68 (0.23)
Number of Lines	27.0 (18.4)	30.4 (13.5)
Size	8.8 (9.4)	9.0 (8.3)
Time per Line	3.9 (3.5)	3.8 (2.0)
Time per Detail	14.9 (11.8)	21.5 (24.6)
Rotation	5.0 (3.4)	7.6 (3.8)

Person		
Number of Details/13	7.8 (2.6)	7.5 (2.6)
Distortion Ratio	0.71 (0.24)	0.78 (0.19)
Size	8.1 (7.1)	11.9 (8.4)
Time per Detail	17.8 (18.3)	10.8 (6.1)

[1]Visual fields could not be determined in one additional subject, due to severe neglect.

*P<.05
***P<.001

diffusely affected than the left-hemisphere group. The incidence of restricted damage in the right-hemisphere neglect group is only 17%, while in the left apraxic group, it is 40%. The occurrence of hemiparesis is very much higher in the right subgroup as well, and the Performance IQ is lower. It would appear that gross neglect on visual searching is typically associated with extensive damage to the right hemisphere, not merely with right parietal damage. Nevertheless, it should be noted that the only 3 patients with restricted damage in this group, had lesions involving at least the parietal lobe.

Despite the extensive damage in the right-hemisphere neglect group, and their lower Performance IQ, the frequency of unidentifiable drawings remains higher in the left apraxic group. However, the frequency of drawings in which there is obvious omission or incompleteness on the side contralateral to the lesion is very much higher in the right-hemisphere group, as would be expected. When one looks at the detailed scoring of the drawings, both groups are very impaired, and there is no significant difference between them on any measure. Probability levels are therefore not shown.

Thus if we select patients with left- or right-hemisphere damage on the basis of striking neuropsychological symptoms, we find some expected differences between them in pertinent drawing characteristics, but no difference in most details of drawing. However, only a small proportion of patients with right-hemisphere damage show severe drawing impairment.

Locus of Lesion

Past studies of constructional disability have emphasized the role of the parietal lobe (Denny-Brown and Chambers, 1958), particularly in the right hemisphere (McFie et al,. 1950). In order to examine the role of locus of damage, three main subdivisions were considered: patients with lesions apparently restricted to an area in front of the Rolandic fissure (anterior group); patients with lesions apparently restricted to an area behind the Rolandic fissure, but including the anterior temporal lobe (posterior group); and a subgroup of the latter, patients with lesions which included a significant sector of the parietal lobe but which might extend beyond the parietal region (parietal+ group). Patients with damage restricted to temporal or occipital lobes would be included in the posterior group, but not in the parietal+ group.

Decisions concerning how such patients should be classified were based on all available medical evidence (see under Patients, above), and were made quite independently of neuropsychological data (with one exception noted below). Where there was any doubt about the extent of the lesions, the radiologist was consulted. Patients were not classified as having a lesion restricted to the anterior region, if they had visual field defects, even if other diagnostic information were compatible with an anterior locus; and patients would not be classified as having a posterior lesion if they had a significant hemiparesis, again despite the lack of other contrary signs. Finally, on only one point was our neuropsychological data used as a classifying criterion, and this was on the somatosensory threshold. Patients were not classified as anterior in locus of damage if the somatosensory 2-point threshold were significantly raised on the contralateral hand, since Corkin et al. (1970) found a strong dependence of somatosensory defect on damage to the postcentral gyrus.

Most patients had damage too extensive to be included in these localized groups. The numbers of patients with restricted lesions in each hemisphere, along with other information about them, is shown in Table 10.

Looking first at crude measures of drawing disability (Table 11), it is clear that by far the largest proportion of unrecognizable drawings is made by patients with left anterior lesions. Patients with parietal-lobe lesions actually produce only a small number of unidentifiable drawings.

When we look in detail at the source of error in the drawings (Table 12), the left anterior group appears to be impaired in almost all aspects of drawing. Thus these patients produce fewer and more distorted details, fewer lines in total, smaller drawings and larger rotations (the latter being matched only by the right anterior group). The next most impaired group is the left parietal+ group, with time per line and per detail being especially poor, across both drawings.

Patients with restricted right-hemisphere lesions appear again to have their primary difficulty with rotation aspects of the drawing (at least in the right anterior group). They draw larger pictures, on average, than patients with restricted left-hemisphere lesions, and they draw more lines, and more quickly. It must be emphasized that we do not have normal persons of this age with which to compare these performances, and it is very probable that many patients with right-hemisphere

TABLE 10. DATA ON PATIENTS WITH RESTRICTED DAMAGE

	Left Ant Lesions	Left Post Lesions	Left Par+ Lesions	Right Ant Lesions	Right Post Lesions	Right Par+ Lesions
Total N =	21	46	13	14	36	16
Mean Age*	52.8 (12.4)	41.2 (15.4)	47.8 (10.4)	52.7 (12.5)	41.6 (17.6)	44.5 (17.5)
WAIS Perf. IQ	89.4 (12.8)	95.4 (14.8)	87.4 (18.4)	96.2 (11.8)	95.4 (14.8)	89.2 (17.7)
# Vascular Pathology	11 (52%)	24 (52%)	8 (62%)	6 (43%)	19 (53%)	7 (44%)
# Tumour Pathology	10 (48%)	20 (43%)	5 (38%)	6 (43%)	14 (39%)	8 (50%)
# Visual Field Defects	0	18 (39%)	6 (46%)	0	18 (50%)	7 (44%)
# Hemiparetic	3 (14%)	0	0	5 (36%)	0	0
# Aphasic	9 (43%)	11 (24%)	5 (39%)	0	1 (3%)	1 (6%)
# Apraxic	8 (38%)	8 (17%)	6 (46%)	0	2 (6%)	1 (6%)

*F ratio (L ant, R ant, L post, R post) = 4.537, P<.01; (L ant, R ant, L par+, R par+) = 1.463, P=.234, NS

TABLE 11. INCIDENCE OF UNRECOGNIZABLE DRAWINGS AND OF CONTRALATERAL OMISSIONS (HOUSE AND PERSON COMBINED) IN PATIENTS WITH RESTRICTED LESIONS

	Left Ant	Left Post	Left Par+	Right Ant	Right Post	Right Par+
# Drawings	39	88	26	26	72	32
# Unrecognizable[1]	8 (21%)	3 (3%)	1 (4%)	0	2 (3%)	2 (6%)
# With Contralateral Omissions	1 (3%)	1 (1%)	0	0	3 (4%)	2 (6%)

[1] X^2 (Anterior Groups) = 6.082, df = 1, P<.02

TABLE 12. MEAN SCORES ON DRAWINGS (LOCALIZED GROUPS)

	Left Ant		Left Par+		Right Ant		Right Par+		Left Post		Right Post	
Total N =	21		13		14		16		46		36	
	Mean	(SD)	Mean	(SD)	Mean	(SD)	Mean	(SD)	Mean	(SD)	Mean	(SD)
House Drawing:												
*Number of Details/17	6.7	(2.4)	7.5	(1.9)	8.5	(1.7)	8.4	(1.4)	8.9	(2.0)	8.9	(1.4)
*Distortion Ratio	0.74	(0.26)	0.62	(0.19)	0.59	(0.27)	0.54	(0.20)	0.49	(0.21)	0.48	(0.20)
*Number of Lines	26.8	(12.9)	34.4	(16.9)	45.0	(32.4)	47.9	(46.4)	46.8	(32.3)	48.4	(40.6)
Size (in^2)	13.0	(13.1)	19.0	(15.0)	18.5	(13.4)	16.2	(14.2)	15.0	(11.8)	16.2	(14.0)
*Time per Line (sec)	3.0	(1.3)	5.9	(6.0)	2.2	(1.1)	1.9	(1.0)	3.1	(3.4)	1.9	(0.9)
Time per Detail (sec)	12.7	(8.3)	19.2	(17.3)	10.8	(8.2)	8.0	(4.6)	12.6	(10.8)	9.1	(6.7)
Rotation (Degrees)	5.6	(2.6)	5.0	(4.7)	5.5	(3.9)	3.3	(1.5)	4.1	(2.9)	3.1	(1.4)
Person Drawing:												
*Number of Details/13	8.0	(3.1)	8.3	(3.4)	8.7	(2.9)	10.0	(4.5)	10.3	(3.6)	10.4	(3.7)
*Distortion Ratio	0.66	(0.30)	0.61	(0.25)	0.57	(0.28)	0.57	(0.32)	0.49	(0.25)	0.43	(0.29)
Size (in^2)	11.4	(15.6)	14.1	(13.3)	21.9	(17.1)	15.8	(11.1)	13.3	(11.7)	14.5	(12.6)
Time per Detail (sec)	16.9	(19.3)	16.6	(18.6)	9.5	(6.7)	10.7	(11.2)	12.8	(11.9)	9.8	(8.0)

*See Table 13 for significance levels

lesions are impaired on several aspects of the drawings. However, it is of interest to note that the most impaired right-hemisphere sub-group is not the right parietal+ group, but the right anterior group. This subgroup of the right-hemisphere-damaged patients has the smaller number of details (person drawing), greater distortion (house drawing), and larger rotation. These patients also make larger drawings and, without comparative data, one can only speculate that the size of some of the person drawings, for example, is perhaps above normal.

There is a significant difference in the ages of the anterior and posterior groups (Table 10), but not between the anterior and parietal+ groups. This is largely due to the deletion, from posterior to parietal+ groups, of patients with temporal-lobe epilepsy, who tend to be younger than patients with tumours or vascular accidents. Therefore, and also because this is the comparison of greatest interest, statistical comparisons are shown in Table 13 for the anterior and parietal+ groups only. Moreover, since some of the patients did not do the person drawings, the N's for these are quite small, and analyses for person-drawing are not shown (though the overall picture is the same as for house drawings). Finally, since only 7 analyses of variance are now being performed, on the 7 house scores, the conventional .05 probability level is shown as significant. For all items in which there is a significant effect for side of lesion (left, right), locus of lesion (anterior, parietal+), or an interaction between these two factors, post-hoc Newman-Keuls tests were done to indicate the source of the effect. They are summarized in Table 13 as well.

Overall, the data confirm the important contribution of the left hemisphere to spontaneous drawing ability. However, it is of particular interest that the right parietal group produces significantly more appropriate details than the left anterior group; and the left parietal group is slower than all other groups, as measured in time per line.

Summary of Results

Left-hemisphere damage results in greater impairment of spontaneous drawing than does right-hemisphere damage. This is true whether a crude measure of identifiability of drawings is employed, or whether finer objective measures are used. Thus, the group of patients with left-hemisphere lesions produces fewer recognizable drawings than the group with right-hemisphere lesions,

138

TABLE 13. F RATIOS FOR 2X2 ANALYSIS OF VARIANCE
(SIDE X LOCUS), ANTERIOR AND PARIETAL+ GROUPS

	Side	Locus	Side X Locus
House Drawing:			
Number of Details[1]	7.077**	.435	.587
Distortion Ratio	4.137*	1.955	.279
Number of Lines	4.466*	.493	.095
Size	.216	.325	1.388
Time per Line[2]	7.821**	2.321	4.330*
Time per Detail	4.361*	.263	2.655
Rotation	.190	.565	.244

*P<.05
**P<.01

[1] Left anterior and right parietal groups differ significantly.
[2] All groups differ significantly from the left parietal group.

and they produce fewer lines and smaller drawings. Patients with right-hemisphere lesions make more contralateral omissions on their drawings, and they rotate them more on a page.

The largest proportion of poor drawings comes from apraxic/aphasic patients, and particularly apraxic patients, with left-hemisphere damage. Such patients are impaired on almost every aspect of drawing, including even a tendency to rotate drawings on the page.

In the group of patients with right-hemisphere damage, drawing disability is greatest in those who show generalized contralateral neglect, most of whom also have a hemiparesis. Drawing disability after right-hemisphere damage thus appears to be related to extensive damage within the hemisphere.

When lesions are restricted in extent, the major contribution of the left hemisphere to spontaneous drawing ability is confirmed. Moreover, the contribution of anterior systems appears to be greater than that of the parietal region, contrary to some previous claims in the literature. However, left parietal damage does significantly slow the drawing performance.

DISCUSSION

Our data indicate that, when patients are selected only on the basis of unilateral damage, lesions of the left hemisphere have a more general detrimental effect on spontaneous drawing than do lesions of the right hemisphere. This contrasts with some prior claims in the literature that right-hemisphere damage causes greater drawing impairment (McFie, Piercy and Zangwill, 1950; Piercy, Hécaen & de Ajuriaguerra, 1960; Hécaen, 1962; Binder, 1982). These earlier conclusions were probably a result of two main factors. First, most of them employed copying tasks, and there is a general consensus in the literature that patients with left-hemisphere damage are assisted in drawing by having a model to copy, whereas patients with right-hemisphere damage are not (see for example, Piercy et al., 1960; Warrington et al., 1966; Gasparrini, Shealy & Walters, 1980). Second, and probably more important, most of the studies reporting the overwhelming constructional deficits in right-hemisphere lesions have come from highly selected cases, rather than consisting of a series of unselected patients. In particular, severe aphasics were probably excluded from consideration, thus artifactually enhancing the defects after right-hemisphere damage relative to left.

The exclusion of severely aphasic cases in much of the literature on constructional apraxia may often have occurred quite inadvertently, through the apparently reasonable aim of assessing only patients who unequivocally understood what was required of them. This degree of assurance can seldom be gained from a severe aphasic, and one must therefore be satisfied with a reasonable estimate of the degree to which comprehension on any one task was achieved. Without information on the performance of such severely impaired patients, one has a very biassed sample of the effects of left-hemisphere damage on any measure. Where care has been taken not to exclude patients with severe aphasia, the paramount contribution of the right hemisphere to drawing has not been confirmed (Colombo, de Renzi & Faglioni, 1976; Arena & Gainotti, 1978).

In the present sample, which was a very large unselected, and so far as we could determine, unbiassed series of unilateral cases, the largest number of impaired drawings came from patients who were aphasic or apraxic (or both). Arena and Gainotti (1978) also reported that aphasic patients were the worst group at copying designs, and found their performance highly correlated with scores on the Token Test, a test of receptive speech function. However, the drawing difficulties in our sample could hardly be attributed to comprehension problems, since the apraxic subgroup was more consistently affected, and drawing disability correlated better with apraxia than with aphasia. Rather, it appeared that the drawings were generally impoverished and poorly executed, sometimes to the point of being unidentifiable unless one knew what the target item was supposed to be. Such impoverishment, or "oversimplification" of drawings has been noted by several previous investigators (Arrigoni and de Renzi, 1964; Gainotti and Tiacci, 1970; Piercy et al., 1960).

Among patients with right-hemisphere damage, difficulties with drawing were seen most frequently in patients who had a hemiparesis and/or who showed contralateral neglect on a visual search task. These variables were not related to drawing difficulties in the left-hemisphere damaged group. The patients impaired on drawing in the right-hemisphere group thus had a high incidence of neurological impairments, and appeared to have more extensive damage within the right hemisphere. These patients still produced fewer unrecognizable drawings than the left apraxic group but they showed more contralateral omissions on their drawings, as would be predicted from their neglect and from the vast majority of studies on this topic. Our data thus indicate quite strongly that drawing disability in the right-hemisphere cases is associated with extensive damage, as suggested by Arrigoni and de Renzi (1964), and that this also gives rise to a higher incidence of neglect. This was coupled, in our cases, with a greater tendency to rotate drawings on the page. It appears that constructional ability is more diffusely organized within the right hemisphere than within the left, a suggestion that has also been made for sensorimotor function (Semmes, 1965).

When damage to more focal regions (anterior, parietal) within a hemisphere was examined for effects on drawing, the most consistently impaired group was the group with left anterior damage, followed by the left parietal group. The right parietal group was, on most measures, the least impaired, although this group did produce a small number of unidentifiable drawings. Our data are clearly at variance with the almost axiomatic position in the literature that the right parietal lobe is of paramount importance for visuoconstructional ability. Again we must emphasize that, had aphasic or apraxic patients been excluded from our sample, the scores for the right-hemisphere group as a whole would have been worse than that for the left. Within the right hemisphere, the right parietal group produced unrecognizable drawings with contralateral omissions at a slightly (but not significantly) higher rate than did restricted lesions in the anterior region of the right hemisphere. Since severe impairment in the right-hemisphere group in our sample was associated with more extensive damage, it may be that previous studies on patients with damage thought to be primarily in the right parietal region, actually had more extensive involvement of that hemisphere. This is supported by the fact that 5 of the 8 cases reported by McFie et al. (1950), and at least 7 of the 10 cases of Ettlinger, Warrington & Zangwill (1957), had a significant left hemiparesis.

We have, unfortunately, no reliable way of equating our own sample of unilaterally damaged patients with others described in the literature. Our expectations of a high incidence of poor drawings after certain focal lesions, based on the previous literature, was not confirmed, but it is difficult to know precisely what the expectations should be. However, there are a number of factors which may have contributed to the apparently small number of poor drawings in our sample of focally damaged cases. One is the degree to which extent of neural damage in living subjects can now be estimated, as compared to the era prior to computerized tomography, when many of the classic papers on the subject were written. That is, some cases formerly assumed to have focal damage might acutally have had more extensive lesions. In addition, the extent and severity of cerebrovascular strokes has been on the decline (Whishnant, 1983), and the availability of antibiotics (Woods & Teuber, 1978) and other inflammation-reducing medications has meant that the actual extent of neural tissue damage has probably been reduced in recent years. The combination in our series of less severe pathology and better localization methods has probably resulted in a more accurate estimate of the effects of frontal or parietal damage on drawing, than was possible in the older literature.

Leaving these historical and selection factors aside, the fact remains that aphasic and manually apraxic patients have a great deal of difficulty in drawing simple objects which they have seen, and probably themselves depicted, many times in their lives. The explanation put forward in the past, that they have an "executive" problem (Piercy et al., 1960), would certainly be consistent with the idea that apraxia consists of a motor selection or programming deficit. One manifestation of such a problem might be the smaller number of lines and details drawn, and in the left parietal group especially, the greater time per detail. One would, however, also expect, not merely that output would be less, but that what output there was would contain errors of execution; particularly since apraxic patients display a variety of errors, especially perseverative and unique (unrelated) errors in the acquisition of a manual skill (Kimura, 1977). If, in the present study, one looks at the number of faulty details (distortion ratio), it is indeed highest in the apraxic group, and one might argue that this is akin to the production of unique errors. There is not, however, much evidence of perseveration of particular lines or configurations in their drawings.

Somewhat more puzzling is the number of very small drawings in the apraxic group. In fact, size of the drawings most sharply differentiated aphasic and apraxic patients. It may be that the smaller drawings permitted execution entirely with the distal musculature, a suggestion which would imply that the proximal musculature is somewhat more under praxic control (see Kimura, 1979, for a discussion of this point).

The objects to be drawn from memory (house, person) were familiar things which could be labelled. It is possible that the spontaneous drawing of such objects is normally controlled, in part, by a implicitly verbal "list" of components. In that case, one would certainly expect aphasic patients to draw fewer details. An additional factor may contribute to the impoverished content of patients with left-hemisphere, and particularly left anterior, damage. Such patients have been shown to have a diminished capacity for generating words beginning with a particular letter (Benton, 1968; Milner, 1964). They are even impaired, though not to the same degree as patients with right frontal lesions, on a task of generating meaningless nonverbal designs (Jones-Gotman and Milner, 1977). This factor has been called "fluency" and it implies an inability to generate items according to some imposed rule. It may well operate in the production of a variety of items, including the spontaneous drawing of familiar things.

The ability to draw familiar objects is of course an extremely complex act. It requires not only spatial synthesis and praxic control, but also some retrieval of the components of an object from long-term store. It has frequently been suggested that different factors are contributed by the left and right hemisphere, and while this is a reasonable position, the objective evidence for it so far is very weak. One major drawback of practically all studies on constructional disability, including our own, is that we typically have available only the final performance on a task, rather than a record of the behaviour that led to that end result. Much of the speculation concerning different mechanisms for impairment after left or right damage can best be confirmed by making detailed behavioural observations. We are carrying out such studies at present, but unfortunately no such information was available for the retrospective analyses reported here.

Another requisite for future research is a reference group of patients who do not have significant CNS damage. "Normal" control groups are often unsatisfactory in that their general health is typically well above that of neurological patients, and hence they may not really elucidate the problems specific to nervous system damage. Instead, what is needed are age- and education-matched controls, who are in comparable hospital settings, etc. Again, it was not feasible to collect data on a retroactive comparison group, but our current research incorporates such controls.

Finally, our study strongly suggests that neuropsychologists must in future exercise great care in the selection of patients for research purposes. Selecting patients because they have a particular symptom, or because they are medical referrals, will not result in a balanced picture of the deficits which occur when either a hemisphere, or a particular sub-system of the brain, is affected. Von Monakow (1914) implicitly recognized this important point when he tallied not only "positive" cases, i.e., instances of a defect after damage to a presumptively critical region, but also "negative" cases. The assumption, of course, is that we must have some mechanism for testing the negative cases.

ACKNOWLEDGMENTS

This research was supported by grants to D. Kimura from the Medical Research Council, and the Natural Sciences and Engineering Research Council, Ottawa. We are grateful to the members of the Clinical Neurological Sciences department, University Hospital, London, Canada for allowing us to see their patients, and to the patients themselves for their co-operation.

REFERENCES

Arena, A. and Gainotti, G. (1978). Constructional apraxia and visuoperceptive disabilities in relation to laterality of cerebral lesions. Cortex, 14, 463-473.

Arrigoni, G. and de Renzi, E. (1964). Constructional apraxia and hemisphere locus of lesion. Cortex, 1, 170-197.

Benton, A. (1968). Differential behavioural effects in frontal lobe disease. Neuropsychologia, 6, 53-60.

Benton, A. (1979). Visuoperceptive, visuospatial, and visuoconstructive disorders. In Clinical Neuropsychology. (eds. K. Heilman and E. Valenstein). New York, Oxford Univ. Press, 186-232.

Binder, L.M. (1982). Constructional strategies on complex figure drawings after unilateral brain damage. Journal of Clinical Neuropsychology, 4, 51-58.

Brodal, A. (1981). Neurological Anatomy. New York, Oxford, 3rd edition.

Colombo, A., de Renzi, E. and Faglioni, P. (1976). The occurrence of visual neglect in patients with unilateral cerebral disease. Cortex, 12, 221-231.

Corkin, S., Milner, B. and Rasmussen, T. (1970). Somato-sensory thresholds. Archives of Neurology, 23, 41-58.

Denny-Brown, D. and Chambers, R.A. (1958). The parietal lobe and behavior. In The Brain and Human Behavior, Proceeds of the Association for Research in Nervous and Mental Disease, 36, 35-117.

Ettlinger, G., Warrington, E. and Zangwill, O.L. (1957). A further study of visualspatial agnosia. Brain, 80, 335-361.

Gainotti, G. and Tiacci, C. (1970). Patterns of drawing disability in right and left hemispheric patients. Neuropsychologia, 8, 379-384.

Gasparrini, B., Shealy, C. and Walters, D. (1980). Differences in size and spatial placement of drawings of left versus right hemisphere brain-damaged patients. Journal of Consulting and Clinical Psychology, 48, 670-672.

Hécaen, H. (1962). Clinical symptomatology in right and left hemispheric lesions. In Interhemispheric Relations and cerebral Dominance. (ed. V.B. Mountcastle). Balti-more, Johns Hopkins Press, 215-243.

Jones-Gotman, M. and Milner, B. (1977). Design fluency: the invention of nonsense drawings after focal cortical lesions. Neuropsychologia, 15, 653-674.

Kimura, D. (1977). Acquisition of a motor skill after left-hemisphere damage. Brain 100, 527-542.

Kimura, D. (1979). Neuromotor mechanisms in the evolu-tion of human communication. In Neurobiology of Social Communication in Primates. (eds. H.D. Steklis and M.J. Raleigh). New York, Academic Press, 197-219.

Kimura, D. (1984a). Neuropsychology Test Procedures, DK Consultants, London, Canada.

Kimura, D. (1984b). Visual search in patients with uni-lateral lesions. Perception Seminar, York University, Toronto, March.

Kimura, D., Barnett, H.J.M. and Burkhart, G. (1981). The psychological test pattern in progressive supranuclear palsy. Neuropsychologia, 19, 301-306.

Kimura, D. and Harshman, R.A. (1984). Sex differences in brain organization for verbal and non-verbal functions.

In Sex Differences in the Brain. Progress in Brain Research. (eds. G.J. de Vries, J.P.C. de Bruin, H.B.M. Uylings and M.A. Corner). Elsevier, Amsterdam, 61, 423-441.

McFie, J., Piercy, M.F. and Zangwill, O.L. (1950). Visual-spatial agnosia associated with lesions of the right cerebral hemisphere. Brain, 73, 167-190.

McFie, J. and Zangwill, O.L. (1960). Visual-constructive disabilities associated with lesions of the left cerebral hemisphere. Brain, 83, 243-260.

Milner, B. (1964). Some effects of frontal lobectomy in man. In The Frontal Granular Cortex and Behavior. (eds. J.M. Warren and K. Akert). New York, McGraw Hill, 313-334.

Monakow, C. von. (1914). Die Localisation im Grosshirn, Wiesbaden, Bergmann Verlag.

Piercy, M., Hécaen, H. and de Ajuriaguerra, J. (1960). Constructional apraxia associated with unilateral cerebral lesions - left and right sided cases compared. Brain, 83, 225-242.

Renzi, E. de and Vignolo, L. (1962). The Token Test: a sensitive test to detect receptive disturbances in aphasics. Brain, 85, 665-678.

Semmes, J. (1965). A non-tactual factor in astereognosis. Neuropsychologia, 3, 295-315.

Warrington, E.K., James, M. and Kinsbourne, M. (1966). Drawing ability in relation to laterality of cerebral lesion. Brain, 89, 53-82.

Whishnant, J.P. (1983). The role of the neurologist in the decline of stroke. Annals of Neurology, 14, 1-7.

Woods, B.T. and Teuber, H-L. (1978). Changing patterns of childhood aphasia. Annals of Neurology, 3, 273-280.

10
Extreme Mathematical Talent: A Hormonally Induced Ability?

Camilla Persson Benbow and Robert Michael Benbow

The Study of Mathematically Precocious Youth (SMPY) was founded in 1971 by Julian C. Stanley to find and help mathematically talented junior high school students. To identify such 12- to 13-year-olds, the concept of a talent search was initiated. Students in the seventh grade (and occasionally in the eighth) were eligible to participate in these talent searches if they had scored in the upper 5% in 1972, 2% in 1973 and 1974, and thereafter in the upper 3% on national norms on the mathematics part of a standardized achievement test. In the beginning these talent searches only spanned the Middle Atlantic states of the U.S. Now they are national in scope.

As part of each talent search, qualified students take the Scholastic Aptitude Test, Mathematics (SAT-M) and Verbal (SAT-V) sections, at a national administration of the SAT in their local area. The SAT-M and SAT-V were designed to measure mathematical and verbal reasoning ability, respectively, of above average 11th and 12th graders. Because the intellectually talented 12-year-olds studied by SMPY have not had much experience with abstract mathematics and have not been exposed to the content of the test, the SAT-M is an especially good measure of mathematical reasoning ability for them (Benbow & Stanley, 1983).

The importance of mathematical reasoning ability is highlighted by the performance of the 1986 U.S. Mathematical Olympiad team. Four of the six students on this Olympiad team were identified several years ago by SMPY as extremely mathematically precocious . Before age 13, each had scored at least 700 on SAT-M. These four males won 2 gold medals and 2 silver medals and were instrumental in the U. S. team's tieing with the USSR for first

*We thank Joel Abramowitz and Michael O'Boyle for helpful comments and suggestions. Support was provided by the National Science Foundation (MDR-8651737) and the Department of Education.

147

place in the 1986 competition (J. Stanley, Personal Communication).

In 1972 when the first talent search was conducted, the staff of SMPY did not anticipate any sex differences in SAT-M scores. Thus, it was quite surprising that a sex difference was found, which has turned out to be one of the largest ever detected. Almost 8% of the 7th-grade boys and 27% of the 8th-grade boys in 1972 scored more than 600 on SAT-M (approximately the 80th percentile of college-bound 12th grade males), while not one single girl did. The mean difference in scores was 37 points for the 7th-graders and 70 points for the 8th-graders. Initially, we believed that the sex difference was most likely an artifact. If not that, it was thought to result from environmental and sociological factors. When a large sex difference was found the following year in the talent search, the staff of SMPY was again surprised. In every talent search subsequently conducted by SMPY, however, a large sex difference has been found (Benbow & Stanley, 1980). For the 9,927 students tested from 1972 through 1979 no statistically significant sex difference in verbal ability, as measured by the SAT-V, was found. Yet there was a consistent sex difference on SAT-M, which was equivalent to about one half of a standard deviation. This was observed even when the boys and girls had been matched on achievement test scores before taking the test.

The results until 1980 were limited, however, to highly motivated mathematically talented students. In 1980 the traditional mathematics talent search expanded into verbal and general ability areas as well. The top 3% in mathematical, verbal, or general ability were now eligible to participate in a talent search. Moreover, as time went on the entire United States was covered by the talent searches conducted at Johns Hopkins, Duke, Northwestern University, and the University of Denver. Two results came from these expansions: Approximately 100,000 students now take the SAT as part of a talent search each year and equal numbers of boys and girls participate. Despite these modifications the sex difference in SAT-M scores has remained stable. For example, the sex difference in the Johns Hopkins talent searches since and including 1980 has ranged from 30 to 36 points. There has been no indication that this difference is changing.

Although this average sex difference may be important, it is not the major finding resulting from the data. Benbow and Stanley (1983) have analyzed in detail the results for the 1980 to 1983 Johns Hopkins talent searches. They determined that sex differences in mathematical reasoning ability are greatest at the top of the scale. For example, among 12-year-old students scoring at least 500 on SAT-M (the approximate average score of college-bound 17-year-old males), there is a ratio of two males (2.1) for every female (based on 5,325 cases); at $\geq$ 600 SAT-M (77th percentile of college-bound 12th grade males) the ratio is 4.1 to 1 (806 such cases); and at $\geq$ 700 SAT-M (94th percentile of college-bound senior males) the ratio is 12.9 to 1 for the 278 cases reported by Benbow and Stanley (1983).

On the basis of results from several 100,000 seventh grade students who have been tested through the various talent searches in the United States, it is quite clear that there are very large sex differences in mathematical reasoning ability by age 12, especially at the highest levels of that ability. Moreover, these findings have been replicated in other countries as well (e.g., Stanley, Huang, & Zu, 1986). Benbow and Stanley (1982) and Benbow and Minor (1986) have shown that the observed sex difference at age 13 can predict subsequent sex differences in mathematics and science participation and achievement in high school. Thus, the sex difference appears to have important predictive value. It is therefore important to determine why it occurs.

Extensive studies conducted over a 14 year period by the staff of SMPY and others, have been unable to find results that are consistent with an exclusively environmental explanation for the sex difference (e.g., Benbow & Stanley, 1982, 1983; Raymond & Benbow, in press). Because we have been unable to find support for an entirely environmental explanation of the sex difference in mathematical reasoning ability in SMPY's high-ability population, we began in 1980 to consider possible biological factors that might influence intellectual ability. It had been shown previously that there was a possible association between mathematical reasoning and spatial ability (which may have a biological basis). In addition, we had tested previously parents of extremely talented students, who turned out to be highly intelligent (Benbow, Zonderman, & Stanley, 1983). Thus, we had some indication that biological factors might be involved in producing extremely high ability.

Little research on possible biological correlates of extremely high reasoning ability has been conducted, possibly because of the social and political ramifications. Nonetheless, three physiological correlates of that ability have been identified to date: left-handedness, symptomatic atopic disease (allergies), and myopia (Benbow, in press). The first two appear to be related to bihemispheric representation of cognitive functions or the influence of fetal testosterone, as will be described below. New data presented in this paper are consistent with that possibility.

The striking findings of Geschwind and Behan (1982) initially prompted our investigations in this area. Geschwind and Behan had reported that left-handers suffer more frequently from immune disorders, learning disabilities, and migraines than right-handers. Their explanation for this unusual finding involved testosterone. They hypothesized that if a developing fetus is exposed to high levels of testosterone or has an increased sensitivity to it, it has two effects. The thymus gland is affected resulting in immune disorders, such as allergies. Moreover, testosterone slows down the development of the left-hemisphere of the brain. As a result, the right-hemisphere compensates and becomes stronger. This increases the likelihood of that individual being left-handed (Geschwind & Behan, 1982). Since mathematical reasoning, in contrast to computation, and spatial ability are considered to be functions more efficiently carried out by the right than the left-hemisphere, we decided to investigate the possibility that extremely mathematically talented students would be more frequently left-handed.

Students representing the top 1 in 10,000 in mathematical reasoning ability (i.e., before age 13, they had an SAT-M score of at least 700) were surveyed using the Oldfield Handedness Inventory. The criterion for left-handedness was LQ < 0. It was found that the rate of left-handedness for this group was about twice that reported for the general population (Benbow, in press). Moreover, the frequency of left-handedness among such students was greater than for their parents or siblings and a comparison group of above average ability (approximately the top 1 in 20) students (Benbow, in press).

Having found the above relationship between left-handedness and mathematical precocity, we decided to investigate the possibility that extremely mathematically talented students would also exhibit an increased frequency of immune disorders (such as, allergies). An allergy questionnaire, which was designed by Dr. Franklin Adkinson of the Johns Hopkins Medical School at Good Samaritan Hospital, was used. Allergies were classified according to type, severity, and duration. This questionnaire was completed by the parents of the students whose mathematical reasoning ability was at the top 1 in 10,000 level. The frequency of allergies found among such students (i.e., approximately 55%) was more than twice the rate found for the general population. This frequency was also greater than that reported for their parents and siblings and the comparison group (Benbow, in press).

Moreover, the frequency of left-handedness and allergies was also investigated among a group of students who represented at least the top 1 in 10,000 in verbal reasoning ability. They also exhibited these traits at a frequency similar to the mathematically talented students. Although this may seem counterintuitive, this finding may be explained by the nature of the ability studied. We studied verbal reasoning. This ability may be more strongly under the influence of the right-hemisphere, especially in contrast to language production (Gardner et al., 1983).

The above findings are, therefore, consistent with the hypothesis that students with extremely high mathematical or verbal reasoning ability are exposed prenatally to higher than normal levels of testosterone. The late Norman Geschwind, upon learning of our results, suggested that we study when such students were born. He predicted that these students would be conceived more frequently in months when daylight occupies more than 12 hours per day. [Daylight affects pineal gland secretion, altering the level of melatonin, which in turn has an inhibitory effect on reproductive hormones (Lewy, 1983; Lewy et al., 1980).] Because of previous controversies in this area, caution was exercised in dealing with this hypothesis. Moreover, the biological basis of Geschwind's hypothesis is not entirely clear. Nevertheless, as is shown below, the data did bear out Geschwind's prediction remarkably well.

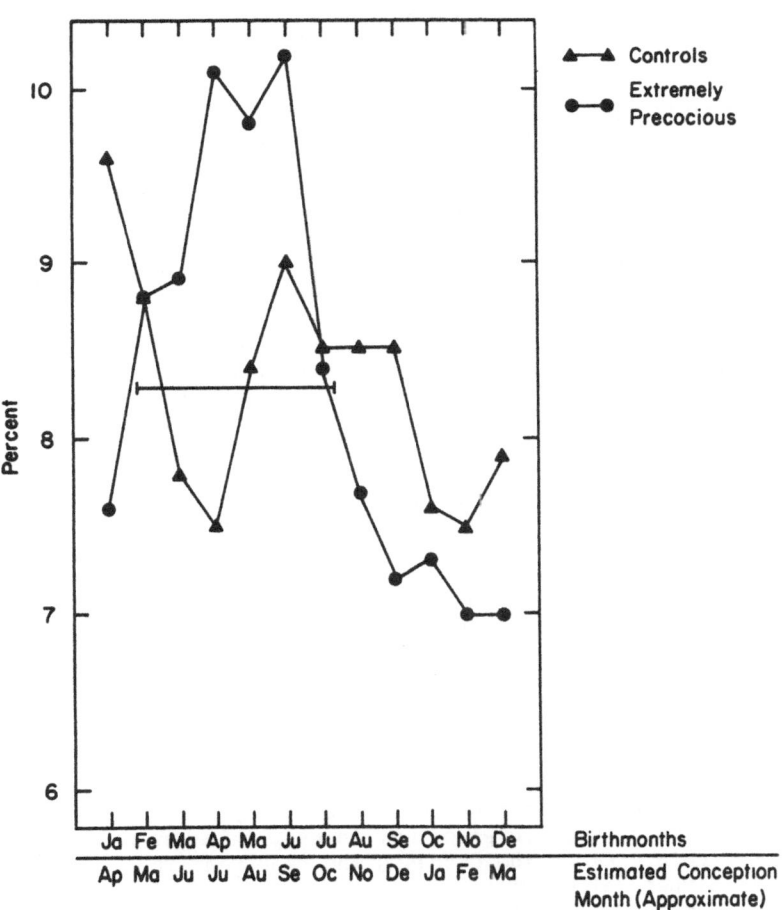

Figure 1

The proportion of extremely talented students born in each
month was computed and adjusted for the length of the month. The
average length of a month is 30.4375 days. The data, which are in
the form of a moving average, are presented in figure 1. This
yielded a smooth curve with the maximum in June and a minimum in
December, exactly six months apart. Moreover, the six months
February through July are all above the mean for the year and the
six months August through January are below. The probability that
the figures for any six consecutive months will all be higher than
the other 6 months is less than .05. The equinoxes are March 21
and September 21; as shown in figure 1, most of the extremely
precocious students were conceived in the months when daylight
occupies more than 12 hours per day. Moreover, birthmonth data for
average ability students born in the same years as the
intellectually precocious students in this sample were reported by
Badian (1984). The data for these students are also depicted in
figure 1. Clearly, there is no seasonal pattern to be found in the
birthmonths of the control group.

An article by Maccoby et al. (1979) suggested another possible
avenue of research. Maccoby et al. (1979) reported that first-born
children are exposed to higher hormone levels (i.e., progesterone
and, for boys only, testosterone) than later-born offspring. Thus,
we decided to investigate what proportion of students in the top 1
in 10,000 in either mathematical or verbal reasoning ability were
first-borns. Moreover, for comparison purposes we also studied
those students whose abilities were in approximately the 95th
percentile and were in the same age range as the extremely
talented. It was found that 62% of the extremely precocious
students and 48% of the comparison group were first-borns, a
difference that was statistically significant at the .05 level.

The data reported above provide several lines of converging
evidence that are consistent with the hypothesis that extremely
precocious students may be exposed to high levels of hormones
during fetal development. This may be important since it has been
shown that progesterone exposure enhances numerical ability and
androgens affect spatial ability (e.g., Reinisch, Gandelman, &
Spiegel, 1979). Moreover, Levy and Gur (1980) proposed that high
levels of fetal sex hormones promoted the maturational rate and
cognitive capacity of the right hemisphere, which has been
implicated as being specialized for non-verbal abilities, such as
spatial and mathematical reasoning. Geschwind and Behan (1982)
made a similar proposal for testosterone as discussed above.

Although the evidence is only circumstantial that high
levels of prenatal hormones or testosterone exposure may be
correlated with extreme intellectual talent, this may not be the
case for myopia. Among the students who represent the top 1 in
10,000 in either mathematical or verbal reasoning ability, we found
that approximately 55% were myopic at this early age. This was
about the same as reported for their parents and much greater than

for their siblings (36%) and the comparison group (22%) (Benbow, in press). The rate of myopia for the extremely precocious students was about four times the frequency found even for high school students.

One last area of our research deserves to be mentioned even though it is preliminary. The above left-handedness findings have broader implications that can be examined directly. Left- and mixed-handers and right-handers with left-handed relatives have been found more frequently to have bilateral or diffuse representation of cognitive functions (Bradshaw & Nettleton, 1983). Since the majority of extremely precocious students fit into one of the above three handedness groupings, this may imply that bilateral representation of cognitive functions per se (rather than greater specialization of the hemisphere) is associated with extremely high mathematical and/or verbal reasoning abilities. Some preliminary results have been obtained that are consistent with this hypothesis.

A subset of students identified by SMPY as being in the top 1 in 10,000 in either mathematical or verbal reasoning ability (N = 72) were tested with a computer simulation of a tachistoscope using a letter matching and a rotation task. In the rotation task the subjects were to decide if the letter "R," which was presented in one of four rotational orientations on the computer screen to the subject's right or left visual field, was in a normal or mirror-image position. In the letter matching task subjects compared two pairs of letters, which were presented sequentially and the second pair (the test stimulus) to either the right or the left visual field, and decided if the letter pairs had the same or a different name. The letter-matching task was developed by Posner et al. (1969) and the experimental procedures for this task were essentially a replication of those utilized by Kroll and Madden (1978).

For both the rotation and the letter-matching task each student was presented with a practice block, which was then followed by three experimental blocks of 32 trials each. Each experimental block consisted of two examples of the 16 possible conditions for the rotation task and four examples of the eight possible conditions in the letter-matching task, all randomly arranged within a block. A beep signalled the beginning of each trial. For the rotation task, 1000 msec later a fixation point appeared and lasted for 1000 msec. Then the screen went blank for 135 msec and the test stimulus was flashed to either the right or the left visual field for 100 msec. A mask was flashed to the opposite side. In the letter-matching task the memory stimulus, which was presented centrally for 150 msec, followed the beep after 1000 msec. The screen went blank for 500 msec and then the fixation point appeared for 1000 msec. The screen went blank for 120 msec and then the test stimulus appeared for 100 msec. On the side opposite the stimulus a mask was flashed. Visual field of presentation was random for both tasks. The accompanying computer program recorded the response time in milliseconds, beginning at the onset of the test stimulus and ending when the subject depressed the response button using their dominant hand. Errors were also recorded.

The mean response times of the extremely talented students for each task and by the right and left visual field are located in Table 1, as are the mean number of errors. On the rotation task no significant difference in response times between the right and left visual field was found by a t-test. On the letter-matching task, however, the left visual field had significantly faster response times than the right (t = 3.94, p < .001). In terms of errors made, the same pattern of results as for response times was found. In this instance, the difference was significant for the letter-matching task at the p < .01 level (t = 3.0). Thus, the results of this study were not in the direction usually reported and indicate that such extremely talented students do have strong right hemispheres.

Table 1. Mean reaction times and number of errors, by visual field, for the rotation and letter-matching task for extremely talented students (N = 72).

Visual Field	Rotation Task		Letter Matching Task	
	Mean Time	Mean Errors	Mean Time	Mean Errors
Left	1.0447	4.8	0.8850	3.8
Right	1.0441	4.3	0.9094	4.8

Because a t-test may "disguise" a possible interaction between handedness and sex, the difference in response times between hemispheres was also subjected to an ANOVA by sex and handedness grouping. The handedness grouping was right-handed with no family history of left-handedness (N = 20) and all others (N = 34). (Because of the cultural bias against left-handedness among Asians, they were studied separately. Results did not differ for them.) Since the ANOVA was nonorthogonal, special procedures for interpretation were followed. Although the differences were for the most part not significant, the trend in the data indicated that the difference in reaction times between the hemispheres in the letter-matching task were much larger if the person was male or had a history of left-handedness, but the difference for all groups consistently favored the right hemisphere. That is, the right hemisphere was faster and more accurate for all sub-groups of subjects, but even more so for left-handers and males. (Right-handed males did not show the typically reported pattern.)

On the rotation task, the relationships were less clear.

Nonetheless, the reversal of the usual pattern of results were stronger for males and individuals with a history of left-handedness. The differences were not very large, however. This was rather surprising. We had expected that this task may have showed the greatest right hemisphere advantage rather than the letter-matching task.

It seems then that extremely intellectually talented students may have strong right hemispheres and exhibit bilateral or diffuse representation of cognitive functions. The interpretability of these results are limited at present by the fact that the hand of response was not randomized or counterbalanced, and by the lack of a control group. A co-worker of Levy (Levy, Personal Communication) has found similar results for a group of extremely intellectually talented students and did have a control group. Moreover, the letter-matching task was a standard task devised by Posner et al. (1969) and was programmed in the same manner as Kroll and Madden (1978). Nonetheless, caution should be used when generalizing from these results.

In conclusion, the Study of Mathematically Precocious Youth, now at Iowa State University, has identified several physiological correlates of extreme intellectual talent. These are left-handedness, allergies, myopia, and possibly prenatal testosterone exposure (or high hormone levels prenatally) and bilateral representation of cognitive functions. Some of these may be related to the large sex difference found in mathematical reasoning ability among intellectually talented students and bear on the etiology of extreme intellectual abilities.

REFERENCES

Badian, N. A. (1984). Reading disability in an epidemiological context: Incidence and environmental correlates. Journal of Learning Disabilities, 17, 129-136.

Benbow, C. P. (In press). Physiological correlates of extreme intellectual precocity. Neuropsychologia.

Benbow, C. P. and Minor, L. L. (1986). Mathematically Talented Students and Achievement in the High School Sciences. American Educational Research Journal, 23.

Benbow, C. P. and Stanley, J. C. (1980). Sex differences in mathematical ability: Fact or artifact? Science, 210, 1262-1264.

Benbow, C. P. and Stanley, J. C. (1982). Consequences in high school and college of sex differences in mathematical reasoning ability: A longitudinal perspective. American Educational Research Journal, 19, 598-622.

156

Benbow, C. P. and Stanley, J. C. (1983). Sex differences in mathematical reasoning ability: More facts. Science, 222, 1029-1031.

Benbow, C. P., Zonderman, A., & Stanley, J. C. (1983). Assortative marriage and the familiarity of cognitive abilities in families of extremely gifted students. Intelligence, 7, 153-161.

Bradshaw, J. L. and Nettleton, N. C. (1983). Human Cerebral Asymmetry. Englewood Cliffs, NJ: Prentice-Hall, Inc.

Gardner, H., Brownell, H. H., Wapner, W., and Michelow, D. (1983). Missing the point: The role of the right hemisphere in the processing of complex linguistic materials. In Cognitive processing in the right hemisphere, E. Perecman (Editor), pp. 169-191. Academic Press, New York.

Geschwind, N. and Behan, P. (1982). Left-handedness: Association with immune disease, migraine, and developmental learning disorder. Proceedings of the National Academy of Sciences, 79, 5097-5100.

Kroll, N. E. A. and Madden, D. J. (1978). Verbal and pictorial processing by hemisphere as a function of the subject's verbal scholastic aptitude score. In Attention and performance VII. New York: Wiley and Sons.

Levy, J. and Gur, R. C. (1980). Individual differences in psycho-neurological organization. In J. Herron (Ed.) Neuropsychology of left-handedness. New York: Academic Press, 199-210.

Lewy, A. J. (1983). The effects of light on human melatonin production and the human circadian system. Prog. Neuro-Psychopharmacology and Biological Psychiatry, 7, 551-556.

Lewy, A. J., Wehr, T. A., Goodwin, F. K., Newsome, D. A., & Markey, S. P. (1980). Light suppresses melatonin secretion in humans. Science, 210, 1267-1269.

Maccoby, E. E., Doering, C. H., Jacklin, C. N., & Kraemer, H. (1979). Concentrations of sex hormones in umbilical-cord blood: Their relation to sex and birth order of infants. Child Development, 50, 632-642.

Posner, M. I., Boies, S. J., Eichelman, W. H., & Taylor, R. L. (1969). Retention of visual and name codes of single letters. Journal of Experimental Psychology, 79, 1-16.

Raymond, C. L. and Benbow, C. P. (In Press). Gender differences in mathematics: A function of parental support and sex-typing? Developmental Psychology.

Reinisch, J. M., Gandelman, R., & Spiegel, F. S. (1979). Prenatal influences on cognitive abilities: Data from experimental animals and human and endocrine syndromes. In Sex-related differences in cognitive functioning: Developmental issues. (Eds. M. Wittig and A. C. Petersen). New York: Academic Press, 215-240.

Stanley, J. C., Huang, J., and Zu, X. (1986). SAT-M scores of highly selected students in Shanghai tested when less than 13 years old. College Board Review.

Session III
Emotion, Attention and Arousal

Chairman: G. Rosen

11
Disorders of Emotional Behaviour and of Autonomic Arousal Resulting from Unilateral Brain Damage

Guido Gainotti

INTRODUCTION

The existence of a link between experience of emotions and autonomic arousal has been widely recognized,although with different formulations in the history of psychology. Thus, James (1890) and Lange (1885-1822) suggested that the arousal of specific patterns of autonomic nervous system activity provides the basis, through feed-back from peripheral visceral responses, for qualitative distinctions among emotions. Successive non-specific arousal theories have argued that a quantitative, but no qualitative relationship exists between emotions and autonomic arousal (Duffy, 1962). This non-specific arousal theory has also provided the basis for theories suggesting that undifferentiated arousal is structured by cognitive factors (Schachter and Singer, 1962; Mandler, 1975) producing the impression that each emotion differs in feeling. Even more recently, Ekman et al.(1983) have revisited the specific autonomic arousal hypothesis, showing that different facial emotional expressions are associated with distinct patterns of autonomic activity.

In recent years, the problem of the relationships between emotions and autonomic arousal has also been taken into account with respect to the question of the hemispheric asymmetry in the processing of emotions and affects. A large series of investigations dealing with the more cognitive components of emotions (namely comprehension and expression of emotions) had, in fact, failed to clarify the meaning of the clinical observations which have prompted, about twenty years ago, the hypothesis of a right hemisphere dominance for emotions and affects. It seemed, therefore, logical to turn to more elementary components of the emotional experience, to see if the clinical differences observed between right and left brain-damaged patients could be better explained in terms of

Acknowledgements: This work was supported in part by a C.N.R. grant.

defective autonomic arousal than in terms of defective communication of emotions. An analysis of data gathered following this
line of research will, therefore, constitute one of the main
purposes of the present survey.

On the other hand, since investigations of possible defects
in autonomic arousal have come relatively late in the study of the
emotional behavior of patients with unilateral brain injury, we
will not by-pass in our presentation the evolution in time of
theories concerning the relationships between emotions and hemispheric specialization. We will, on the contrary, remind how facts
and theories about this problem have emerged and have evolved
during the last 20 years. The general plan of our presentation
will, therefore, include:

a short survey of the contribution of those classical authors
who have made important pioneeristic observations on this topic;

a more analytical account of the clinical investigations which
have described a contrasting emotional behavior in right and left
brain-damaged patients;

a schematic review of the experimental investigations which
have studied, both in normal subjects and in brain-damaged patients,
possible hemispheric asymmetries in comprehension and expression
of emotions;

a discussion of the different interpretations that have been
advanced to explain the results of clinical and experimental
investigations;

a description of new research strategies (including the study
of the autonomic arousal in brain-damaged patients) proposed to
choose among these alternative interpretations;

a short account of some results recently obtained by our
research group, following these new research strategies.

THE PIONEERS

More than a century ago, Jackson (1880) observed that when an
epileptic seizure begins with an emotional experience, such as
"dreamy states", "reminiscences", or "emotions of fear", the first
spasm is usually on the left side of the body.

Some decades later, Babinski (1914) drew attention to the phenomenon whereby a patient affected by organic hemiplegia was sometimes seemingly unaware of his disability (anosognosia) or treated
his paralysis with undue unconcern and cheerful acceptance (anosodiaphoria) and raised the question of the relationships between
these patterns of behavior and lesions of the right hemisphere.

These Babinski's observations were developed by Schilder
(1935) who quoted the hypothesis, advanced by Hirschl and by Potzl
of a possible dominance of the right hemisphere for vegetative
functions and by Critchley (1955) who suggested that right brain-
damaged patients might present a prevalence not only of anosognosia
and of anosodiaphoria, but also of other abnormal attitudes toward
the disability, such as the personification of the paralyzed limbs,
or misoplegia, defined as the tendency to use expressions of hatred
toward the paralyzed limbs.

The intriguing observations of these authors did not modify,
however, the prevailing views considering the right and left hemisphere as equivalent as for emotional behavior. This viewpoint was
stressed, for example, by Hécaen (1962) who, reviewing the clinical
work on cerebral dominance, explicitly claimed: "When there is a
disturbance of the activities of synthesis (such as personality
disturbances derived from cerebral lesions) no difference can be
found between left and right lesions".

These claims, however, were at variance with results obtained
almost in the same period by Terzian and Ceccotto (1959), Alema
and Donini (1960) and Perria et al. (1961), observing the emotional
behavior of patients submitted to pharmacological inactivation of
the right and left hemisphere. Injection of sodium amytal in the
left carotid artery produced a "depressive-catastrophic" reaction,
whereas barbiturization of the right hemisphere produced an opposite "euphoric" reaction.

Although observations reported by these authors have not been
afterwards systematically confirmed by other authors who have used
the same technique (negative results have been obtained, for ex.
by Milner, 1967 and more recently by Kolb and Milner, 1981) they
have the merit of having actually introduced the problem of the
relationship between emotions and hemispheric specialization.

ANALYSIS OF EMOTIONAL BEHAVIOR ASSOCIATED WITH HEMISPHERIC LESIONS

Clinical observations incidentally made in patients with unilateral brain lesions seemed consonant with the finding of a different emotional reaction following pharmacological inactivation of
the right and left hemisphere. In order to submit these clinical
impressions to a more rigorous control, we undertook in 1969 a
first systematic observation of the emotional behavior associated
with injury of either hemisphere. We distinguished (Gainotti, 1969)
three types of emotional reactions:
- catastrophic reactions, with anxiety and sudden bursts of tears
- well controlled emotional reactions;
- indifference reaction with cheerful acceptance of the disability and indifference toward failures.

A double dissociation was observed between right and left brain damaged patients, since catastrophic reactions prevailed in patients with left-sided lesions, whereas indifference reactions were more frequent in subjects with right hemisphere injury.

An analysis of the clinical context of these contrasting forms of emotional behavior gave some hints about the nature of catastrophic, but not of indifference reactions. Catastrophic reactions, being closely related to severe language and sensori-motor disorders and being usually elicited by repeated failures in verbal communication, could be considered as a dramatic but psychologically appropriate form of emotional reaction. Indifference reactions, on the other hand, could not be considered, as originally suggested by Hécaen (1962) as a sort of demential lack of awareness, since they were significantly associated with manifestations of neglect for the left half of the body and of extra-personal space, but not with mental deterioration.

We tried, therefore, to clarify the meaning of this abnormal form of emotional reaction, by undertaking a more analytical study of the patterns of emotional behavior shown by right and left brain-damaged patients (Gainotti, 1972). Results of this new study showed that anxiety reactions, bursts of tears, swearing and sharp or depressed refusals to go on with the examination are significantly associated with left hemisphere injury, whereas anosognosia, minimization of the disability, indifference to failures and tendency to joke are significantly more frequent in right-sided patients. Interestingly, patients with right-sided lesions also showed a prevalence of expressions of hatred toward the paralyzed limbs, thus confirming Critchley's (1955) suggestion that not only anosodiaphoria, but also misoplegia could be more frequent in patients with right hemisphere lesions.

Our finding of a double dissociation between right and left brain-damaged patients, with a prevalence of catastrophic reactions in patients with left hemisphere injury and of indifference reactions in subjects with right-sided damage has been afterwards confirmed by many other authors (e.g. Cutting, 1978; Robinson and Price, 1982; Denes et al., 1982; Robinson et al., 1984). This clinical observation can, therefore, be considered as one of the most firmly established facts in the study of the relationships between emotions and hemispheric specialization. On the other hand, since the clinical observation cannot explain the meaning of the contrasting emotional reaction of right and left brain-damaged patients (and in particular cannot clarify the meaning of the abnormal indifference reaction of patients with right-sided lesions) a large series of experimental investigations were undertaken both in normal subjects and in in patients with unilateral brain injury to more deeply explore the nature of the link between emotions and hemispheric specialization.

LATERALITY EFFECTS IN COMPREHENSION AND EXPRESSION OF EMOTIONS:
EXPERIMENTAL STUDIES IN NORMALS AND IN BRAIN-DAMAGED PATIENTS.

Although several different aspects of the emotional behavior
have retained the interest of neuropsychologists and experimental
psychologists, attention has been focused above all on the most
cognitive components of emotions, namely comprehension of emotionally laden stimuli and expression of emotions. This was due, on
one hand, to the tendency to assume that inter-hemispheric differences should emerge on late rather than on earlier stages of
processing (Moscovitch, 1979), that is to say in the more complex
rather than in the more elementary components of emotions, and, on
the other hand, to the development of a series of sophisticated
techniques of analysis of non-verbal communication (e.g. Ekman and
Friesen, 1975; Rosenthal et al., 1979) which have attracted the
attention of various authors toward the communicative aspects of
emotions. The present review of the experimental investigations of
emotional behavior will, therefore, be focused on studies dealing
with comprehension and expression of emotions, both in normal
subjects and in brain-damaged patients.

Comprehension and expression of emotions in patients affected
by unilateral brain lesions: Heilman et al. (1975) Tucker et al.
(1977) Ross (1981 and 1984) Ruckdeschel-Hibbard et al. (1984) and
Heilman et al., (1984) have shown that right brain-damaged patients
although perfectly able to understand the meaning of a sentence,
are more impaired than patients with left-sided injury in recognizing the emotion conveyed through prosody.

Results obtained with emotional visual stimuli have mirrored
those obtained with auditory affective material. De Kosky et al.
(1980), Cicone et al.(1980), Goldblum (1981), Benowitz et al.(1983)
and Etcoff (1984) have shown, in fact, that patients suffering
from unilateral right lesions are significantly more impaired than
subjects with left-sided damage in recognizing the emotion displayed
by an emotional face.

A quite similar situation can be observed if we pass from the
receptive to the expressive level, since Tucker et al.(1977), Ross
and Mesulam (1979), Ross (1981 and 1984), Hughes et al.(1983) and
Borod et al.(1984) have shown that right hemisphere lesions significantly impair the ability to convey emotion through the tone of
voice, whereas Buck and Duffy (1980) and Borod et al.(1984) have
demonstrated that right brain-damaged patients are less facially
expressive than subjects with left-sided injury.

These findings, however, have not been confirmed by all authors
since Schlanger et al.(1976) have observed no laterality effects
in the perception of emotional prosody and Kolb and Milner (1981)
have observed no difference between right and left brain-damaged
patients in emotional facial expressiveness.

Asymmetries in recognition of emotional cues and in expression of emotions in normal subjects: Several authors have shown that ear differences exist in evaluating the emotional tone and respectively the verbal content of sentences. Safer and Leventhal (1977) and Ley and Bryden (1982) have shown, for ex., that when the emotional tone of the speaker contrasts with the linguistic meaning of a sentence, the left ear (right hemisphere) pays attention above all to the emotional tone of voice, whereas the right ear (left hemisphere) bases its judgement on the verbal content of the sentence. Further- more, King and Kimura (1972) and Carmon and Nachson (1973) have shown a left ear superiority in identification of emotional non- speech sounds (such as outbursts of weeping or of laughter) and a similar left ear advantage has been demonstrated by Bryden et al. (1982) for identification of the emotional (positive or negative) quality of short tonal sequences.

Results very similar to those obtained in the auditory moda- lity have been obtained in the visual modality by presenting faces expressing a given emotion to the right or left of a fixation point and asking the subjects to identify the emotion displayed by the emotional face. Suberi and Mc Keever (1977), Campbell (1978), Landis et al. (1979), Ley and Bryden (1979), Ladavas et al. (1980), Hansch and Pirozzolo (1980), Safer (1981), Strauss and Moscovitch (1981) and Natale et al. (1983) have, in fact, consistently shown that the right hemisphere is faster and more accurate in identifying facial emotional expressions than is the left hemisphere. It must be acknowledged, however, that laterality effects were, in general, rather small and were sometimes limited only to subsets (e.g. males or females) of the whole experimental sample.

A last group of investigations has studied possible asymme- tries in facial expression of emotions, either by subjectively rating the expressiveness of the two sides of faces or by comparing the emotional intensity of mirror-image composites, formed by two right or two left sides of the same face. Results obtained with these methods by Campbell (1978), Sackeim and Gur (1978), Borod and Caron (1980), Heller and Levy (1981), Moscovitch and Olds (1982) and Dopson et al. (1984) have shown that asymmetries in facial expression of emotions often exist and that the left half face is generally rated as the more expressive one. Now, since the motor control of the left half face is accomplished by the right motor cortex, these findings suggest a right hemisphere superiority not only in processing emotional informations, but also in expressing emotions.

An important caution on this subject has been, however, suggested by Ekman,et al. (1981), who have made a careful distin- ction between "spontaneous" and "posed" facial emotional expression. Only the former should, in fact, be considered as emotional in nature, whereas the latter should be viewed as deliberate attempts to simulate an emotional experience. Now, since most photographs used to evaluate asymmetries of emotional expression did not reproduce spontaneous but posed emotional expressions, asymmetries observed on these photographs do not point to a greater right hemisphere involvement in the production of emotional expressions, but rather to a greater involvement of the right motor cortex "in any cortically directed facial mouvements" (Ekman et al., 1981).

So, if we try to summarize results obtained studying emotional communication in normal subjects and in brain-damaged patients, we can say that these studies have generally shown an advantage of the right hemisphere both in comprehending and in expressing emotions, but that the meaning of this right hemisphere superiority is obscured by at least three reasons:
(1) significant interactions have been sometimes observed between the laterality factor and other factors, such as: -the (positive or negative) type of emotion;-the linguistic or visual-perceptual material used to communicate emotions, - the task demands and so on;
(2) laterality effects are generally small;
(3) some negative results, concerning both recognition of emotionally laden stimuli and facial expression of emotions have been obtained.

Interpretations Advanced to Explain Hemispheric Asymmetries in Comprehension and Expression of Emotions

Several alternative interpretations have been advanced to explain results obtained studying expressive and receptive components of emotional communication in normal subjects and in brain-damaged patients.

The first interpretation assumes that the biological evolution of verbal and non-verbal communicative systems may have propted a different lateralization of language functions in the left hemisphere and of emotional communicative systems in the right hemisphere. Asymmetries in comprehension and expression of emotions should be due to the major involvement of the right hemisphere in non-verbal (emotional) communicative functions.

The second interpretation maintains that the right hemisphere advantage in comprehension and expression of emotions might be but an aspect of the more general superiority of this hemisphere for various aspects of emotional behavior (emotional appropriateness, comprehension and expression of emotions, emotional arousal). These two theories, despite some differences, agree in attributing to the right hemisphere a general superiority in the expression and comprehension of emotional stimuli.

This statement, however, is not accepted by supporters of the third theory, who, instead of a general superiority of the right hemisphere for the processing of emotional stimuli, suggest the existence of an interaction between hemisphere and type of emotion,

with a superiority of the right hemisphere in the treatment of negative emotions and an advantage of the left hemisphere in the
processing of positive emotions.

Right hemisphere and non-verbal communication: Ross (1981 and
1984) and Ruckdeschel-Hibbard et al. (1984) have proposed that a
parallelism might exist between the anatomo-functional organization
of language at the level of the left hemisphere and the corresponding organization of emotional communication at the level of the
right hemisphere. According to these authors, disruption of emotional communication should be considered as the primary disorder
resulting from right hemisphere injury and other emotional disturbances observed in these patients should be viewed as a consequence
of their basic inability to comprehend and express emotions.

Thus, the emotional indifference often described in these
patients should not be considered as a true indifference but as
the consequence of their primary inability to express an otherwise
intact experience of emotions. By the same token, these authors
maintain that the autonomic arousal should be normal in right
brain-damaged patients, their basic inability consisting in a
defect in emotional communication and not in a disorder of the more
basic components of the experience of emotions.

Hemispheric asymmetries and the distinction between "positive"
and "negative" emotions: Reuter-Lorenz and Davidson (1981), Natale
et al. (1983) and Reuter-Lorenz et al. (1983) have shown, that
negative emotions are better recognized by the left visual field
(right hemisphere) but positive emotions are better identified by
the right visual field. Analogously, Sackeim and Gur (1978) and
Schwarts et al. (1979) have shown that negative emotions are expressed more intensely by the left half face, whereas the emotion of
joy is expressed more intensely by the right half face. This dissociation between "positive" and "negative" emotions has been considered as a proof supporting the thesis of a different specialization of the right and left hemisphere for opposite aspects of the
mood.

This general theory will not be discussed in detail here for
obvious time/space reasons (see Sackheim et al., 1982 and Gainotti,
1983 for different viewpoints on this problem). I will, rather,
limit myself to note that the great majority of results obtained
studying comprehension and expression of emotions in normal subjects
and in brain-damaged patients do not support the hypothesis of an
opposite specialization of the right hemisphere for negative emotions and of the left hemisphere for positive affects. Campbell
(1978), Ladavas et al.(1980), Strauss and Kaplan (1980), Strauss
and Moscovitch (1981), Hirschman and Safer (1982), Duda and Brown
(1984) and Gage and Safer (1985) have shown, in fact, that in
normal subjects the left visual field is superior to the right one
in processing both negative and positive emotions, whereas Chaurasia
and Gaswami (1975), Campbell (1978), Heller and Levy (1981) and

Moscovitch and Olds (1982) have shown that in normals the left half face is more expressive than the right half for both negative and positive emotions.

The influence of sensori-motor asymmetries on tasks of emotional expression and comprehension: We have already said that Ekman et al.(1981) have questioned the hypothesis of a right hemisphere superiority in the production of facial emotional expressions suggesting that asymmetries observed on tasks of facial emotional expression could be due to a general advantage of the right hemisphere in the production of intentional facial movements, rather than to a specific superiority of this hemisphere in the production of properly emotional facial expressions.

This line of reasoning can also be applied to asymmetries observed in comprehension of emotional stimuli, since all these tasks require a fine grained analysis of complex perceptual material, for which the right hemisphere is more endowed than the left one. The general question raised by these methodological considerations consists, therefore, in evaluating if results obtained by normal subject and brain-damaged patients on tasks of emotional comprehension and expression point to a specific ability of the right hemisphere in the processing of emotional material, or can simply be explained on the ground of a more general superiority of the right hemisphere for various kinds of perceptual and motor functions.

Different strategies have been used by various authors to disentangle emotional from non-emotional factors. Some authors have contrasted, for ex., in the same subjects the recognition of emotional and non-emotional faces (e.g. Suberi and Mc Keever, 1977; Ley and Bryden, 1979; Hansch and Pirozzolo, 1980; De Kosky et al., 1980; Strauss and Moscovitch, 1981) or of emotional and non-emotional prosody (e.g. Weintraub,et al., 1981; Hughes et al., 1983; Heilman et al., 1984) to see if one and the same perceptual ability subsumes performances obtained on emotional and non-emotional tasks.

Other authors (e.g. Mc Keever and Dixon, 1981; Bryden and Ley, 1983; Gage and Safer, 1985) have tried to separate even more clearly emotional from non-emotional factors, by holding the stimulus material (e.g. the faces to be memorized) constant and varying only the emotional factor (i.e. the affective context in which memorization of emotionally relevant stimuli took place). Results obtained in all these research, although not conclusive, seem to indicate that both the physical characteristics and the emotional nature of the stimuli are in part responsible for the right hemisphere advantage.

TOWARD NEW RESEARCH STRATEGIES

Since results obtained studying hemispheric asymmetries in comprehension and expression of emotions have led themselved to diverging interpretations, new research strategies have been recently developed to approach from new directions the problem of the relationships between emotions and hemispheric specialization. Two main characteristics differentiate the new research strategies from those followed in experimental investigations that we have just summarized:
- the complexity of the experimental design
- the component of the emotional behavior on which attention is directed.

As for the first point, previous investigations had generally taken into account only one aspect at a time of the emotional behavior of brain-damaged patients, whereas more recent research tend to consider simultaneously various aspects of the emotional behavior of right and left brain-damaged patients.

As for the second point, previous investigations had payed attention to the most cognitive components of emotions (i.e. comprehension and expression of emotional situations) whereas more recent research tend to focus attention on more elementary components of the emotional experience (namely the autonomic arousal response to emotional stimuli, observed in patients with unilateral brain injury).

The advantage of studying in the same subjects various aspects of the emotional behavior consists in the fact that this research strategy could allow to control whether an internal consistency exists between the various aspects of the emotional behavior disrupted by brain damage or if disorders of emotional comprehension and expression can be better explained as a derivative of a more general perceptual or executive disorder produced by brain injury.

The interest of shifting attention from comprehension and expression of emotions to the study of autonomic arousal in unilateral brain-damaged patients derives from at least three sources: (a) the unquestionable importance of emotional arousal in the experience of emotions; (b) the relevance of data concerning the autonomic arousal to evaluate the meaning of the strange "indifference reaction" of right brain-damaged patients, since supporters of the "emotional communication defect" hypothesis have explicitly claimed that autonomic arousal is normal in right brain-damaged patients and that the apparent indifference of these subjects is a consequence of their basic inability to express in otherwise intact emotional experience, and, on the other hand; (c) the existence of data strongly suggesting that one of the main disorders of right brain-damaged patients does, indeed, consist in a defect of autonomic arousal. Heilman et al. (1978) have, in fact, shown that right brain-damaged patients present a defective autonomic response to painful stimuli, whereas Morrow et al. (1981) and Zoccolotti et al. (1982) have made similar observations during presentation of emotionally laden visual stimuli.

We have, therefore, recently developed, in cooperation with Pizzamiglio, Caltagirone, Zoccolotti, Mammucari and Guariglia and with the assistance of Ekman and Friesen, a new research program, designed to study at the same time the following components of the emotional behavior of right and left brain-damaged patients:
- emotional reaction to the disability;
- comprehension of emotional stimuli;
- expression of emotions;
- autonomic response to emotional stimuli.

Comprehension of emotional stimuli was assessed by asking patients to identify the emotion displayed by an emotional face. Stimuli were 36 photographs taken from the set of Ekman and Friesen (1975) of actors expressing six different emotions (happiness, disgust, suprise, anger, fear and sadness).

In order to control if performances obtained on this task were really due to the emotional nature of the stimuli (or to their physical characteristics), patients also received the "Face recognition test" (Benton and Van Allen, 1968), in which facial identity and not facial expression has to be recognized.

Facial expression of emotions was studied by showing the patients through a rear projector, movies devised to elicit pleasant, neutral and intensely unpleasant emotional reactions and videotaping their facial expressions.

At the end of the stimuli presentation, subjects were asked to briefly describe each movie and only patients who had understood the general meaning of each movie were included in the study. Emotional facial expressions were analyzed by means of the EMFACS procedure, which is an objective scoring technique, developed by Ekman and Friesen (1975), and were considered as congruent or not with the emotion elicited by each movie.

In order to control if possible differences in facial expression were due to the emotional nature of the task or to non specific asymmetries in facial mobility, each patient also received a test of imitation of non-emotional facial movements, based on Ekman and Friesen (1978) Facial Action Coding System (FACS).

Autonomic response to emotional stimuli: The same procedure used to provoke spontaneous facial emotional expressions was also used to study the autonomic responsivity to positive, neutral and negative emotional stimuli. Galvanic skin response was taken as an index of sympathetic arousal, whereas heart rate deceleration was retained as an index of para-sympathetic arousal.

Preliminary Results of Our New Research Program

Results of the above mentioned line of research are still preliminary, but are nevertheless, in part unexpected. As a matter of fact, no significant difference was found between right and left brain-damaged patients on tasks of emotional comprehension and expression, but right-sided patients showed behavioral and psychophysiological evidence of insensitivity to emotional stimuli.

As for the first point, Guariglia (1986) has found no significant difference between right and left brain-damaged patients on tests of recognition of facial emotions and facial identity, although showing that normal controls score much better than the brain-damaged groups on both tests.

In a similar manner, Mammucari et al. (1986) have failed to observe any significant difference between patients with right-sided and left-sided injury, in the frequency with which they displayed congruent facial expressions to emotional movies. In this case too, normal controls produced a number of congruent facial emotional expressions significantly higher than both brain damaged groups.

In the same experimental situation, however, an interesting difference was observed between right-sided patients and normal controls or left brain-damaged patients, during the negative emotional movie. In this situation, normal controls and left-sided patients showed the tendency to move the eyes away from the most unpleasant scene (which consisted in a bloody presentation of a surgical toilette). Right brain-damaged patients, however, did not show, during the projection of this film, an incidence of avoidance eye movements greater than that shown during the projection of the neutral and of the pleasant movies.

This finding suggests an emotional indifference of right brain-damaged patients in front of stressing situations, which is confirmed by results obtained studying the autonomic response to emotional stimuli.

Analysis of the galvanic skin response to emotional and non-emotional movies showed, indeed, that the sympathetic arousal of right brain-damaged patients was significantly different from that shown by normal controls and by left brain-damaged patients. The former presented a very weak response, which was poorly influenced by the emotional or non-emotional nature of the stimuli, whereas the latter consistently showed the highest level of response while viewing emotional (and above all unpleasant) visual scenes.

Analysis of heart rate deceleration (considered as an index

of para-sympathetic activation) gave very similar results, since in this case too normal controls and left brain-damaged patients showed the greatest heart rate change while viewing the unpleasant emotional film, whereas right brain-damaged patients did not significantly change their heart rate as a function of the (emotional or non-emotional) and of the (positive or negative) connotation of the visual stimuli (Zoccolotti et al., 1986).

CONCLUSIONS

In conclusion, the preliminary results of our present line of research do not support the theory considering disruption of emotional communication as the primary disorder of right brain-damaged patients, since no significant difference was found between right and left-sided patients on tasks of emotional faces comprehension and of spontaneous emotional facial expression.

The same results are also at variance with the hypothesis of a different specialization of the right and left hemisphere respectively for negative and positive emotions, since no interaction was observed between laterality of lesion and type of emotions.

Our findings seem to show, on the other hand, that the emotional reaction of right brain-damaged patients is inappropriate to the situation, since these patients do not move the eyes away from very unpleasant visual scenes and do not present, in front of emotional stimuli, the autonomic responses presented by normal controls and by left brain-damaged patients.

Thus, the safest interpretation of these findings could consist in maintaining that a specific relationship between right hemisphere damage and impairment of emotional behavior does indeed exist, but lies more at the level of the autonomic arousal, intimately linked with the experience of emotions, than at interface between the emotional experience and the most cognitive and communicative aspects of emotion.

REFERENCES

Alemà, G. and Donini, G. (1960). Sulle modificazioni cliniche ed elettroencefalografiche da introduzione intracarotidea di iso-amil-etil-barbiturato di sodio nell'uomo. Boll. Soc. Ital. Biol. Sper., 36, 900-904.

Babinski, J. (1914). Contribution à l'étude des troubles mentaux dans l'hémiplégie organique cérébrale (Anosognosie). Rev. Neurol. (Paris), 27, 845-848.

174

Benowitz, L.I., Bear, D.M., Mesulam, M.M., Rosenthal, R., Zaidel, E. and Sperry, R.W. (1983). Nonverbal sensitivity following lateralized cerebral injury. Cortex, 19, 5-12.

Benton, A.L. and Van Allen, M.W. (1968). Impairment in facial recognition in patients with cerebral disease. Cortex, 4, 344-358.

Borod, J.C. and Caron, H. (1980). Facedness and emotion related to lateral dominance, sex, and expression type. Neuropsychologia, 18, 237-241.

Borod, J.C., Koff, E., Perlman, M. and Nicholas, M. (1985). Channels of emotional expression in patients with unilateral brain damage. Archives of Neurology, 42, 345-348.

Bryden, M.P., Ley, R.G. and Sugarman, J.H. (1982). A left ear advantage for identifying the emotional quality of tonal sequences. Neuropsychologia, 20, 83-87.

Bryden, M.P. and Ley, R.G. (1983). Right Hemispheric Involvement in Imagery and Affect. In Cognitive Processing in the Right Hemisphere. (Ed. E. Perecman). Academic Press, New York.

Buck, R. and Duffy, R.J. (1980). Nonverbal communication of affect in brain-damaged patients. Cortex, 16, 351-362.

Campbell, R. (1978). Asymmetries in interpreting and expressing a posed facial expression. Cortex, 14, 327-342.

Carmon, A. and Nachson, I. (1973). Ear asymmetry in perception of emotional non-verbal stimuli. Acta Psychol. (Amst.), 37, 351-357.

Chaurasia, B.D. and Goswami, H.K. (1975). Functional asymmetry in the face. Acta Anatomica, 91, 154-160.

Cicone, M., Wapner, W. and Gardner, H. (1980). Sensitivity to emotional expressions and situations in organic patients. Cortex, 16, 145-158.

Critchley, M. (1955). Personification of paralysed limbs in hemiplegics. Br. Med. J., 30, 284-286.

Cutting, J. (1978). Study of Anosognosia. Journal of Neurol. Neurosurg. and Psychiat., 41, 548-555.

Dekosky, S.T., Heilman, K.M., Bowers, D. and Valenstein, E. (1980). Recognition and discrimination of emotional faces and pictures. Brain Lang., 9, 206-214.

Denes, G.F., Semenza, C., Stoppa, E. and Lis, A. (1982). Unilateral spatial neglect and recovery from hemiplegia. A follow-up study. Brain, 105, 543-552.

Dopson, W.G., Beckwith, B.E., Tucker, D.M. and Bullard-Bates, P.C. (1984). Asymmetry of facial expression in spontaneous emotion. Cortex, 29, 243-251.

Duda, P.D. and Brown, J. (1984). Lateral asymmetry of positive and negative emotions. Cortex, 20, 253-261.

Duffy, E. (1962). Activation and Behavior. Wiley, New York.

Ekman, P. and Friesen, W.V. (1975). Unmasking the face   N.J. Englewood Cliffs . Prentice Hall.

Ekman, P. and Friesen, W.V. (1978) The facial action coding system. Consulting Psychologists Press, Palo Alto, CA.

Ekman, P., Hager, J. and Friesen, W.V. (1981). The symmetry of emotional and deliberate facial action. Psychophysiology, 18, 101-106.

Ekman, P., Levenson, R.W. and Friesen, W.V. (1983). Autonomic nervous system activity distinguishes among emotions. Science, 221, 1208-1210.

Etcoff, N.L. (1984). Perceptual and conceptual organization of facial emotions: Hemispheric differences. Brain and Cognition, 3, 385-412.

Gage, D.F. and Safer, M.A. (1985). Hemisphere differences in the mood state-dependent effect for recognition of emotional faces. Journal of Experimental Psychology (in press).

Gainotti, G. (1969). Réactions "catastrophiques" et manifestations d'indifférence au cours des atteintes cérébrales. Neuropsychologia, 7, 195-204.

Gainotti, G. (1972). Emotional behavior and hemispheric side of lesion. Cortex, 8, 41-55.

Gainotti, G. (1983). Laterality of affect: the emotional behavior of right and left brain-damaged patients. In Hemisyndromes: Psychobiology, Neurology, Psychiatry. (ed. M.S. Myslobodsky). Academic Press, New York.

Goldblum, M.D. (1980). La reconnaissance des expressions faciales émotionnelles et conventionnelles au cours de lésions corticales. Rev. Neurol. (Paris), 136, 711-719.

Guariglia, C. (1986) Riconoscimento delle espressioni facciali emozionali in soggetti cerebrolesi (unpublished thesis).

Hansch, E.C. and Pirozzolo, F.J. (1980). Task relevant effects on
the assessment of cerebral specialization for facial emotion. Brain
Lang., 10, 51-59.

Hécaen H. (1962). Clinical symptomatology in right and left hemisphere lesions. In Interhemispheric relations and cerebral dominance. (ed. V.B. Mountcastle). Johns Hopkins, publ., Baltimore.

Heilman, K.M., Bowers, D., Speedie, L. and Coslett, H.B. (1984).
Comprehension of affective and nonaffective prosody. Neurology, 34,
917-921.

Heilman, K.M., Scholes, R. and Watson, R.T. (1975). Auditory affecti
ve agnosia: Disturbed comprehension of affective speech. J. Neurol.
Neurosurg. Psychiat., 38, 69-72.

Heilman, K.M., Schwartz, H.D. and Watson, R.T. (1978). Hypoarousal
in patients with neglect and emotional indifference. Neurology, 28,
229-232.

Heller, W. and Levy, J. (1981). Perception and expression of emotion in right handers and left handers. Neuropsychologia, 19, 263-
272.

Hirschman, R.S. and Safer, M.A. (1982). Hemisphere differences in
perceiving positive and negative emotions. Cortex, 18, 569-580.

Hugher, C.P., Chan, J.L. and Su, M.S. (1982). Aprosodia in Chinese
patients with right cerebral hemisphere lesions. Arch. Neurol.,
40, 732-736.

Jackson, J.H. (1980). On right- or left-sided spasm at the onset
of epileptic paroxysms, and on crude sensation warnings, and elaborate mental states. Brain, 3, 192-206.

James, W. (1890). The principles of psychology. Holt, New York.

King, F.L. and Kimura, D. (1972). Left-ear superiority in dichotic
perception of vocal non-verbal sounds. Can. J. Psychol., 26, 111-
116.

Kolb, B. and Milner, B. (1981). Observations on spontaneous facial
expression after focal cerebral and after intracarotid injection
of sodium amytal. Neuropsychologia, 19, 505-514.

Ladavas, E., Umilta, C. and Ricci-Bitti, P.E. (1980). Evidence for
sex differences in right-hemisphere dominance for emotions.
Neuropsychologia, 18, 361-366.

Landis, T., Assal, G. and Perret, E. (1979). Opposite cerebral hemispheric superiorities for visual associative processing of emotional facial expressions and objects. Nature, 278, 739-740.

Lange, C. (1922). The emotions. In The emotions (ed. K. Dunlap) (I. Haupt, Trans.).Williams and Wilkins, Baltimore, Md. (Originally published in 1885).

Ley, R.G. and Bryden, M.P. (1979). Hemispheric difference in processing emotions and faces. Brain Lang., 7, 127-138.

Ley, R.G. and Bryden, M.P. (1982). A dissociation of right and left hemispheric effects for recognizing emotional tone and verbal content. Brain Cogn., 1, 3-9.

Mammucari, A., Caltagirone, C., Gainotti, G., Pizzamiglio, L. and Zoccolotti, P.(1986). Emotional Facial Expressions in Brain-Damaged Patients. Paper presented at the 4th European Meeting of Cognitive Neuropsychology (Bressanone, January).

Mandler, G. (1980). The generation of emotion: a psychological theory. In Emotion: theory, Research Experience, Vol. I. (eds. R. Plutchik and H. Kellerman).Academic Press, New York.

McKeever,W.F. and Dixon, M.S. (1981). Right-hemisphere superiority for discriminating memorized from nonmemorized faces: Affective imagery, sex, and perceived emotionality effects. Brain Lang., 12, 246-260.

Milner, B. (1967). Brain mechanisms suggested by studies of the temporal lobes. In Brain mechanisms underlying speech and language. (eds. C.H. Milikan and F.L. Darley). Grune & Stratton, New York.

Morrow, L., Vrtunski, B., Kim, Y. and Boller, F. (1981). Arousal responses to emotional stimuli and laterality of lesion. Neuropsychologia, 19,65-71.

Moscovitch, M. (1979). Information processing and cerebral hemispheres. In Handbook of Behavioral Neurobiology Vol. 2. (ed. M.S. Gazzaniga). Plenum Press, New York.

Moscovitch, M. and Olds, J. (1982). Asymmetries in spontaneous facial expressions and their possible relation to hemispheric specialization Neuropsychologia, 20, 71-81.

Natale, M., Gur, R.E. and Gur, R.C. (1983). Hemispheric asymmetries in processing emotional expressions. Neuropsychologia, 21, 55-565.

Perria, L., Rosadini, G. and Rossi, G.F. (1961). Determination of side of cerebral dominance with Amobarbitol. Arch. Neurol., 4, 173-181.

Reuter-Lorenz, P. and Davidson, R.J. (1981). Differential contribution of the two cerebral hemispheres to the perception of happy and sad faces. Neuropsychologia, 19, 609-613.

Reuter-Lorenz, P., Givis, R. and Moscovitch, M. (1983). Hemispheric specialization and the perception of emotion: evidence from
right-handers and from inverted and non-inverted left handers.
Neuropsychologia, 21, 687-692.

Robinson, R.G., Kubos, K.L., Starr, L.B., Rao, K. and Price, T.R.
(1984). Mood disorders in stroke patients. importance of lesion
location. Brain, 107, 81-93.

Robinson, R.G. and Price, T.R. (1982). Post stroke depressive
disorders: A follow up study of 103 patients. Stroke, 13(5), 635-
640.

Rosenthal, R., Hall, J.A., Archer, D., Di Matteo, M. and Rogers,
P.L. (1979). The PONS test: Measuring sensitivity to nonverbal
cues. In Nonverbal Communication. (ed. S. Weitz). Oxford University Press, New York, pp. 357-370.

Ross, E.D. (1981). The aprosodias. Arch. Neurol., 38, 561-569.

Ross, E.D. and Mesulam, M.M. (1979). Dominant language functions
of the right hemisphere. Prosody and emotional gesturing. Arch.
Neurol., 36, 144-148.

Ruckdeschel-Hibbard, M., Gordon, W.A. and Diller, L. (1984). Affe-
ctive Disturbances Associated with Brain Damage. In Handbook of
Neuropsychology, vol. 2 (eds. S. Filskov and T. Boll). John
Wiley & Sons, Inc.

Sackeim, A., Greenberg, M.S., Weiman, L., Gur, R.C., Hungerbuhler,
J.P. and Geschwind, N. (1982). Hemispheric asymmetry in the
expression of positive and negative emotions. Arch. Neurol.,
39, 210-218.

Safer, M.A. (1981). Sex and hemisphere differences in access to
codes for processing emotional expressions and faces. J. Exp.
Psychol.: Gen., 1, 86-100.

Safer, M.A. and Loventhal, H. (1977). Ear differences in evaluating
emotional tones of voice and verbal content. J. Exp. Psychol.:
Hum. Percept. Peform., 3, 75-82.

Schwartz, G.E., Ahern, G.L. and Brown, S.L. (1979). Lateralized
facial muscle response to positive and negative emotional stimuli.
Psychophysiology, 16, 561-571.

Strauss, E. and Kaplan, E. (1980). Lateralized asymmetries in
self-perception. Cortex, 5, 283-293.

Strauss, E. and Moscovitch, M. (1981). Perception of Facial Expressions. Brain Lang., 13, 308-332.

Suberi, M. and McKeever, W.F. (1979). Differential right hemispheric memory storage of emotional and non-emotional faces. Neuropsychologia, 15, 757-768.

Terzian, H. and Cecotto, C. (1959). Determinazione e studio della dominanza emisferica mediante iniezione intracarotide di Amytal sodico nell'uomo. I. Modificazioni cliniche. Boll. Soc. Ital. Biol. Sper., 35, 1623-1626.

Tucker, D.M., Watson, R.T. and Heilman, K.M. (1977). Discrimination and evocation of affectively intoned speech in patients with right parietal disease. Neurology, 27, 947-950.

Weintraub, S., Mesulam, M.M. and Kramer, L. (1981). Disturbances in prosody: A right hemisphere contribution to language. Arch. Neurol., 38, 742-744.

Zoccolotti, P., Caltagirone, C., Benedetti, N. and Gainotti, G. (1986). Perturbation des réponses végétatives aux stimuli émotionnels au cours des lésions hémisphériques unilatérales. L'Encéphale, in press.

Zoccolotti, P., Scabini, D. and Violani, V. (1982). Electrodermal responses in patients with unilateral brain damage. J. Clin. Neuropsychol., 4, 143-150.

12
Hemisphere Specialization: A Mechanism for Unifying Anterior and Posterior Brain Regions

Don M. Tucker

Hemispheric specialization in the human brain was recognized largely through the study of hemispheric contributions to cognitive processes. In each of the research approaches to studying asymmetries of brain function, including studies of unilateral brain damage, commisurotomy studies and work with lateralized stimulus presentation with normals, the emphasis has been on describing hemispheric cognitive functions, rather than emotional aspects of hemispheric specialization. As described by Gainotti (this volume), there is a growing body of evidence which indicates that the hemispheres differ in their contributions to the understanding and expressing of emotion, and furthermore, may differ in their characteristic emotional valences. A fruitful approach to explaining these hemispheric differences in emotional functioning may be to consider how lateralized cognitive processes may afford different advantages in the cognitive handling of emotional experience and behavior (Buck, 1985, Safer and Levanthal, 1977; Tucker, 1981). In the present paper, however, we will take the reverse approach, proposing that hemispheric specialization for elementary emotional and motivational mechanisms has been the primary factor in the evolution of brain lateralization, and that many aspects of asymmetric cognitive function can be explained by these asymmetric emotional and motivational controls. This paper begins by reviewing concepts of lateralized activation and arousal systems proposed by Tucker and Williamson (1984). Next, it then attempts to develop this theory further, in an effort to avoid the simplistic division of brain operations into those which are exclusively motor versus those which are exclusively perceptual. The new theoretical proposal is that hemispheric specialization evolved because it allowed the higher-order elaboration of basic activation and arousal mechanisms, each within its own hemisphere. Thus, new phenomena were created: the activation system modulating not only motor systems, but perceptual mechanisms within the posterior left hemisphere; the arousal system modulating not only perceptual operations but motor

180

organization within the anterior right hemisphere. These complementary aspects of hemispheric specialization afford interesting new theoretical possibilities not only for the attentional control of cognition, but also for the motivational and emotional control of experience and behavior.

ACTIVATION AND AROUSAL THEORY

Perhaps the most interesting and controversial observations on hemispheric function and emotion have been the suggestions that each hemisphere is sensitive to different kinds of emotional arousal. Right hemisphere cognitive processes seem to be particularly impaired in depressed persons, but when the depressed mood is improved through treatment, right hemisphere cognition becomes normal. Further evidence that this represents a dynamic influence of depression on right hemisphere function comes from studies of the effects of depression in normal persons; both trait (Schaffer, Davidson, and Saron, 1983) and state (Tucker, Stenslie, Roth, and Shearer, 1981) depression seems to alter right hemisphere cognition and EEG. Although the evidence is less clear, we have observed in our laboratory that left hemisphere activity and function is not immune to the effects of emotional arousal, but may be especially sensitive to the influence of anxiety. High anxious persons show a right ear attentional bias (Tucker, Antes, Stenslie, and Barnhardt, 1978); they may show a strongly analytic approach to perception (Tyler and Tucker, 1982), and highly anxious subjects show an impairment of right visual half field performance for both verbal and non-verbal stimuli (Tucker et al, 1978).

What could it mean that each hemisphere seems to become engaged by, or even over-activated by, a different kind of emotional excitement? In order to address this question, Tucker and Williamson (1984) examined the literature on neural systems maintaining and regulating the activity of the brain. An An important formulation in this literature is that of Pribram and McGuinness (1975), who describe an _activation_ system which emerges from the motor readiness mechanisms of the basal ganglia to produce a tonic increment in brain activity in preparation for action. This contrasts with an _arousal_ system which produces a more phasic increment in brain activity in response to a stimulus and which supports orienting to perceptual input. Tucker and Williamson elaborated on these elementary attentional mechanisms by considering the effects of altering the activity of neurotransmitter pathways that seem integral to these attentional systems. Rather than simply increasing brain activity in a quantitative fashion, as has been assumed for reticular activating mechanisms, these neurotransmitter regulatory mechanisms appear to alter brain function qualitatively. The dopaminergic system seems to not only support motor activity, consistent with a major role in the activation system, but it also restricts the range of activity; with high dopaminergic modulation, behavior becomes routinized and stereotyped. In contrast, norepinephrine modulation seems to increase the tendency of the brain to habituate to constant input, consistent with an arousal system that produces a phasic increase in brain activity in response to novelty.

By elaborating on the Pribram and McGuinness framework, Tucker and Williamson found some interesting new conceptual tools to use in describing the neural regulation of attention and cognitive processing. They suggested that, consistent with the left hemisphere's specialization for complex and sequential motor operations, activation is particularly important to left hemisphere attention and cognition. Several important aspects of left hemisphere cognition might be explained to arise from the operational characteristics of an attentional control that increases the redundancy of mental operations.

Assume the brain has a limited capacity for short-term or working memory. The redundancy bias of the activation system and the habituation or novelty selection bias of the arousal system will produce opposite ways of allocating this working memory. The redundancy bias of activation causes the contents of current processing to be restricted to a few perceptual and motor elements, which are, because of the redundancy of representation, thoroughly represented in working memory and able to be manipulated repeatedly. In contrast, the effect of the novelty bias of the arousal system is to continually input new information, such that the capacity of current working memory becomes saturated with unique perceptual elements.

The effects of these opposite elementary control modes are to produce opposite forms of attention. The restriction of unique elements produced by the activation system seems to yield a focal form of attention; Tucker and Williamson propose that this focal attentional mode is the reason for the left hemisphere's superiority in analytic perception and cognition. On the other hand, when the arousal system is dominant, the saturation of working memory with a diverse range of new information produces an expansive attentional mode; this may form the basis for the right hemisphere's skill in holistic perception and concept-formation.

Thus, it may be possible to reason from elementary attentional controls to describe the basis for hemispheric cognitive skills. Pribram and McGuinness (1975) point out that the difference between tonic activity and phasic responsiveness as neural control mechanisms can be seen in the most simple nervous systems and in the vertebrate spinal cord; the most elementary neural circuits responsible for modulating activity in perceptual and motor systems either sensitize or habituate to repetitive stimulation. It is an interesting theoretical possibility that the highest level of specialized cognitive function in the human brain retains a continuity with the most primitive cybernetics of neural networks.

The Tucker and Williamson (1984) paper certainly over-simplifies the neurotransmitter mechanisms that serve as the mid-brain sources of activation and arousal systems. However, this theorizing may provide some new, and testable, ideas on emotional and motivational aspects of attentional control because it does link a functional model to neurochemical mechanisms. It seems fairly easy to consider how the redundancy bias of the activation system could lead to the motor stereotypies of obsessive-compulsive disorders or schizophrenia. In both these conditions, dopaminergic blockade abolishes the motor stereotypies and produces clinical improvement (Matthysse, 1977). What is perhaps a new theoretical possibility is to appreciate the attentional and cognitive function that may also be explained by excessive activation: the obsessive-compulsive's thinking is not only routinized but also intellectualized and overly analytic. These symptoms of anxiety may represent the excessive operation of the activation system which is particularly important to the left hemisphere.

It may be that other signs of the influence of anxiety on perception and cognition may be explained by an excessive redundancy bias. For example, a well-documented effect in cognitive psychology is a decreased utilization of peripheral cues in high anxiety (Easterbrook, 1959). This seems consistent with an overly focal attention produced by excessive activation. A similar effect may have been observed in the analytic perceptual strategy adopted by anxious subjects in our research on visual perception (Tyler and Tucker, 1982).

The important aspect of these findings is not just the suggestion of neural mechanisms for psychopathology, but rather these effects may be extreme examples of how elementary attentional controls are important to regulating normal, adaptive cognition. If a certain degree of anxiety is required for focused attention and for sequential, deliberate cognitive operations, it may be that anxiety is a requisite motivational substrate for left hemisphere information processing. From the elementary control systems for routinizing motor operations have emerged attentional mechanisms that produce the consistency and deliberate focus of motivated action.

If we are correct in identifying norepinepherine pathways as the primary substrate of the arousal system, it is possible then to consider the psychopathological exaggeration of arousal in the affective disorders, which are thought to be produced by abnormal norepinepherine and serotonin functioning. If excessive arousal has characteristics of mania, while low arousal is associated with depression, it would seem that the arousal system is important not only to general responsivity to novelty in the environment, but also to hedonic processes as well.

These hedonic processes would seem to require substantial involvement from limbic systems. Bear (1983) suggests that the right hemisphere is particularly specialized for the parietal organization of orientation in space, and that the parietal visual pathway has important connections to cingulate cortex. With substantial input from locus coeruleus and integral to the limbic circuitry, the cingulate cortex has neurons that fire only when a stimulus has hedonic significance (Mesulam, 1981).

It would make sense if the organism's capacity to orient to stimuli in space were regulated by a monitoring system that kept track of the adaptive significance of the stimulus. If so, the

adaptive meaning and the representational qualities of perception
may be handled in parallel. Impairment of the spatial orienting
functions of this right-lateralized system may help explain some
of the attentional and orienting deficits of depressives. There
have been reports suggesting that some depressives may show
left-sided neglect not unlike that seen with right parietal
lesions (Freeman, Galaburdn, Cabal and Geschwind, 1975; Liotti, et
al. 1986). In normals during a depressed mood, the relative
decrease in judged loudness of left versus right ear tones (Tucker
et al, 1981) and the slow performance to simple visual stimuli in
the left visual field (Liotti et al., 1986) may suggest that a
transient depressed mood preferentially influences this right
lateralized orienting system as well. At high levels of the
operation of this arousal system, we find not only an overly
expansive attentional mode but also a markedly positive hedonic
tone, suggesting that many kinds of stimulation are perceived by
the manic as affectively pleasurable. Could it be that the
regulation of attentional and hedonic function by the arousal
system could help to explain the cognitive bias in affective
disorder patients, whereby the manic is inappropriately optimistic
in making life plans, and the depressive is unrealistically
pessimistic?

A NEW KIND OF COGNITIVE CONTROL AFFORDED BY HEMISPHERIC
SPECIALIZATION
        Hemispheric specialization may have evolved to allow the
cybernetics of activation and arousal to be elaborated, each
within an entire hemisphere. Prior to hemispheric specialization,
control by activation was restricted to the operations of the
motor system of the anterior brain. With left hemisphere
specialization for this cybernetic mode, the advantages of this
form of regulating neural and mental functions could be actualized
within a whole hemisphere, including both input and output
functions integrated in a tight network. Similarly, right
hemisphere specialization may have allowed control by the
cybernetics of arousal to be elaborated within that side of the
brain, such that now it was possible to develop motor and sensory
circuits that operated on the same control principle. This
lateral specialization may have allowed new levels of integration
between anterior and posterior systems. When motor organization
and perceptual organization each had its own cybernetics, only
fairly limited integrations could be achieved across this boundary
between front and back brain systems. With hemispheric
specialization, the more unified functioning of front and back
systems may have allowed the emergence of higher-order, perhaps
more abstract, sensorimotor integrations. By integrating across
the front-back boundary, of course, hemispheric specialization
opened up a new division between the way attention was regulated
on the left and the right, and it is the problem of the
unification of this new differentiation that led us to this
conference.
        Tucker and Williamson (1984) emphasized left hemisphere

specialization for the cybernetics of the motor system and right hemisphere specialization for the cybernetics of the perceptual orienting system. Although this may be the major alignment of left/right with front/back, it is especially interesting to consider the new forms of neuropsychological functioning that hemisphere specialization created. What may have been most important from an evolutionary point of view was the capacity for the posterior left hemisphere's perceptual operations to draw on the advantages of self-regulation by activation (the primitive motor control). And, in parallel to this, what was new from an evolutionary point of view was the right hemisphere motor system's ability to rely more extensively upon control by arousal mechanisms.

Front-Back Dynamics In The Left Hemisphere

Certainly even amphibian brains are capable of focusing their perceptual operations on a motivationally relevant target, and the basal ganglia of the activation system carry out important functions of sensorimotor integration in the rat brain (Bunney and Aghajanian, 1978) that would allow a motivated attention to be focused. However, it may have been that prior to left hemisphere specialization the majority of the perceptual operations of the posterior cortex were regulated by the basic control mode of the arousal system. The novelty-selection by the arousal system would have been well-suited to most of the functions of the posterior cortex, maintaining an internal representation of the external context and its changes. This would seem quite descriptive of the participatory mode of functioning that Pribram (1981) attributes to posterior cortex; the brain through its perceptual representation of surrounding events "participates" in the immediate context.

But this would not allow a brain that would be capable of the regularized, routinized conceptual operations that are required for language. For as control by arousal seems well-suited to analogical perception (where the internal representation is a continuous mirror of the external input), the comprehension of language requires the ability to parse a continuous auditory stream into discrete coded units; this is a more active, propositional form of representation.

It may have been that the capacity for phonetic decoding required the attentional substrate of sequential cognitive operations provided by the activation system. A cognitive psychology model of language (Neisser, 1961), suggests that language perception may be achieved through an analysis by synthesis; the auditory input stream is parsed into meaningful phonetic units by paralleling the input data with one's own speech generation mechanisms. Although Wernicke's area of the posterior left hemisphere is a primary center for language comprehension, data from both blood flow (Risberg, this volume) and electrical stimulation studies during surgery (Ojemann, this volume) have suggested that anterior left hemisphere regions are active and apparently functional during speech comprehension. The advantage

of left hemisphere specialization was not only that it allowed the
motor system's unique attentional control (the redundancy bias) to
be applied to speech perception, but it allowed a new level of
integration between input and output functions within the left
hemisphere. The anterior and posterior centers were now regulated
by the same cybernetic mode.

Non-speech complex motor behavior coordinated by the left
hemisphere (Kolb and Milner, 1981) is impaired by left parietal as
well as frontal lesions. This may suggest that not only is
motoric involvement important to the left hemisphere's perceptual
operations, but motor organization may be tightly regulated by the
criteria for actions represented in the posterior left hemisphere.
An important example of this is the loss of speech organization
capacity that often follows posterior left hemisphere damage. In
certain forms of Wernicke's aphasia, the patient cannot speak
correctly, but produces jargon and disorganized speech (Benson and
Geschwind, 1975). This suggests that the production of
syntactically and semantically correct speech requires continuous
control from the speech perception regions, a critical
self-monitoring of one's own output as it is generated. With its
differentiation from the right hemisphere, and reliance on a
common cybernetic mode, the left hemisphere seems to have become a
highly integrated machine.

When the anterior and posterior regions of the left
hemisphere become uncoupled, there are very interesting changes
not only in cognitive self-regulation but emotional functioning as
well. As demonstrated by Gainotti's (1972) important study, the
classic effect of lesions to the left hemisphere is to produce a
depressive-catastrophic reaction. This contrasts with the greater
denial of problems and inappropriate cheerfulness following right
hemisphere lesions. These observations suggest a fundamental
difference in the hemispheric representation of the valence of
emotion, but interpreting what this difference means for normal
hemisphere function has been difficult. Sackeim, Greenberg, Gur,
Hungubuhler, and Geschwind (1982) have proposed that the
depressive-catastrophic reaction with left damage represents a
release of the right hemisphere's normal affective negativity.
Tucker (1981), on the other hand, has argued that this
interpretation is not consistent with studies of normals and
psychiatric patients, and proposes that left hemisphere damage may
produce a disinhibition of subcortical mechanisms, thus
exaggerating the left hemisphere's characteristic emotional
orientation of anxiety. Recently, this disinhibition
interpretation seems to have been supported by evidence from
Robinson, Kubos, Rao, and Price (1984) who found that a more
affectively negative reaction follows a more anterior lesion of
the left hemisphere, but an inappropriately cheerful reaction
follows a more anterior lesion of the right hemisphere. Given the
disinhibition that follows frontal lesions (Hecaen, 1964;
Flor-Henry, 1977), the Robinson et al. findings are consistent
with the interpretation that the depressive-catastrophic reaction
represents a disinhibition of left hemisphere's characteristic

affective orientation.

The emotional changes that follow different kinds of aphasia provide an even more specific model of the balance between anterior and posterior mechanisms within the left hemisphere. Consistent with the Robinson findings, aphasics with anterior left hemisphere lesions and speech production deficits often show a more negative emotional orientation (Benson and Geschwind, 1975). In contrast, some of Wernicke's aphasics may be inappropriately cheerful. There may be an important lesson here on the emotional self-regulation that is intrinsic to the cognitive self-regulation provided by these anterior and posterior left hemisphere regions. Just as these regions are in a tensile balance in regulating speech, they may maintain a similar balance in their roles in personality self-regulation. The posterior left hemisphere regions may provide a critical self-monitoring of behavior according to a perceptual model of appropriate behavior. When this function does not develop or is damaged, the individual may show unconstrained and disorganized behavior. In contrast, when this self-monitoring function becomes exaggerated, from loss of frontal inhibition or perhaps from developmental or situational factors, the person may operate under a highly critical self-consciousness. Thus the posterior systems of the left hemisphere, drawing on the cybernetics of activation, provide a tight self-monitoring that may not only explain the excessive self-consciousness in anxiety disorders, but which may represent an essential component of self-regulation in normal personality.

Front-Back Dynamics in the Right Hemisphere

Right hemisphere specialization for the control mode of the arousal system may have integrated its anterior and posterior functions as well. Although the evidence on movement disorders following brain lesions suggests the importance of left hemisphere control for sequential and fine motor operations, more continuous and fluid motor processes may require a primary contribution from the right hemisphere.

Recent observations with a pursuit motor task by Rizzolotti and associates suggest the right hemisphere may be particularly skilled at integrating continuous visual feedback with an ongoing motor operation (Liotti, personal communication, 1986). One way to explain these would be to propose that the right hemisphere is particularly adept at an analogical representation of the motor activity. In contrast to what may be a more propositional, discrete representation of desired motor outcomes by the left posterior regions, the right parietal lobe may be more suited for a fluid and continuous reflection of ongoing motor action. With the front-back integration that right hemisphere specialization seems to have achieved, the contribution of right anterior regions to motor control may be especially synchronized with this ongoing mirroring of the motor outcome by perceptual operations.

It is possible that this integration of input and output applies to the right hemisphere's role in understanding and expressing emotion as well. Since Darwin (1872), emotion theorists

have emphasized the importance of emotional communication in
synchronizing the internal states of social animals. The right
hemisphere's holistic and analogical representations seem to
provide a rich elaboration of internal subjective responses to
ongoing events. A close synchrony of right hemisphere motor
control with this internal affective monitoring function would
cause the expression of emotion to closely mirror the person's
subjective response to current events. It may be that
self-regulation through arousal is important not only to a fluid
and richly elaborated affective response, but also to the close
translation of this response into social media such as vocal
intonations or facial displays.

The right hemisphere's elaboration of more complex and
subtle aspects of emotional experience may be linked with its
specialization for maintenance of arousal in response to
environmental events. Heilman et al (Heilman 1979; Heilman and
Van Den Able, 1980) have examined the mechanisms underlying the
attentional neglect that is particularly severe with right
parietal lesions. They have observed that not only does the right
parietal lobe play a major role in orienting attention, right
parietal lesions produce bilateral deficits in electrodermal
responsiveness to environmental events. Tucker and Williamson
(1984) proposed that it is not all attentional mechanisms for
which the right hemisphere is specialized, but specifically a
receptive orienting to novel environmental events that emerges
from the arousal system. Bowers and Heilman (1980) showed that
in normal subjects a left visual field warning stimulus was more
effective in improving performance than a right visual field
stimulus. That it is specifically the maintenance of alertness
and responsiveness to environmental stimuli for which the right
parietal region is specialized has been shown in research by
Posner and associates on the effect of parietal lesions on
specific components of the attentional system (Posner, Friedrich,
Walker, and Rafal, 1983). They have found that both left and
right parietal lesions impair the capacity to direct attention to
the contralateral field, and there was no strong asymmetry to this
impairment in their patients. However, there was a clear right-
lateralized effect for experimental conditions which influence the
subject's capacity to maintain alertness from one trial to the
next (such as the lack of a warning cue, and longer intertrial
intervals).

A particularly interesting demonstration of a right
hemisphere role in alerting has been provided by Levy, Wagner,
Luh, Heller, and Levine (1986). They studied a syllable
identification task with visual half field presentation and found
that if the previous trial was presented to the left visual field,
current performance in either left or right visual field is
facilitated. Apparently, engaging the right hemisphere is
effective in allocating arousal for both hemispheres' performance.
This effect was obtained in two replications. The particular
effectiveness of left visual field pre-stimulation was also shown
in a face-identification task, but here the facilitation of

current trial performance was only for left visual field
presentation. Levy et al proposed that the right hemisphere is
responsible for allocating arousal for the entire brain, and that
it allocates arousal more specifically to itself when it is
specialized for the task at hand.

The integration of this right-lateralized arousal mechanism
with emotional functioning is emphasized by Bear's (1983) model of
emotional and attentional pathways. He suggests that the right
hemisphere is particularly interconnected with limbic system
mechanisms. This would be consistent with the observation by Bear
and Fedio (1977) that right temporal lobe epileptics, who seem to
show a "hyper-connection" of limbic with cortical processes, show
unusual emotional expressiveness, whereas left temporal lobe
epileptics show an unusual concern with the significance of ideas.

If the right posterior mechanisms are specialized to
facilitate an affective responsiveness, we might expect that right
frontal systems may be important in balancing this responsivity
through inhibitory control. Luria (1973) observed that the
classic frontal lobe disinhibition syndrome, which includes social
and sexual inappropriateness, is more frequent with right than
left frontal lesions. This would be consistent with the unusual
cheerfulness and denial of problems observed with right anterior
lesions in recent studies (Robinson et al, 1984). Consistent with
the idea that the right hemisphere is particularly involved with
the depression-elation dimension, Perris and Monakhov (1978)
observed that those patients for whom a depressed mood and not
anxiety was the major symptom, this depressed mood covaried with
the EEG activity of the right frontal region. In an important
dissociation that supports the current alignment of left and right
hemisphere mechanisms with the anxiety and depression-elation
dimensions, Perris and Monakhov found that the patients for whom
anxiety is a major symptom showed a covariance of left anterior
EEG activity with degree of the anxiety.

We have interpreted the frontal activity observed during a
depressed mood in normal subjects (Tucker, Slenslie, Roth and
Shearer, 1981) as reflecting frontal inhibitory control over the
right hemisphere's emotional responsiveness. It seems quite
likely that when this frontal inhibitory control is inadequate, we
may find not only an extraversion that represents the posterior
right hemisphere's strong responsivity to the emotional
significance of environmental events, but also increased
impulsivity of the individual's behavior. This impulsivity in
behavioral self-control, occurring not only with frontal lesions
but with extraverted personalities, may reflect the modulation of
the motor system by the right-lateralized cybernetics of arousal.
Rather than the restricted and tightly organized control by
activation, modulation of behavior by the arousal system would
prime motor action for quick release in synchrony with the
responsive right posterior orienting mechanisms.

The more loose control over generative mechanisms that
occurred as arousal-regulation became important to right
hemisphere motor functions may have led to important cognitive

advantages. In contrast to the more sequential, highly determined cognitive processes regulated by the redundancy bias of the tonic activation system, cognitive processes regulated by arousal would be more responsive to change as a function of new connections that appear, and arousal-regulation would specifically favor novel constructions. By integrating anterior and posterior right hemisphere systems to capitalize on the advantages of this kind of control, hemispheric specialization may have led to a new level of creativity in human imagination.

SUMMARY AND CONCLUSIONS

In attempting to understand the motivational and emotional mechanisms underlying hemispheric specialization, Williamson and I found it was necessary to recognize the alignment of lateral specialization with the more fundamental functional differentiation of the brain that has developed motor organization in anterior regions and perceptual organization in posterior regions (Tucker and Williamson, 1984). This alignment is consistent with Hughlings Jackson's formulation at the end of the last century that the left hemisphere is specialized for expressive functions, and the right for receptive functions (Taylor, 1958). What seems to have happened in evolution is that hemispheric specialization occurred through an elaboration of the elementary regulatory algorithms of the motor and perceptual control systems: the left hemisphere's analytic attentional capacities emerged from the tonic activation system with its redundancy bias; the right hemisphere's more holistic attention emerged from the novelty selection bias of the phasic arousal and orienting system. Within this framework, each hemisphere's affective characteristics can be understood as intrinsic to these attentional control systems. The anxiety that seems to accompany left hemisphere function may represent the phenomenological manifestation of the tonic motor readiness system. The variations along the depression-elation dimension that seem to accompany changes in right hemisphere function may reflect the intrinsic hedonic features of the phasic perceptual arousal system.

If we assume that higher cognitive processing skills could only emerge through drawing on the cybernetics of more primitive control systems, then it may have been that the advantage of hemispheric specialization was that it allowed an entire hemisphere to share a common cybernetic mode, the left hemisphere specializing posterior as well as anterior regions for the analytic, sequential, and routinized cognitive operations that are consistent with self-regulation by tonic activation, the right hemisphere specializing anterior as well as posterior regions for the more holistic, parallel, and novel cognitive operations that are consistent with self-regulation by phasic arousal.

In this paper I have tried to consider the new kinds of neuropsychological organization that were created by hemispheric specialization. The activation-regulated perceptual operations of the posterior left hemisphere may have been integral to developing the perceptual competence to process the codified vocalizations of language. Because this form of control is inherently linked to the motivational functions of the motor readiness mechanisms, the new activation-regulation of the posterior left hemisphere was not without important adaptive implications. It may have allowed new abilities in critical self-monitoring. The arousal-regulated generative capacities of the anterior right hemisphere may have allowed a new flexibility in imagination. The increased anterior-posterior integration in the right hemisphere may have facilitated a close modulation of expressive communication by the right hemisphere's emotional responses. In addition, the greater arousal-regulation of the motor processes mediated by the anterior right hemisphere may have led to the impulsivity that is associated with extraversion.

For each hemisphere, reliance on a common cybernetic mode may have afforded new capacities for abstraction, for developing representational forms that transcend the sensory-motor boundary. If this view is correct, then this new capacity for abstraction would be not only a cognitive but an emotional abstraction. Because of the essential motivational and emotional features of the control mechanisms that support each hemisphere, the powerful and complex new options for organizing intelligence that arose with hemispheric specialization were paralleled by equally powerful and complex processes of motivation and emotion.

References

Bear, D. M. (1983). Hemispheric specialization and the neurology of emotion. Archives of Neurology, 40, 195-202.

Bear, D.M., and Fedio, P. (1977). Quantitative analysis of interictal behavior in temporal lobe epilepsy. Archives of Neurology, 34, 454-467.

Benson, D. F. & Geschwind, N. (1975). The aphasias and related disturbances. In A. B. Baker & L. D. Baker (Eds.), Clinical Neurology. New York: Harper & Row.

Bowers, D., and Heilman, K. M. (1980). Material specific hemispheric activation. Neuropsychologia, 18, 309-320.

Buck, R. (1985). Prime theory: An integrated view of motivation and emotion. Psychological Review, 92, 389-413.

Bunney, B. S., Aghajanian, G. K. (1978). Mesolimbic and mesocortical dopaminergic systems: Physiology and pharmacology. Psychopharmacology: A Generation of Progress, 159-169.

Darwin, C. (1872). Expressions of the emotions in man and animals. London: John Murray.

Easterbrook, J. A. (1959). The effect of emotion on cue utilization and the organization of behavior. Psychological Review, 66, 183-201.

Flor-Henry, P. (1977). Progress and problems in psychosurgery. In J. H. Maserman (Ed.), Current psychiatric therapies. New York: Grune & Stratton.

Freeman, R. L., Galaburda, A. M., Cabal, R. D., and Geschwind, N. (1985). The neurology of depression. Archives of Neurology, 42, 289-291.

Gainotti, G. (1972). Emotional behavior and hemispheric side of the lesion. Cortex, 8, 41-55.

Hecaen, H. (1964). Mental symptoms associated with tumors of the frontal lobe. In J. M. Warren & K. Akert (Eds.), The frontal granular cortex and behavior. New York: McGraw-Hill.

Heilman, K. M. (1979). Neglect and related disorders. In K. M. Heilman & E. Valenstein (Eds.), Clinical Neuropsychology. New York: Oxford University Press.

Heilman, K., & Van Den Abell, T. (1980). Right hemisphere dominance for attention. The mechanism underlying hemispheric asymmetry of innattention. Neurology,30, 327-330.

Kolb, B. & Milner, B. (1981). Performance of complex arm and facial movements after focal brain lesions. Neuropsychologia, 19, 491-503.

Levy, J., Wagner, N., Luh, K. E., Heller, W., and Levine, S. C. (1986; in preparation). Asymmetric and adaptive regulation of cerebral arousal.

Liotti, M., Sava, D., Caffarra, P., and Rizzolatti, G. (1986; in preparation). Hemispheric differences in patients affected by depression and anxiety.

Liotti, Mario (1986). Personal communication.

Luria, A. R. (1973). The working brain; an introduction to neuropsychology. New York: Basic Books.

Matthysse, S. (1977). Dopamine and selective attention. Advances in Biochemical Psychopharmacology, 16, 667-669.

Mesulam, M. M. (1981). A cortical network for directed attention and unilateral neglect. Annals of Neurology, 10, 309-325.

Neisser, U. (1967). Cognitive Psychology. New York: Appleton-Century-Crofts.

Perris, C., Monakhov, K., VonKnorring, L., Botskarev, V. and Nikiforov, A. (1978). Systemic structural analysis of the EEG of depressed patients Neuropsycholbiology, 4, 207-228

Posner, M. I., Friedrich, F. J., Walker, J. and Rafal, R. (1983). Neural control of the direction of covert visual orienting. Paper presented at the meetings of the Psychonomic Society, November.

Pribram, K.H. (1981). Emotions. In, S.K. Filskov and T.J. Boll (Eds.), Handbook of clinical neuropsychology. New York: Wiley-Interscience, 102-134.

Pribram, K. H., and McGuinness, D. (1975). Arousal, activation, and effort in the control of attention. Psychological Review, 82, 116-149.

Robinson, R. G., Kubos, K. L., Rao, K., and Price, T. R. (1984). Mood disorders in stroke patients: Importance of location of lesion. Brain, 107, 81-93.

Sackeim, H. A. Greenberg, M.S., Weiman, A.L., Gur, R.C., Hungerbuhler, J.P., & Geschwind, N. (1982). Hemisphereic asymmetry in the expression of positive and negative emotions: Neurologic evidence. Archives of Neurology, 39, 210-218.

Schaffer, C. E., Davidson, R. J. and Saron, C. (1983). Frontal and parietal electroencephalogram asymmetry in depressed and nondepressed subjects. Biological Psychiatry, 18, 753-762.

Taylor, J. (Ed.) (1958). Selected writings of John Hughlings Jackson. New York: Basic Books.

Tucker, D. M. (1981). Lateral brain function, emotion, and conceptualization. Psychological Bulletin, 89, 19-46.

Tucker, D. M., Antes, J. R., Stenslie, C. E., and Barnhardt, T. N. (1978). Anxiety and lateral cerebral function. Journal of Abnormal Psychology, 87, 380-383.

Tucker, D. M., Stenslie, C. E., Roth, R. S., and Shearer, S. (1981). Right frontal lobe activation and right hemisphere performance decrement during a depressed mood. Archives of General Psychiatry, 38, 169-174.

Tucker, D. M., & Williamson, P. A. (1984). Asymmetric Neural Control Systems in Human Self-Regulation. Psychological Review, 91, 185-215.

Tyler, S.K., and Tucker, D.M. (1982). Anxiety and perceptual structure: Individual differences in neuropsychological function. Journal of Abnormal Psychology, 91, 210-220.

13
Hemispheric Asymmetry and Positive–Negative Affect

Howard Ehrlichman

For more than a century, functional asymmetry of the human brain was considered a matter of the lateralization of cognitive processes. Yet today most investigators would probably agree that emotional and affective processes, as well as cognitive processes, are lateralized in the brain. The inclusion of affect within the domain of cerebral laterality represents one of the most important and far-reaching consequences of the large body of research in functional brain asymmetry that has developed during the past two decades. Nevertheless, despite the evidence from many studies using a variety of laterality paradigms that affective processes are in some way lateralized, there remain major controversies as to the nature of emotional laterality. Most important, there continue to be fundamental disagreements as to exactly what is lateralized where.

To some extent, these disagreements reflect interpretive problems relating to the complex, dynamic nature of affective processes and to the crucial role of subcortical structures in emotional functions. When one considers the possibilities of inhibitory and/or excitatory connections along the anterior-posterior, left-right, and cortical-subcortical planes one is left with so many plausible interpretations that almost any result can be assimmilated to a particular perspective. For example, Robinson and his colleagues (1985) interpret their finding that depression in stroke patients is highly related to locus of lesion in the left hemisphere as reflecting a left hemisphere basis for depressed affect. In contrast, Sackeim and his colleagues (1982) see destructive lesions in the left hemisphere as disinhibiting dysphoric emotional processess in the right hemisphere. A similar difference of interpretation also exists between Davidson (1984) and Tucker (1984) with respect to findings with EEG asymmetry.

*Portions of the research reported in this paper were supported by a grant from the Fragrance Research Fund, Ltd.

194

Both investigators have reported finding relatively greater right hemisphere EEG activation associated with depressed affect (Schaffer et al., 1983; Tucker et al., 1981). Davidson views this finding as consistent with the view that the right hemisphere mediates negative mood. In contrast, Tucker views the results as indicating inhibitory processes suppressing the right hemisphere's positive mood.

Perhaps the most fundamental disagreement, however, is whether the two sides of the brian differ in their degree of involvement in emotion and affect in general of whether the two sides of the brain are involved in qualitatively different emotions. The predominant point of view has been that the right hemisphere plays a special role in all emotion and affect, or that it "subserves" emotional functions. However, some investigators have challenged this position, maintaining that different emotions are lateralized differently and that the two sides of the brain differ with regard to the kinds of emotions they subserve. Most of these writers hold the view that the right hemisphere has a special involvement in "negative" affect and the left in "positive" affect; however, others have taken the reverse position, placing negative affect on the left side of the brain and positive affect on the right (see Gainotti, this volume).

The view that hemispheric lateralization for affect can be described without distinguishing among different affects, and particularly between the broad classes of positive and negative affect, is supported by a large number of studies which seem to indicate that the expression and perception of affect are mediated by the right hemisphere, regardless of hedonic valence (Bryden, 1982). However, it is not clear that this necessarily implies that the right hemisphere is dominant for all emotional experience. For example, a number of writers have noted that the inner affective experience of brain-damaged patients may be quite different from their outward emotional displays (Gainotti, 1983; Ross, 1984; Sackeim et al., 1982). Similarly, it is one thing to perceive or judge the affective qualities of a stimulus and quite another to have that stimulus change one's own affective state. Davidson (1984) has suggested that in studies involving the perception of affect, a right hemisphere superiority will generally be obtained regardless of the affective valence of the stimuli; however, when the stimuli produce affective feelings in the subject, differences between the left and right sides will be found as a function of affective valence, that is, whether the feelings are positive or negative.

Although a comprehensive review of the relevant literature is beyond the scope of this paper, I think it is fair to draw a distinction similar to Davidson's between the perceptual and expressive aspects of emotion and affect on one hand, and the experiential aspects on the other. If we accept that a distinction between affective decoding/encoding and affective experience may at

least have heuristic value, the choice of research strategies
becomes an important issue. Although it is relatively simple to
create situations involving perception, judgment, or expression of
affect, studying affective experience in controlled settings
presents serious difficulties. This has been a major
methodological problem for many areas of research on affect. For
example, studies of mood and memory have relied primarily on verbal
induction techniques such as the Velten procedure (e.g., Teasdale &
Fogarty, 1979) or hyponosis (Bower, 1981) to induce mood in the
laboratory. Not only do such techniques run the risk of providing
strong demand cues (Polivy & Doyle, 1980), they also obviously
involve complex mixes of cognitive, motivational and affective
processes. When we come to studying laterality, the problem is
compounded by the fact that many of the techniques used for
lateralizing stimuli seem to require rapid or degraded input with
limited potential for producing affective experience. Moreover,
even when we use monitoring techniques such as EEG, inducing affect
through methods such as funny or frightening films obviously
involves complex cognitive components that may themselves have
lateralized effects on such measures. This is not to say that
cognitive factors play no role in affective experience; they surely
do. However, there may be some value in attempting to isolate
those qualities of experience that are most basic to affect.

Although they do not state it in just these terms, a number of
writers have suggested that affective differences between the two
sides of the brain may reflect differences in more biologically
fundamental motivational tendencies involving approach and
withdrawal (Ahern & Schwartz, 1979; Davidson, 1983; Flor-Henry,
1983). A modification of this idea which I would like to suggest
is that, at base, affect can be described as an experiential
representation or felt sense of what is good or bad for the
organism. It is felt in that it is a quality of experience not a
cognitive label or judgment, although in humans it is obviously
often influenced by these (cf. Lazarus, 1984; Leventhal, 1984;
Zajonc, 1984). Note that this general approach presupposes that it
is sensible to talk about broad categories of positive and negative
affect. I think this can be justified on numerous grounds, despite
the popularity of so-called discrete emotion theory in recent years
(Ekman, 1971; Izard, 1977; Tomkins, 1984). First, there is
overwhelming evidence that the judgment of pleasant versus
unpleasant, or good versus bad, is the most fundamental aspect of
meaning. I refer here, of course, to Osgood's (1962) well-known
finding of an evaluative dimension in factor analyses and
multidimensional scaling of words and other stimuli using the
semantic differential. Second, even discrete emotion theorists
classify their various basic emotions as positive or negative.
Third, recent data on self-reported affective experience suggest
that emotions of similar hedonic valence tend to co-occur (Diener &
Iran-Nejad, 1986; Gotlib & Meyer, 1986). Indeed, one can
appreciate that underlying the varieties of good and bad
experiences that we have is something that allows us to tell if

they are in fact good or bad, despite the other differences between them. Fear is not sadness, and joy isn't satisfaction. Neither is pain the same as the smell of rotten eggs, or laughter the same as a tasty meal, yet in all these cases we know and directly feel which is good and which is bad. What I am suggesting, then, is that in some basic way affect can be defined as the evaluative component of experience, regardless of whether the experience is self-referenced (e.g., "I feel happy") or felt as a quality of an external stimulus (e.g., "That tastes good"). This, in turn, may be related to approach and withdrawal. Indeed, linguistically the concepts of positive and negative affect and approach-withdrawal are closely connected. Although we use the terms "positive" and "negative" as if they are synonymous with "good" and "bad" or "pleasant" and "unpleasant," this is not part of the dictionary definition of these words (World Book Dictionary, 1983). In fact the closest definition relevant to this usage is the biological meaning of "moving or turning toward or away from stimuli." For its part, "affective" means "of or having to do with feelings," and "feelings" is defined as "a pleasant or painful mental state." Thus, the terms positive and negative affect at root seem to refer to the felt component of approach and withdrawal tendencies.

This way of thinking about affect has led me to begin a research program using odor as a prototype of affective experience. Odor has a number of properties that recommend it for this purpose. First, the stimuli are less complex and multifaceted than other affect-inducing stimuli. Second, odor experience is known to be highly saturated with the pleasant/unpleasant dimension. Indeed, unlike visual or auditory stimuli, the experience of liking or disliking is a salient part of most olfactory experiences (Engen, 1982). Third, on both phylogenetic and physiological grounds odor seems to be directly tied to the basic motivational approach and withdrawal systems. In many species, including our own, even tiny amounts of odorant substances can cause powerful approach and/or withdrawal reactions. Although very little research has been done using odor as a means of inducing affective experience, what little has been done supports the idea of a connection between odor experience and affect (e.g., Kraut, 1982). In a study recently completed in my lab (Halpern & Ehrlichman, in preparation), we found that the happiness of memories that subjects associated to neutral words varied as a function of whether they were exposed to mildly pleasant or unpleasant odorants during recall. Since these are the same effects that have been found using verbal emotion-induction procedures (Teasdale & Fogarty, 1979), we feel these results support the notion that odor experience and affect are closely related.

Our first study using odors to examine brain laterality was designed to see if ratings of pleasantness or unpleasantness would differ as a function of whether the odor was presented to the left or right nostril. The rationale for this procedure is that the primary olfactory projections to the olfactory bulb and limbic system are lateralized, although unlike other sensory systems the connections are ipsilateral rather than contralateral. Moreover, although the nasal passages do eventuate in a common pathway at the back of the throat, at the level of the olfactory epithelium they are completely isolated by the septum. Evidence that unilateral olfactory stimulation can also produce lateralized responding in the brain is provided by work with commissurotomized patients in which errors were made in naming or verbally detecting odorants presented to the right nostril (Gordon & Sperry, 1969). Furthermore, there have been reports that unilateral brain damage may impair olfatory function in the ipsilateral nostril while sparing the contralateral (Potter & Butters, 1980).

The study was modeled in part on the experiments carried out by Dimond and his colleagues (Dimond & Farington, 1977; Dimond et al., 1976) in which films were presented to the left and right hemiretinas using occluding contact lenses. In those studies, subjects' affective and physiological responses differed depending on whether the stimuli were presented to the left or right hemispheres. Along the same lines, Bryden, Ley and Sugarman (1982) obtained different ratings of the affective quality of musical stimuli depending on ear of presentation. Following these studies, we reasoned that if the two sides of the brain differ in their roles in positive and negative affect in the way Davidson and others have suggested, we should expect a difference between the nostrils along a pleasant-unpleasant dimension, with odorants presented to the left nostril rated as more pleasant or less unpleasant then odorants presented to the right nostril. However, if the right side mediates affect regardless of valence, we would predict that the hedonic experience of both pleasant and unpleasant odors would be amplified by presentation to the right nostril, and would not differ from each other as a function of which nostril was stimulated.

Three experiments were run to explore different methods of presenting the odors unilaterally. In these experiments, one nostril received an odorant while the other received no odorant. Interestingly, in pilot work we found that subjects were unable to discriminate which nostril received the odorant, unless the substance included a trigeminal stimulant. In fact, to assure that we were dealing exclusively with olfaction, we did not use any odorant substance in which subjects could detect which nostril was being stimulated. This was necessary because the trigeminal connections are contralateral. Each experiment involved 15 right-handed men sniffing two pleasant odors (almond extract and orange extract) and two unpleasant odors (thiophine and pyridine). The unpleasant odors were diluted to the point at which they produced hedonic ratings equivalent to the ratings of the pleasant odors but with a negative valence. In the first study, an active sniffing procedure was used, in which subjects sniffed through two glass bulbs connected to individual flasks, one flask containing an odorant substance, the other distilled water. On each trial, subjects rated the odor on a scale from -100 (extremely unpleasant) to +100 (extremely pleasant). In the second study a passive sniffing procedure was used, in which subjects inserted the tips of plastic bottles into their nostrils and then pressed a lever which delivered an equivalent puff of air to each nostril. Again, one bottle contained an odorant, the other only water and subjects rated the odor on the -100 to +100 scale. The third experiment used the same presentation technique as the second but with fewer trials (144 in study 2, 48 in study 3).

The results of these studies provide partial support for the view that positive and negative experiences are lateralized differently. In all three studies, negative odors were rated more unpleasant when presented to the right nostril than when presented to the left nostril. Although this nostril difference was not significant in any of the three studies taken singly, when the data for the 45 subjects were analyzed together, the result was significant at the .05 level, two-tailed. The mean ratings for left and right nostril presentations of negative odors were -29.1 and -35.3, respectively. In contrast, the ratings for positive odors were identical for left and right nostril presentation (M = +31.8).

To the extent that these results reflect central rather than peripheral asymmetries, they suggest that negative olfactory experience involves a lateralized component, whereas positive olfactory experience does not. Such generalizations, however, are extrememly precarious when based on a single method, and, indeed, it is one of the strengths of laterality research that many different approaches exist for testing theoretical propositions. Accordingly, in our next study we shifted from assessing behavioral responses to lateralized stimuli, to assessing asymmetrical hemispheric activation in response to bilaterally-presented stimuli.

In this study, lateral eye movements (LEMs) were used as a measure of asymmetrical hemispheric activation as subjects sniffed pleasant and unpleasant odors presented to both nostrils. Although the validity of LEMs as a measure of asymmetrical hemispheric activation has been uncertain (Ehrlichman & Weinberger, 1978), recent research has supported the underlying model linking LEM direction to asymmetrical hemispheric activation (Ehrlichman, 1984). Moreover, some of the major problems that have plagued LEM research were avoided in this study. In almost all previous studies, verbal questions have been used to engage both left and right hemispheres. In addition, verbal responses have most often been required from subjects. The use of verbal stimuli and responses has obviously been less than optimal for producing strong differences between the hemispheres in activation given the specialization of the left hemisphere for verbal processes. In the present study, we had a non-verbal stimulus (the odor) and a 5-second delay before the subject was required to respond.

Moreover, the response itself, although verbal, was minimal, simply a rating ranging from -10 to +10.

Nineteen right-handed subjects participated in this study. Each subject sniffed five pleasant and five unpleasant odors randomly presented in six blocks of ten trials while eye movements were recorded on video tape. In order to insure that subjects were looking straight ahead at the time they sniffed, they were told to signal the experimenter that they were ready by looking into the camera. The experimenter stood behind the subjects, observing their eye movements on a video monitor. Lateral eye movements were scored subsequently from the video tapes with the sound off so that the judgements of LEM direction were made without knowledge of the odor or the subject's rating. Possibly because of the instruction to look into the camera at the start of each trial, there were a large number of trials on which no discernible eye movement occurred. Six of the 19 subjects had fewer than 15 scorable lateral eye movements. For the remaining 13 subjects, the mean number of scorable lateral eye movements was 42. In order to see if eye movement direction and experienced pleasantness and unpleasantness of the odors were related, the algebraic mean rating over trials on which a left LEM occurred and the algebraic mean rating over trials on which a right LEM occurred were computed for each subject who met the criterion of a least 10 LEMs in each direction. The mean ratings over subjects for left and right LEM trials were -2.27 and -0.22, respectively (p=.023, two-tailed). All seven subjects who met the criterion had more negative ratings when the odors elicited left as compared to right eye movements. To further explore the nature of the relationship between LEMs and odor experience, trials were selected in which the odor ratings fell within the categories of very pleasant (+6 to +10), very unpleasant (-6 to -10), or relatively neutral (-2 to 2). Although an insufficient number of scorable trials remained to permit statistical analysis with individuals as the units of observation, it was possible to evaluate all of the trials classified in this way across the 13 subjects. This analysis showed that the lateral eye movement pattern for trials which subjects rated as neutral was very similar to the pattern for trials rated as very pleasant, the neutral trials eliciting equal numbers of left and right eye movements and the very pleasant trials eliciting slightly more (51%) left eye movements. In contrast, trials rated as very unpleasant elicited 62% left eye movements. Although these results are obviously tentative, they are interesting in the light of the previous study where only ratings of unpleasant odors differed according to which nostril was stimulated; here, it appeared that only the negative, and not the positive odors, showed a lateral eye movement pattern that differed from the lateral eye movement pattern for neutral odors.

I believe that the use of odors has great potential for experimental studies of affect. However, research on affective experience induced in the lab, whether by odors of other means, must be complemented by research on naturally-occurring affect in order to achieve adequate external validity. Correlational studies are useful in this respect (Cronbach, 1957; Underwood, 1975). I would like to describe a study in which individual differences in hemispheric activation as measured by lateral eye movements were correlated with self-report measures of characteristic affective tendencies. Like EEG, lateral eye movements have been used as indicators of transient, situation-specific arousal asymmetries and as indicators of stable, trait-like arousal asymmetries. Most of the research on personality correlates of people's preferred direction of lateral eye movements has focussed on characteristics presumably linked to the information processing differences between the two hemispheres (Ehrlichman & Weinberger, 1978). However, if brain asymmetry also involves emotional processes, it seems reasonable to expect personality differences in characteristic mood, or what some have called "affectivity" (Watson & Clark, 1984), to correlate with the tendency to make left or right eye movements.

In order to see if lateral eye movements and characteristic affect were related, we used a set of mood adjective scales that has been employed by Diener and Emmons (1984) in a program of research on the structure and correlates of long-term affective tendencies. Diener and others (Watson & Tellegen, 1985) have been claiming that positive and negative affect are not the endpoints of a single dimension, but are separate dimensions with different characteristics and correlates. In part, this claim is based on factor analytic studies in which separate positive and negative factors have been found.

We administered the 14 mood adjective scales to 168 female undergraduates and found the same orthogonal positive and negative factors Diener has reported. We then computed factor scores in order to ask whether lateral eye movements are related to positive affect alone, negative affect alone, both or neither. Twenty-four volunteers from the larger sample were run individually to assess their preferred direction of lateral eye movements. Lateral eye movement indexes (R-L/R+L) were computed and correlated with the two factor scores. We found that the negative affect factor score correlated significantly at .35 with the lateral eye movement index, such that higher levels of negative affect were associated with a tendency to make left eye movements. We also obtained scores on two personality scales that have been characterized as measures of positive and negative affectivity, the Well-being scale from Tellegen's (1982) Differential Personality Questionnaire and the Neuroticism scale from the Eysenck Personality Inventory (Watson & Clark, 1984). These scales correlated with each other and with both of the factor pure mood adjective scales, suggesting that each of the personality scales is actually a complex mix of both negative and positive affective tendencies. Interestingly, neither scale correlated significantly with lateral eye movements. The present findings suggest, then, that the tendency to make left or right lateral eye movements is related to one's score on the negative affect dimension, but not to one's score on the positive affect dimension.

Although these studies I have described used relatively small samples and clearly need to be replicated, they do appear to have a common pattern. In the study comparing left and right nostrils, only negative odors showed a laterality effect. In the study of lateral eye movements in response to odors, only the odors rated as unpleasant elicited directional eye movements which appeared to differ from those elicited by the odors rated as neutral. In the correlational study of lateral eye movements and affectivity, only variation along the negative affect dimension was related to eye movement direction. And in all three studies, the direction of the effects was toward an association between negative affective experience and greater right hemisphere activation or involvement.

Earlier I pointed out that there is a fundamental question as to whether affective laterality differs for positive and negative affect. These studies suggest that while positive and negative valence do need to be separated when describing differential hemispheric involvement in affect, it may not be simply a matter of negative affect on one side and positive affect on the other. Rather, it may be that while negative affective experience is lateralized (probably to the right side of the brain), positive affective experience may be less strongly tied to lateralized neural systems. Although it is not obvious why this might be, it is worth noting that a number of authors have pointed out that positive and negative affect differ in ways other than in their experiential qualities. Natale et al. (1983) have suggested that there may be a preponderance of negative emotions in humans. In support of this view they cite studies indicating that there are more forms of dysphoric than euphoric reactions to brain damage, more English words descriptive of negative than of positive experience, more distinguishable negative facial expressions, and a greater prevalence of negative affect in psychiatric disorders. In discussing differences in the effects of negative and positive mood on memory, Isen (1984) has noted that humans employ cognitive and behavioral strategies which serve to produce and maintain positive affect. Diener and Larsen (1984) have reported that changes in levels of positive affect are more situationally determined than are changes in levels of negative affect, whereas negative affect is more likely to show cross-situational consistency for individuals. One interpretation of these various observations is that negative affect may be more endogenously or automatically triggered than is positive affect. In contrast, positive affect may be more embedded in cognitive, purposive and social processes. If positive affect (and perhaps approach activity) tends to occur in a complex psychosocial matrix, it may be understandable why it could be more diffusely linked to neural systems on both sides of the brain, while negative affect (and perhaps withdrawal/avoidance activity) may be more tightly connected to neural systems on one side of the brain.

REFERENCES

Ahern, G.L. and Schwartz, G.E. (1979). Differential Lateralization
for Positive and Negative Emotions. Neuropsychologia, 17, 693-698.

Bower, G. (1981). Mood and Memory. Am. Psychol., 36, 129-148.

Bryden, M.P. (1982). Laterality: Functional Asymmetry in the Intact
Brain. Academic Press, New York.

Bryden, M.P., Ley, R.G. and Sugarman, J.H. (1982). A Left-ear
Advantage for Identifying the Emotional Quality of Tonal Sequences.
Neuropsychologia, 20, 83-88.

Cronbach, L.J. (1957). The Two Disciplines of Scientific Psychology.
Am. Psychol., 12, 671-684.

Davidson, R.J. (1983). Affect, Cognition, and Hemispheric
Specialization. In Emotions, Cognition and Behavior. (eds. C.E.
Izard, J. Kagan and R. Zajonc). Cambridge University Press, New
York.

Davidson, R.J. (1984). Hemispheric Asymmetry and Emotion. In
Approaches to Emotion. (eds. K.R. Scherer and P. Ekman). Erlbaum,
Hillsdale, N.J.

Diener, E. and Emmons, R.A. (1984). The Independence of Positive
and Negative Affect. J. Pers. Soc. Psychol., 47, 1105-1117.

Diener, E and Iran-Negad, A. (1986). The Relationship in Experience
Between Various Types of Affect. J. Pers. Soc. Psychol., 50, 1031-
1038.

Diener, E. and Larsen, R.J. (1984). Temporal Stability and Cross-
situational Consistency of Affective, Behavioral, and Cognitive
Responses. J. Pers. Soc. Psychol., 47, 871-883.

Dimond, S.J. and Farrington, L. (1977). Emotional Response to Films
Shown to the Right or Left Hemisphere of the Brain as Measured by
Heart Rate. Acta Psychol., 41, 255-260.

Dimond, S.J., Farrington, L. and Johnson, P. (1976). Differing
Emotional Response from the Right and Left Hemispheres. Nature,
261, 690-692.

Ehrlichman, H. (1984). Methodological Issues in Lateral Eye
Movement Research. Paper presented at the meeting of the American
Psychological Association, Toronto.

Ehrlichman, H. and Weinberger, A. (1978). Lateral Eye Movements
and Hemispheric Asymmetry: A Critical Review. Psychol. Bull, 85,
1080-1101.

Ekman, P. (1971). Universal and Cultural Differences in Facial Expression of Emotion. Neb. Symp. Motiv., 19, 207-283.

Engen, T. (1982). The Perception of Odors. Academic Press, New York.

Flor-Henry, P. (1983). Hemisyndromes of Temporal Lobe Epilepsy: Review of Evidence Relating Psychopathological Manifestations in Epilepsy to Right- and Left-sided Epilepsy. In Hemisyndromes: Psychobiology, Neurology, Psychiatry. (ed. M.S. Myslobodsky). Academic Press, New York.

Gainotti, G. (1983). Laterality of Affect: The Emotional Behavior of Right- and Left-brain-damaged Patients. In Hemisyndromes: Psychobiology, Neurology, Psychiatry. (ed. M.S. Myslobdsky). Academic Press, New York.

Gordon, H.W. and Sperry, R.W. (1969). Lateralization of Olfactory Perception in the Surgically Separated Hemispheres of Man. Neuropsychologia, 7, 111-120.

Gotlib, H.W. and Meyer, J.P. (1986). Factor Analysis of the Multiple Affect Adjective Check List: A Separation of Positive and Negative Affect. J. Pers. Soc. Psychol., 50, 1161-1165.

Halpern, J.N. and Ehrlichman, H. (in preparation). Affect and Memory: Effects of Pleasant and Unpleasant Odors on Retrieval of Happy and Unhappy Events. CUNY Graduate Center, New York.

Isen, A.M. (1984). Toward Understanding the Role of Affect in Cognition. In Handbook of Social Cognition. (eds. R.S. Wyer, Jr. and T.K. Srull). Erlbaum, Hillsdale, N.J.

Izard, C.E. (1977). Human Emotions. Plenum, New York.

Kraut, R.E. (1982). Social Presence, Facial Feedback, and Emotion. J. Pers. Soc. Psychol., 42, 853-863.

Lazarus, R.S. (1984). Thoughts on the Relations Between Emotion and Cognition. In Approaches to Emotion. (eds. K.R. Scherer and P. Ekman). Erlbaum, Hillsdale, N.J.

Leventhal, H. (1984). A Perceptual Motor Theory of Emotion. In Approaches to Emotion. (eds. K.R. Scherer and P. Ekman). Erlbaum, Hillsdale, N.J.

Natale, M., Gur, R.E. and Gur, R.C. (1983). Hemispheric Asymmetries in Processing Emotional Expressions. Neuropsychologia, 21, 555-565.

Osgood, C.E. (1962). Studies on the Generality of Affective Meaning Systems. Am. Psychol., 17, 10-28.

Polivy, J. and Doyle, C. (1980). Laboratory Induction of Mood
States Through the Reading of Self-referent Mood Statements:
Affective Changes or Demand Characteristics? J. Abnorm. Psychol.,
89, 286-290.

Potter, H. and Butters, N. (1980). An Assessment of Olfactory
Deficits in Patients with Damage to the Prefrontal Cortex.
Neuropsychologia, 18, 621-628.

Robinson, R.G., Lipsey, J.R., Bolla-Wilson, K., Bolduc, P.L.,
Pearlson, G.D., Rao, K. and Price, T.R. (1985). Mood Disorders in
Left-handed Stroke Patients. Am. J. Psychiat., 142, 1424-1429.

Ross, E. (1984). Right Hemisphere's Role in Language, Affective
Behavior and Emotion. Trends Neurosci., 7, 342-346.

Sackeim, H.A., Weinman, A.L., Gur, R.C., Greenberg, M.,
Hungerbuhler, J.P. and Geschwind, N. (1982). Pathological Laughing
and Crying: Functional Brain Asymmetry in the Experience of Positive
and Negative Emotions. Arch. Neurol., 39, 210-218.

Schaffer, C.E., Davidson, R.J. and Saron, C. (1983). Frontal and
Parietal Electroencephalogram Asymmetry in Depressed and Nondepressed Subjects. Biol. Psychol., 13, 753-762.

Teasdale, J.D. and Fogarty, S.J. (1979). Differential Effects of
Induced Mood on Retrieval of Pleasant and Unpleasant Events from
Episodic Memory. J. Abnorm. Psychol., 88, 248-257.

Tellegen, A. (1982). Brief Manual for the Differential Personality
Questionnaire. Unpublished manuscript, University of Minnesota.

Tomkins, S.S. (1984). Affect Theory. In Approaches to Emotion.
(eds. K.R. Scherer and P. Ekman). Erlbaum, Hillsdale, N.J.

Tucker, D.M. (1984). Lateral Brain Function in Normal and
Disordered Emotion: Interpreting Electroencephalographic Evidence.
Biol. Psychol., 19, 219-235.

Tucker, D.M., Stenslie, C.E., Roth, R.S. and Shearer, S.L. (1981).
Right Frontal Lobe Activation and Right Hemisphere Performance
Decrement During a Depressed Mood. Arch. Gen. Psychiat., 38,
169-174.

Underwood, B.J. (1975). Individual Differences as a Crucible in
Theory Construction. Am. Psychol., 30, 128-139.

Watson, D. and Clark, L.A. (1984). Negative Affectivity: The
Disposition to Experience Aversive Emotional States. Psychol.
Bull., 96, 465-490.

206

Watson, D. and Tellegen, A. (1985). Toward a Consensual Structure of Mood. Psychol. Bull., 98, 219-235.

World Book Dictionary. (1983). World Book, Chicago.

Zajonc, R.B. (1984). On Primacy of Affect. In Approaches to Emotion. (eds. K.R. Scherer and P. Ekman). Erlbaum, Hillsdale, N.J.

14
Variations in Arousal Asymmetry: Implications for Face Processing

Susan C. Levine, Marie T. Banich and Hongkeun Kim

INTRODUCTION

A variety of findings suggest a highly consistent pattern of hemispheric specialization in dextrals, characterized by left hemisphere superiority for processes utilized in language tasks and right hemisphere superiority for processes utilized in spatial tasks as well as in the perception of non-verbal auditory stimuli. Evidence for this pattern comes from studies of the cognitive abilities of the isolated hemispheres of commissurotomy patients (e.g., Levy et al., 1972; Sperry, 1974), from the rarity of aphasic symptoms following right hemisphere damage in dextrals (e.g., Weisenburg & McBride, 1935; Kimura, 1983), as well as from sodium amytal tests which indicate that 95-99% of dextrals have language output functions lateralized to the left hemisphere (e.g., Wada & Rasmussen, 1960; Rasmussen & Milner, 1977).

In contrast to these results, perceptual asymmetries of normal dextral adults as measured by standard laterality tasks (e.g., tachistoscopic presentation of stimuli to lateral view and dichotic listening), are extremely variable in both magnitude and direction. Explanations for this apparent discrepancy include random error in measurement (e.g., Teng, 1981) and noise introduced by subjects' perceptual biases on standard laterality tasks (e.g., Kirsner & Schwartz, 1986; White, 1969). However, perceptual biases such as reading scanning habits fail to adequately account for the large variations among dextrals in the direction and degree of lateral asymmetries on a wide range of verbal and spatial laterality tasks. Moreover, random error in the laterality measurements is inconsistent with the stability of subjects' scores on re-test (e.g., Blumstein et al., 1975; Chiarello et al., 1984; Wexler et al., 1981), and across different tasks (e.g., Burton, 1985; Levy et al., 1983a).

*We are grateful to Biomedical Research Grant #5-24010 and to the Spencer Foundation for support of this research.

The data reported in this paper support an alternative explanation for variations among dextrals' asymmetry scores on standard laterality tasks: in particular, the explanation proposed by Levy et al. (1983a) that a subject's perceptual asymmetry reflects not only his/her pattern of hemispheric specialization which is stable both within and across dextrals, but also his/her hemispheric arousal asymmetry, which is stable within an individual, but variable across dextrals. The existence of individual variations in hemispheric arousal asymmetry among dextrals is supported by numerous findings, obtained using diverse measures (asymmetric hemispheric blood flow; EEG; direction of lateralized eye movements to reflective questions). Moreover, these findings suggest that hemispheric arousal asymmetries are characteristic of individuals as they are related to a variety of stable cognitive and personality measures (e.g., Gur & Gur, 1975; Gur & Reivich, 1980; Tyler & Tucker, 1982).

Levy et al.'s (1983a) arousal hypothesis has certain implications concerning subjects' asymmetry scores on different tasks. First, it suggests that there should be a significant relation between subjects' asymmetry scores on "non-lateralized" tasks and tasks that differentially engage the specialized processes of the cerebral hemispheres, since the arousal factor should operate similarly on lateralized and non-lateralized tasks. Thus, individual differences in perceptual asymmetries on tasks that are non-lateralized for the group of dextrals as a whole should not merely reflect random fluctuations around zero. In fact, Levine et al. (1984) report that under conditions of bilateral stimulus presentation, subjects with a left visual field (LVF) advantage on a task that was non-lateralized across the group of dextrals tested (recognition of chairs) had a significantly greater LVF advantage on a right hemisphere specialized face recognition task than subjects with a right visual field (RVF) advantage on the non-lateralized chair task. In a second study with a different group of dextrals, Levine et al. (1984) found that subjects with a RVF advantage on another non-lateralized task (recognition of line drawings of objects) had a significantly greater RVF advantage on a left hemisphere specialized word recognition task than subjects with a LVF advantage on the non-lateralized task.

The arousal hypothesis also predicts that there should be a significant relation between subjects' asymmetry scores on left and right hemisphere specialized tasks. Although the mean laterality score for a group of dextrals shifts depending on whether the task is left hemisphere specialized, non-lateralized, or right hemisphere specialized, a strong version of the arousal hypothesis predicts that individual subjects will maintain the same position relative to other subjects on the distributions of asymmetry scores for each of these tasks (see Figure 1 for hypothetical results predicted by the arousal hypothesis). In fact, consistent with this prediction, Levy et al. (1983a) report that subjects with a strong leftward (right hemisphere) bias on a free-vision chimeric face task that is right hemisphere specialized across dextrals have only a small non-signi-

Figure 1

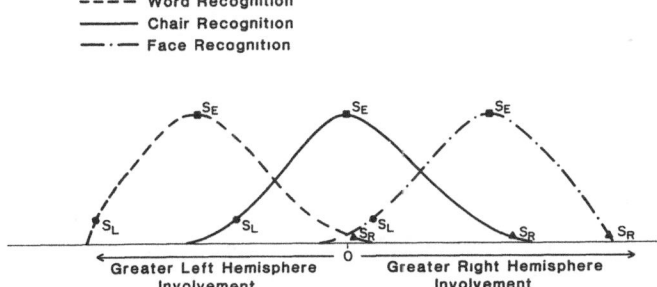

Asymmetry scores predicted by strong version of arousal hypothesis of a left hemisphere aroused subject (S_L), a symmetrically aroused subject (S_E) and a right hemisphere aroused subject (S_R) on a left hemisphere specialized word recognition task, a non-lateralized chair recognition task and a right hemisphere specialized face recognition task (note that the curves are truncated as the degree of asymmetry measurable on a task is limited by the finite number of trials).

ficant RVF advantage on a tachistoscopic CVC-syllable task that is left hemisphere specialized across dextrals. In contrast, subjects with a weak leftward bias on the free-vision chimeric face task have a large and significant RVF advantage on the CVC-task.

It should be noted that these findings are inconsistent with the hypothesis that individual dextrals vary in degree of hemispheric specialization ("specialization strength hypothesis"). According to this hypothesis, certain dextrals are "strongly" lateralized, showing large RVF advantages on left hemisphere specialized tasks and large LVF advantages on right hemisphere specialized tasks. Other dextrals are "weakly" lateralized and show small asymmetries on both left and right hemisphere specialized tasks. The arousal and specialization strength hypotheses make specific and contradictory predictions about the direction of the correlation between subjects' asymmetry scores on left and right hemisphere specialized tasks. If visual field asymmetry scores on left and right hemisphere tasks are computed by subtracting the number of stimuli correctly recognized in one visual field from the other (RVF - LVF), the arousal hypothesis predicts a positive correlation between subjects' asymmetry scores on two such tasks whereas the specialization strength hypothesis predicts a negative correlation.

Two additional predictions of the arousal hypothesis have to do with the variance and covariance relations of subjects' responses in the two visual fields. First, Levy et al. (1983a) predict (where

s_L^2 and s_R^2 are the variances of unilateral correct responses to stimuli presented in the LVF and RVF, respectively) that $s_L^2 > s_R^2$ for a right hemisphere specialized task and $s_R^2 > s_L^2$ for a left hemisphere specialized task. Levy et al.'s (1983a) explanation is as follows for a right hemisphere specialized task such as face recognition. For LVF trials, arousal asymmetry in favor of the right hemisphere will both increase attention to the LVF and promote face processing as the right hemisphere is specialized for this task. In contrast, arousal asymmetry in favor of the non-specialized left hemisphere will both decrease attention to the LVF and interfere with face processing. Thus, the arousal factor will tend to spread subjects apart in LVF performance. In contrast, for RVF performance, arousal in favor of the right hemisphere will tend to promote face processing but· diminish attention to the RVF, whereas arousal in favor of the left hemisphere will tend to interfere with face processing but increase attention to the RVF. Thus, relative to LVF variance, RVF variance will be reduced, due to the arousal factor. (For a left hemisphere specialized task such as word recognition the argument is the same in form, but left and right must be reversed).

Levy et al. (1983a) also predict (where B is the proportion of bilaterally correct responses and L and R are the proportions of unilaterally correct responses in the LVF and RVF, respectively) that COV $_{B,L}$ < COV $_{B,R}$ for a right hemisphere specialized task whereas COV $_{B,R}$ < COV $_{B,L}$ for a left hemisphere specialized task. They argue that characteristic hemispheric arousal asymmetry concordant with the hemisphere specialized for a task results in an attentional bias to the visual field contralateral to the specialized and aroused hemisphere and a relative neglect of information in the ipsilateral visual field. This attentional biasing decreases the probability of a bilaterally correct response. On the other hand, characteristic arousal asymmetry discordant with the hemisphere that is specialized for a task increases the probability of bilaterally correct responses as attention to the left and right visual fields is more equally distributed.

Consistent with these two predictions of the arousal hypothesis, Levine et al. (1984) found the following variance and covariance values (see Table 1).

Table 1

Variance and Covariance Values
(from Levine et al., 1984)

	s_L^2	s_R^2	COV $_{B,L}$	COV $_{B,R}$
FACES	7.62	3.65	−1.61	−0.51
WORDS	8.24	21.72	0.53	−1.27

In the present paper, two studies are reported that further
investigate predictions of the arousal hypothesis. The first study
is concerned with the relation of dextrals' asymmetry scores across
left hemisphere specialized, right hemisphere specialized and non-
lateralized tasks. Each subject is administered multiple laterality
tasks and the relation of subjects' asymmetry scores on these tasks
is investigated in light of the predictions of the arousal hypothesis. The second study is concerned with whether differences in
hemispheric arousal asymmetry are associated with differences in
cognitive processing in a particular domain, the recognition of
familiar classes of visuo-spatial stimuli such as faces.

STUDY 1: VARIATIONS AMONG DEXTRALS IN PERCEPTUAL ASYMMETRY ON RIGHT
HEMISPHERE SPECIALIZED, LEFT HEMISPHERE SPECIALIZED AND NON-LATERAL-
IZED TASKS.

This study tests several specific predictions of the arousal
hypothesis. First, it tests the prediction that subjects' asymmetry
scores on "non-lateralized" tasks are significantly related to their
asymmetry scores on tasks that differentially engage the specialized
processes of the left or right cerebral hemisphere. Second, it
tests the prediction of a positive correlation between subjects'
asymmetry scores on tachistoscopically presented left and right
hemisphere specialized tasks. Third, it tests whether subjects'
asymmetry scores on a free vision task that differentially engages
the right hemisphere (chimeric faces) are significantly related to
their asymmetry scores on tachistoscopically presented left and
right hemisphere specialized tasks. Finally, it tests whether the
variance and covariance relations between subjects' performance in
the two visual fields proposed by Levy et al. (1983a) are met.

Methods

Subjects. Sixteen male and sixteen female students from the
University of Chicago community were tested. All were right-handed
with right-handed parents and had normal or fully corrected vision,
according to self-report.

Stimuli and Procedure. Four different stimulus types were
tachistoscopically presented to bilateral binocular view in a Gerbrands two-channel tachistoscope: line drawings of common objects,
words, and black and white front view photographs of both previously
unfamiliar faces and chairs. Subjects were presented with 18 stimulus pairs of each type in blocked presentation. The non-lateralized
tasks (chairs and line drawings) were presented first, with the
order of presentation counterbalanced across subjects. The lateralized tasks (faces and words) were presented next, again with presentation order counterbalanced across subjects. For each stimulus
type, 8 practice and 18 test pairs were presented.

Subjects began each trial by viewing a pre-exposure field consisting of the outline of a small black rectangle at the center of the visual field. The stimulus card appeared 500 msec. after the subject initiated a trial by depressing a telegraph key. On each trial, to monitor central fixation, the subject first identified a symbol (+, =, ▽, ∽, *, o) which appeared at the center of the stimulus card and then reported the lateralized stimuli. For face and chair stimuli, subjects responded by selecting the presented stimuli from a 12-item array, whereas for words and line drawings subjects verbally reported the stimuli.

Because degree of asymmetry is sensitive to overall accuracy, exposure duration was varied from trial to trial in an attempt to equate performance level across both subjects and stimulus types, but it was never allowed to exceed 200 msec. The starting exposure duration on each block of test trials was the minimum time at which the subject responded correctly to one but not both items on the practice trials. Exposure duration on test trials remained the same if a subject responded correctly to one item of a pair, was increased by 10 msec. if both items were missed, and was decreased by 10 msec. if both items were correct.

Following the tachistoscopic tasks, subjects were presented with Levy et al.'s (1983b) free vision chimeric face task. Each chimera consists of one-half of an individual's face with a smiling pose and the other with a neutral pose. The subject's task on each of 36 pairs was to decide which of two mirror-imaged chimeras looked happier, the one with the smile to the left or the one with the smile to the right.

Results

A repeated-measures analysis of variance on visual field difference scores for the group as a whole revealed a main effect of Stimulus Type (F=12.73, df 3,93, p<.01). The RVF-LVF difference score was significantly different from zero for words (p<.005), and faces (p<.001), but not for line drawings or chairs (two-tailed t-tests).

In order to test the prediction that subjects' asymmetry scores on non-lateralized tasks are related to their asymmetry scores on left and right hemisphere specialized tasks, separate analyses dividing subjects into left and right hemisphere arousal groups were performed. For one of these, arousal groups were defined on the basis of a median split on asymmetry scores on the chair task, and for the other, arousal groups were defined on the basis of a median split on asymmetry scores on the line drawing task. For each of these, a repeated measures analysis of variance was performed on visual field difference scores (RVF-LVF) with Arousal Group (Right Hemisphere, Left Hemisphere) as a between subjects factor and Stimulus Type (Words, Faces) as a within subjects factor. Both analyses

Figure 2

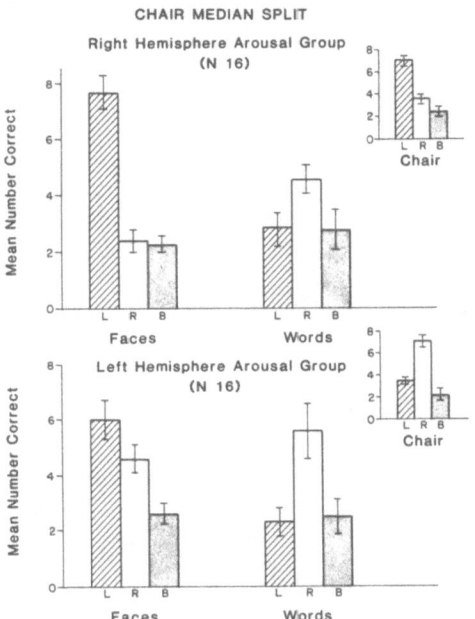

a) Right Hemisphere Arousal Group (as defined by median split on chair asymmetry scores): mean number of unilateral left correct responses (L), unilateral right correct responses (R), and bilateral correct responses (B), for faces and words. b) Same for Left Hemisphere Arousal Group.

showed significant effects of Stimulus Type (p<.001 for both chairs and line drawings). The effect of Group was significant for the chair median split (p<.025), but not for the line drawing median split, although results were in the expected direction. For the chair median split, t-tests (two-tailed) reveal that the LVF advantage for faces significantly differed from zero for the Right Hemisphere Arousal Group (p<.002) but not for the Left Hemisphere Arousal Group, whereas the RVF advantage for words significantly differed from zero for the Left Hemisphere Arousal Group (p<.05) but not the Right Hemisphere Arousal Group. It should be noted that there was no Group by Stimulus Type interaction in either analysis as the offset in visual field asymmetry scores between left and right hemisphere arousal groups was similar in magnitude and direction for words and faces (see Figure 2 for results on Chair median split).

As predicted by the arousal hypothesis, the correlation of subjects' asymmetry scores on the right hemisphere specialized face

Figure 3

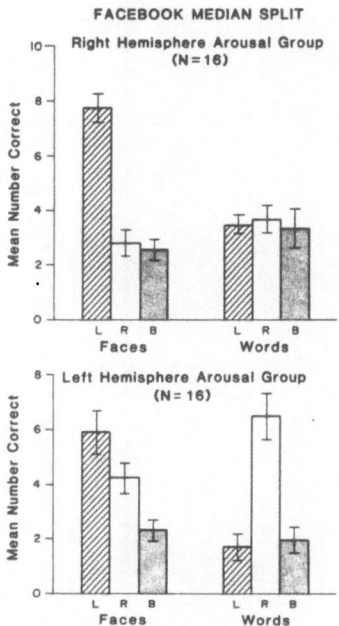

a) Right Hemisphere Arousal Group (as defined by median split on chimeric facebook asymmetry): mean number of unilateral left correct responses (L), unilateral right correct responses (R), and bilateral correct responses (B) for faces and words. b) Same for Left Hemisphere Arousal Group.

task and the left hemisphere specialized word task was positive (r=.21), although it was not significant. An additional analysis was performed to test the prediction that subjects' asymmetry scores on the right hemisphere specialized free vision chimeric face task are related to their asymmetry scores on the tachistoscopic word and face recognition tasks. Subjects were divided by median split into two arousal groups on the basis of facebook asymmetry. A repeated measures analysis of variance was performed on visual field asymmetry scores (RVF-LVF) with Arousal Group (Right Hemisphere, Left Hemisphere) as a between subjects factor and Stimulus Type (Words, Faces) as a within subjects factor. Analogous to the results when arousal groups were defined by median splits on the non-lateralized tasks, there were significant main effects of Group (p<.001) and Stimulus Type (p<.001), but no Group by Stimulus Type interaction. Again, t-tests (two-tailed) reveal that the LVF advantage for faces was significantly different from zero for the Right Hemisphere Arousal Group (p<.002) but not the Left Hemisphere Arousal Group

Table 2

Variance and Covariance Values (from Study 1)

	s_L^2	s_R^2	$COV_{B,L}$	$COV_{B,R}$
FACES	7.51	4.45	−0.45	−0.43
WORDS	3.53	9.30	0.72	−1.71

whereas the RVF advantage for words was significantly different from zero ($p < .002$) for the Left Hemisphere Arousal Group but not the Right Hemisphere Arousal Group (see Figure 3 for results on Chimeric Facebook median split).

Variance and covariance values needed to test Levy et al.'s (1983a) predictions were calculated (see Table 2). All the predictions were supported except that for faces $COV_{B,L}$ was approximately equal to rather than less than $COV_{B,R}$. This may be related to face recognition being a right hemisphere specialized task, and the finding that the right hemisphere activates both itself and the left hemisphere, better than the left hemisphere can activate even itself (Heilman & Van den Abell, 1979). Consistent with this possibility, $COV_{B,L}$ and $COV_{B,R}$ were closer for faces than for words in Levine et al.'s (1984) study (see Table 1).

Conclusion

The overall pattern of results in this study provides support for the hypothesis that asymmetry scores on standard laterality tasks not only reflect patterns of hemispheric specialization but also individual differences in hemispheric arousal asymmetry. First, dividing subjects by a median split on both lateralized and non-lateralized tasks predicts their asymmetry scores on other tasks. Second, the correlation between word and face laterality is positive, as predicted by the arousal hypothesis rather than negative as predicted by the specialization strength hypothesis. (However, the non-significance of this correlation can be viewed as support for Bryden's (1982) hypothesis that specialization of one hemisphere for a particular function implies nothing about the specialization of the other hemisphere.) Finally, the covariance and variance predictions of the arousal hypothesis are upheld for the most part.

The support that the present findings give to the arousal hypothesis should not be regarded as implying that hemispheric specialization and arousal asymmetry are the only factors reflected by subjects' perceptual asymmetries on laterality tasks. Other factors that may be independent of arousal asymmetry and/or hemispheric specialization, notably individual variations in task strategies, may also affect perceptual asymmetries.

STUDY 2: INDIVIDUAL DIFFERENCES IN RIGHT HEMISPHERE ORIENTATION
SPECIFICITY

Study 2 investigates whether differences among dextrals in
hemispheric arousal asymmetry are related to differences in the
manner in which faces are processed. Given the evidence that the
right hemisphere plays an important role in the recognition of
faces, and that individuals vary in the degree to which they engage
left and right hemisphere processes during face recognition tasks
(Ross & Turkewitz, 1981; Ross-Kossak & Turkewitz, in press), it
seems plausible that good face recognition ability may be associated
with arousal asymmetries in favor of the right cerebral hemisphere.

In a prior study (Leehey et al., 1978) dextrals as a group
showed a LVF-right hemisphere advantage for upright but not inverted
faces, which are equivalent in physical complexity. This difference
was attributable solely to better recognition of upright than inverted faces in the LVF, as RVF performance for upright and inverted
faces was equivalent. This finding is viewed as support for an
orientation specific face recognition schema which differentially
involves right hemisphere processes. Study 2 investigates whether
this orientation specific schema is utilized by subjects with right
hemisphere but not left hemisphere arousal asymmetries. In addition, Study 2 investigates whether right hemisphere processing of
other classes of mono-oriented stimuli is orientation specific by
presenting both upright and inverted faces and houses to lateralized
tachistoscopic view.

Methods

Subjects. Sixteen male and sixteen female subjects from the
University of Chicago community were tested. All were right-handed
with right-handed parents.

Stimuli and Procedure. The tachistoscopic stimuli consisted of
black and white front view photographs of chairs, houses and faces,
used in Study 1. Five test sets of 18 trials each were tachistoscopically presented to each subject: chairs (CH), upright faces
(UF), upright houses (UH), inverted faces (IF), and inverted houses
(IH). The same arrays and stimulus cards were used for both the
upright and the inverted orientations, but a different random order
of stimulus cards was used for each orientation. The chair stimuli
were always presented first. For the face and house stimuli, order
of orientation and stimulus type conditions were counterbalanced
across subjects. The free vision chimeric face task developed by
Levy et al. (1983b) was also administered. Other aspects of the
procedure were essentially the same as in Study 1.

Results

A repeated measures analysis of variance on visual field asymmetry scores with the within subjects factors of Stimulus Type (Faces, Houses) and Orientation (Upright, Inverted) revealed no main effects or interactions. Although there was a larger LVF advantage for upright than inverted stimuli for both faces and houses, the effect of Orientation did not reach significance (F=1.90, df 1,30, p=.18).

For the next analysis, subjects were divided on the basis of a median split on their asymmetry scores on the chimeric face task. An analysis of variance on visual field difference scores with Arousal Group (Right Hemisphere, Left Hemisphere) as a between subjects factor and Orientation (Upright, Inverted) and Stimulus Type (Faces, Houses) as within subjects factors was performed (see Figure 4). The main effect of Group (F=7.45, df 1,30, p<.01) and

Figure 4

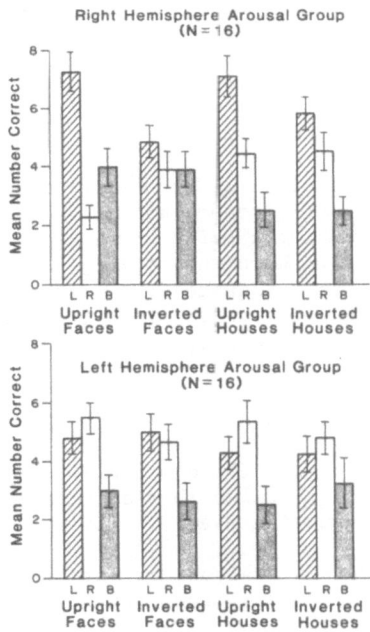

a) Right Hemisphere Arousal Group (as defined by median split on chimeric facebook asymmetry scores): mean number of unilateral left correct responses (L), unilateral right correct responses (R), and bilateral correct responses for upright and inverted faces and houses. b) Same for Left Hemisphere Arousal Group.

Table 3

Processing Efficiency
(Percentage Information Processed Per Unit Time)

	Strong Asymmetry Group	Weak Asymmetry Group
Upright Faces	29.9	17.3
Upright Houses	9.7	9.4
Inverted Faces	7.0	8.7
Inverted Houses	7.2	6.4

the Group by Orientation interaction (F=6.28, df 1,30, p<.025) were significant. For the Right Hemisphere Arousal Group t-tests (two-tailed) showed that the LVF advantage for upright (p<.01) but not inverted stimuli significantly differed from zero. For the Left Hemisphere Arousal Group visual field asymmetries did not significantly differ from zero for either upright or inverted stimuli. Moreover, tests of simple effects revealed that the two groups significantly differed in their asymmetry scores for upright stimuli (p<.001) but not inverted stimuli (F<1). There were no main effects or interactions involving Stimulus Type (see Figure 4).

The absence of a stimulus type effect in the present study may be due to the fact that exposure duration was varied in an attempt to equate performance level for houses and faces in both upright and inverted orientations. When the ability to recognize upright faces exceeds the ability to recognize other stimulus types, as is the case in everyday life (e.g., Yin, 1969), the right hemisphere may play a greater role in face recognition. In fact, the exposure duration data in the present study are consistent with this possibility. A measure of each subject's processing efficiency for each stimulus type was computed as follows:

Processing Efficiency (Percentage Information Processed per Unit Time) = [Total number of stimuli recognized/ Mean exposure time (in msec)] x 100

A repeated measures analysis of variance on subjects' processing efficiency with Arousal Group (Right Hemisphere, Left Hemisphere) (based on facebook median split) as a between subjects factor and Stimulus Type (Faces, Houses) and Orientation (Upright, Inverted) as within subjects factors revealed a significant Arousal Group by Orientation by Stimulus Type interaction (F=5.36, df 1,30, p<.05). Tests of simple effects showed that the arousal groups differed in processing efficiency for upright faces (p<.001) but not for the other stimulus types (see Table 3). Results were almost identical to those in Table 3 when subjects were divided by median split on the chair task.

Conclusion

Dextrals with an arousal asymmetry in favor of the right hemisphere differentially engaged that hemisphere in the processing of familiar classes of stimuli such as faces and houses in their canonical upright orientation. In contrast to this pattern of results, dextrals with an arousal asymmetry in favor of the left hemisphere showed no significant asymmetry in the processing of either upright or inverted stimuli. It is noteworthy that the left and right hemisphere arousal groups show different asymmetry patterns for upright but not inverted stimuli. Thus, arousal asymmetry apparently does not result in equivalent shifts in visual field asymmetry for all stimulus types, but rather may interact with hemispheric specialization for a task. Moreover, the finding that left and right hemisphere arousal groups differed in their processing efficiency for upright faces but not upright houses suggests that orientation specific processing may be particularly advantageous to stimuli that we are expert at recognizing.

GENERAL DISCUSSION

Results of the two studies reported in this paper suggest that the variability of dextrals' asymmetry scores on standard laterality tasks is partially accounted for by individual variations in hemispheric arousal asymmetry. The results of Study 1 show that subjects' asymmetry scores on both lateralized and non-lateralized tasks predict their asymmetry scores on other tasks. Moreover, the finding that subjects' asymmetry scores on the free vision chimeric face task also predict their asymmetry scores on tachistoscopically presented left and right hemisphere specialized tasks suggests that these results are not based on task strategies specific to tachistoscopic presentation. Even if one were to posit that subjects' asymmetry scores across different tasks reflect leftward or rightward attentional biases, one should not lose sight of the fact that numerous studies have shown that greater activation of one hemisphere than the other is related to contralateral attentional and motoric biases (e.g., Kinsbourne, 1972; Trevarthen, 1972).

In addition, the results of Study 2 suggest that differences in hemispheric arousal asymmetry are related to differences in processing efficiency of upright faces. Such findings make it extremely unlikely that our measures of hemispheric arousal asymmetry are indexing something so trivial as reading scanning biases. Moreover, were the results simply attributable to scanning or attentional biases, the mean asymmetry score of the Left Hemisphere Arousal Group should be displaced from the mean asymmetry score of the Right Hemisphere Arousal Group by the same amount across all stimulus types. Results of Study 2 clearly show that the magnitude of the displacement of mean asymmetry scores for left and right hemisphere arousal groups is greater for upright than inverted stimuli. This finding suggests that hemispheric specialization and arousal asymmetry are not simply additive. Note that this finding departs somewhat from the hypothetical results presented in Figure 1.

In summary, administering multiple laterality tasks to subjects makes it possible to begin to disentangle the various factors that affect perceptual asymmetry and task performance. Our results suggest that subjects' asymmetry scores on laterality tasks reflect both hemispheric specialization for a task and individual variation in subjects' hemispheric arousal asymmetry. Moreover, these individual variations in hemispheric arousal asymmetry appear to be related to at least some cognitive processing differences.

REFERENCES

Blumstein, S., Goodglass, H. and Tarttar, V. (1975). The reliability of ear asymmetry in dichotic listening. Brain and Language, 2, 226-236.

Bryden, M.P. (1982). Laterality: Functional Asymmetry in the Intact Brain. Academic Press, New York.

Burton, L. (1985). Sex Differences in Cerebral Asymmetry. Ph.D. Dissertation, University of Chicago.

Chiarello, C., Dronkers, N.F. and Hardyck, C. (1984). Choosing sides: on the variability of language lateralization in normal subjects. Neuropsychologia, 22, 363-374.

Gur, R.E. and Gur, R.C. (1975). Defense mechanisms, psychosomatic symtomatology, and conjugate lateral eye movements. Journal of Consulting and Clinical Psychology, 43, 416-420.

Gur, R.E. and Reivich, M. (1980). Cognitive task effects on hemispheric blood flow in humans: Evidence for individual differences in hemispheric activation. Brain and Language, 9, 78-92.

Heilman, K.M. and Van Den Abell, T. (1979). Right hemispheric dominance for mediating cerebral activation. Neuropsychologia, 17, 315-321.

Kimura, D. (1983). Speech representation in an unbiased sample of left-handers. Human Neurobiology, 2, 147-154.

Kinsbourne, M. (1972). Eye and head turning indicates cerebral lateralization. Science, 176, 539-541.

Kirsner, K. and Schwartz, S. (1986). Words and hemifields: Do the hemispheres enjoy equal opportunity? Brain and Cognition, 5, 354-361.

Leehey, S.C., Carey, S., Diamond, R. and Cahn, A. (1978). Upright and inverted faces: The right hemisphere knows the difference. Cortex, 14, 411-419.

Levine, S.C., Banich, M.T. and Koch-Weser, M. (1984). Variations in patterns of lateral asymmetry among dextrals. Brain and Cognition, 3, 317-344.

Levy, J., Heller, W., Banich, M.T. and Burton, L. (1983a). Are variations among right-handers in perceptual asymmetries caused by characteristic arousal differences between hemispheres? Journal of Experimental Psychology: Human Perception and Performance, 9, 329-359.

Levy, J., Heller, W., Banich, M.T. and Burton, L. (1983b). Asymmetry of perception in free viewing of chimeric faces. Brain and Cognition, 2, 404-419.

Levy, J., Trevarthen, C. and Sperry, R.W. (1972). Perception of bilateral chimeric figures following hemispheric deconnexion. Brain, 100, 105-118.

Rasmussen, T. and Milner, B. (1977). The role of early left-brain injury in determining lateralization of cerebral speech functions. In Evolution and Lateralization of the Brain. (eds. S.J. Diamond and D.A. Blizard). New York Academy of Science, New York.

Ross, P. and Turkewitz, G. (1981). Individual differences in cerebral asymmetries for facial recognition. Cortex, 17, 199-214.

Ross-Kossak, P. and Turkewitz, G. (in press). A micro and macro developmental view of the nature of changes in complex information processing: A consideration of changes in hemispheric advantage during familiarization. To appear in Neuropsychology of facial expression. (ed. R. Bruyer). Lawrence Erlbaum, Hillsdale, NJ.

Sperry, R.W. (1974). Lateral specialization in the surgically separated hemispheres. In The neurosciences: Third study program. (eds. F. Schmitt and F. Worden). MIT Press, Cambridge, MA.

Teng, E.L. (1981). Dichotic ear difference is a poor indicator for the functional asymmetry between the cerebral hemispheres. Neuropsychologia, 19, 235-240.

Trevarthen, C. (1972). Brain bisymmetry and the role of the corpus callosum in behavior and conscious experience. In Cerebral hemisphere relations. (eds. J. Cernacek and F. Podivinsky). Slovak Academy of Science, Bratislava, Czechoslovakia.

Tyler, S.K. and Tucker, D.M. (1982). Anxiety and perceptual structure: Individual differences in neuropsychological function. Journal of Abnormal Psychology, 91, 210-220.

Wada and Rasmussen (1960). Intracarotid injection of sodium amytal and the lateralization of cerebral speech dominance. Experimental and clinical observations. Journal of Neurosurgery, 17, 266-282.

Weisenburg, T. and McBride, K.E. (1935). Aphasia, a clinical and psychological study. Commonwealth Fund, New York.

Wexler, B.E., Halwes, T. and Heniger, G.R. (1981). Use of statistical significance criterion in drawing inferences about hemispheric dominance for language function from dichotic listening data. Brain and Language, 13, 13-18.

White, M. (1969). Laterality differences in perception: A review. Psychological Bulletin, 72, 387-405.

Yin, R.K. (1969). Looking at upside-down faces. Journal of Experimental Psychology, 81, 141-145.

15
Lateralization of Associative Processes: Human Conditioning Studies

Kenneth Hugdahl

INTRODUCTION

Looking at research on hemispheric asymmetry in the intact brain it is probably not an exaggeration to conclude that lateralized effects for perceptual and emotional processes have been revealed (see Bryden, 1982; Bradshaw & Nettleton, 1983 for reviews). Considering the role played by perception and emotionality in psychological theories in general, one may thus wonder whether it is symmetry rather than asymmetry which needs explaining in future studies (cf. Rogers, 1986). One implication of this is that it is probably as important for experimental psychologists to study how hemispheric asymmetry modifies psychological processes, as it is to study how psychological processes modify hemispheric asymmetry. To give an example, although most students of hemispheric asymmetry are aware of the role played by selective attention in dichotic listening (e.g. Bryden, Munhall & Allard, 1983; Hugdahl & Anderson, in press), it is more seldom that students of cognitive psychology incorporate effects of hemispheric asymmetry when postulating general models of attention. Thus, it could be argued that hemispheric asymmetry is critical to most psychological processes and not only to perception and emotion.

One area of experimental psychology where asymmetry may be thought of to have an influence concerns associative processes in terms of conditioning and learning (c.f. Hugdahl, in press; Hugdahl & Brobeck, in press). Apart from the pioneering studies by Hellige in the mid seventies (Hellige, 1975) it is surprising that this area has attracted so little interest from both students of hemispheric asymmetry and from students of general processes of conditioning. This is the more surprising considering the close connection between associative conditioning on the one hand and perceptual, emotional, and attentional processes on the other hand.

A lateralized perspective on associative processes is perhaps most intriguing concerning human conditioning, and especially conditioning of electrodermal responses and of evoked potentials. Starting from recent views of human conditioning as involving higher-order information processing (Dawson, Schell, Bears & Kelly, 1982), stressing such concepts as "attention", "information extraction", and "expectancy", it is a short step to infer that conditioning to a hemisphere specific conditioned stimulus (CS) should be dependent on which hemisphere the CS is initially fed to. Second, from recent cognitive models of associative conditioning (Rescorla, 1972), emphasizing the role of the CS in providing information about the unconditioned stimulus (UCS), it could be argued that the cerebral hemispheres are differentially capable of extracting such information depending on the nature of the CS, and/or on the nature of the processing requirements inherent in the conditioning task (cf. Moscovitch, 1979; Hugdahl, in press).

A third line of evidence linking conditioning and hemispheric asymmetry is the recent demonstration by Rogers (1986) that both habituation and discrimination learning are retarded in chickens after selective injections of monosodium glutamate into the left but not the right forebrain. Moreover, the discovery of the role played by noradrenaline in avoidance conditioning (Archer, 1982), and the demonstration of asymmetry of noradrenaline concentrations in the brain (Oke, Keller, Mefford, & Adams, 1978) further attests to a functional relationship between hemispheric asymmetry and conditioning.

A final example of how hemispheric asymmetry may influence associative conditioning concerns the issue of biologically prepared learning. In 1970, Seligman proposed that the general laws of learning may not be valid for all kinds of stimuli, and for all kinds of responses. Instead he argued that organisms had evolved a biologically determined preparedness to more easily associate certain stimuli with certain reinforces, and with the so called Garcia-phenomenon of taste aversion as the prototype of a prepared contingency (Garcia, McGowan & Green, 1972). In 1972, Seligman and Hager defined prepared learning in terms of how degraded the input could be before that output reliably occurred which meant that learning had taken place. Following this, it could now be hypothesized that the left cerebral hemisphere should be more "prepared" to form an association to a degraded verbal stimulus, whereas the right hemisphere should be more "prepared" to form an association to a degraded emotional stimulus, like a picture of an angry face. Lateralized effects of stimulus degradation have previously been reported for perceptual processes (e.g. Sergent, 1985), and it is hypothesized also for associative processes in the present paper.

Empirical evidence

In this section, I will review a series of experiments from our laboratories at the Universities of Uppsala and Bergen performed together with C.G. Brobeck, G. Kvale, H. Nordby, and J.B. Overmier on electrodermal conditioning to lateralized CS presentations.

The CSs used are either dichotic presentations of consonant-vowel (CV) syllables, or half-field presentations of slides of color-words written in incongruent colors. The UCS is always an aversive white noise presented binaurally. The experimental paradigms are called the "dichotic extinction paradigm", and the "dichoptic acquisition paradigm", respectively. Both paradigms are developed from the standard dichotic listening (Bradshaw, Burden & Nettleton, 1986) and visual half-field (Beaumont, 1982) techniques. Bilateral skin conductance responses (SCRs), and evoked potentials are used as dependent measurements. Only the SCR data are however reported in the present paper.

A review of bilateral skin conductance responses and hemispheric asymmetry is to be found in Hugdahl (1984; see also Hugdahl, Broman & Franzon, 1983). Skin conductance responses have been found to be under the influence of cortical hemisphere functions (Kimble, Bagshaw & Pribram, 1965). Furthermore, several authors have reported differences in response-amplitudes between the left and right hand recordings to supposedly hemisphere specific stimuli (Lacroix & Comper, 1979; see however Hugdahl, 1984 for a critical discussion). Finally, both skin conductance orienting activity and habituation have been related to both stimulus contents in terms of verbal vs. non-verbal attributes, and to the hemisphere initially stimulated (Lacroix & Comper, 1979; Hugdahl et al., 1983).

It may thus be hypothesized that conditioning to a verbal CS should be primarily regulated by the left hemisphere, whereas conditioning to a non-verbal CS should be primarily regulated by the right hemisphere.

The dichotic extinction paradigm. The rationale behind the dichotic extinction paradigm is that when an auditory verbal stimulus (CS+) previously paired with an aversive UCS, is presented in the right ear with another verbal stimulus (CS-) not previously paired with the UCS, simultaneously presented in the left ear, then greater resistance to extinction will occur as compared to when the CS+ and CS- inputs are reversed. This prediction is based on the following assumptions: Since the contralateral auditory projections are predominant (Brodal, 1981), and since the left hemisphere is language specialized, then the verbal CS+ dichotically presented to the right ear during extinction should overshadow a verbal CS- simultaneously presented to the left ear. Thus, responding should decrease in the typical asymptotic manner. However, if the headphones are reversed, then the subject will receive the CS+ in the left ear and the CS- in the right ear. In this case responding is predicted to cease almost immediately after the very first trials.

Figure 1 shows the essential features of the dichotic extinction paradigm.

The Tape

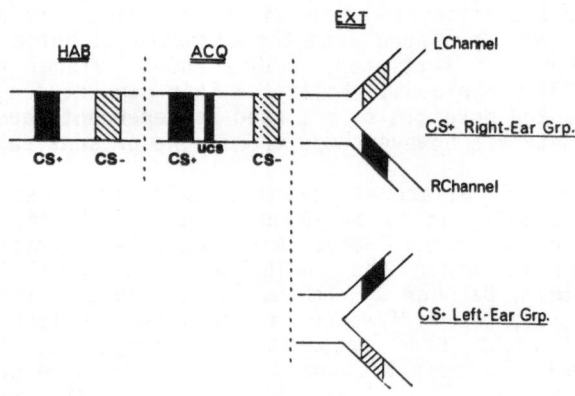

A general outline and illustration of the dichotic extinction paradigm. Note that the illustration is made from the point of view of how the tape is played to the subject. Order of presentation of stimuli counter-balanced across subjects during the habituation and acquisition phases.

Experimentation with the paradigm consists of three phases. The first phase is a habituation phase where the CS+ and the CS- are presented separately in time, and randomized. Each stimulus is presented monotically, i.e. the same stimulus is presented to both ears. In the second, acquisition phase, the CS+ is always followed by the UCS whereas the CS- is never presented together with the UCS. Thus, the CS+ is made into a conditioned signal for the occurrence of the aversively significant UCS, with the CS- as a control stimulus. Each stimulus is presented separately in time and monotically also during the acquisition phase.

During both the habituation and the acquisition phases all subjects are treated in an identical manner. In the third, dichotic extinction phase, the CS+ and CS- are presented dichotically, i.e. simultaneously one in each ear. The only experimental manipulation in the whole experiment is thus that half of the subjects receive the CS+ in the right ear, and the CS- simultaneously in the left ear. The other half of the subjects have the headphones reversed, i.e. they receive the CS+ in the left ear and the CS- in the right ear (see also Figure 1). Thus, one of the unique attributes of the paradigm is that reversal of the headphones is the only difference between the two groups.

From traditional conditioning theory, there is no theoretical model that can predict a difference in extinction performance between the CS+ Right ear and CS+ Left ear groups. This is true for both cognitive (e.g. Rescorla, 1982) and biological (e.g. Seligman, 1970) theories of conditioning.

Figures 2 and 3 show the results from the Hugdahl and Brobeck (in press) study. The CS+ and CS- cues were the CV-syllables "Ba" and "Pa". Half of the subjects had " Ba" as the CS+ and "Pa" as CS-. The other half had "Pa" as CS+ and "Ba" as CS-. Apart from the CS+ Right ear and CS+ Left ear groups a third, control group, was included with no UCS presentations during the acquisition phase. For the sake of simplicity, only the acquisition and dichotic extinction phases are shown in Figures 2 and 3.

Figure 2

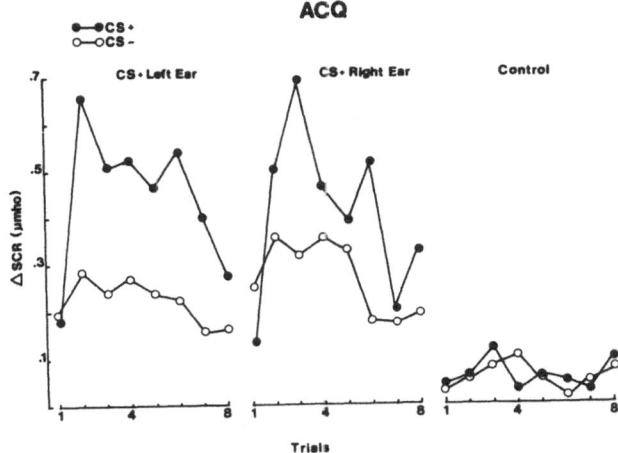

Mean skin conductance responses (SCRs) across hands as a function of trials during the <u>acquisition</u> phase for each of the three groups in the experiment.

Figure 3

EXT

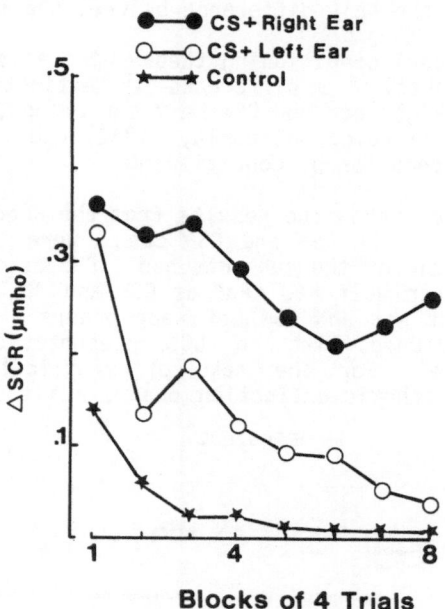

Mean skin conductance responses (SCRs) across hands as a function of trial blocks during the <u>dichotic extinction</u> phase for each of the three groups in the experiment.

As can be seen in Figure 2, there is a marked increase in responding to the CS+ from Trial 2 as compared to the CS- in the two conditioned groups. Thus, it is obvious that conditioned acquisition was demonstrated in the subjects who received the UCS. No difference between the CS+ and CS- was observed for the control group, which was also not expected since this group never received any UCS presentations. The effect was further equal in magnitude for all subjects in the two conditioned groups. Thus, any difference between the CS+ Right ear and CS+ Left ear groups during the later dichotic phase of the experiment cannot be attributed to differences already existing during the acquisition phase. Note that the denotation CS+ Right ear and CS+ Left ear groups does not refer to actual groups during the habituation and acquisition phases, since all subjects were treated similarly. The denotation means those subjects who had the CS+ in the rght and left ears, respectively, during the <u>dichotic extinction phase.</u>

As can be seen in Figure 3, there is a significant difference during the dichotic extinction phase between the two conditioned groups depending on whether the CS+ was presented in the right or in the left ear. Moreover, the CS+ Right ear group is superior on all 32 trials during the extinction period. The conclusion is therefore that effects of hemispheric asymmetry to auditory CSs presented in a dichotic extinction paradigm were demonstrated despite the minimal experimental manipulation. This is a finding not previously reported in the literature.

<u>The dichoptic acquisition paradigm.</u> Turning to the dichoptic acquisition paradigm and the use of visual CSs, the general features of the paradigm are seen in Figure 4.

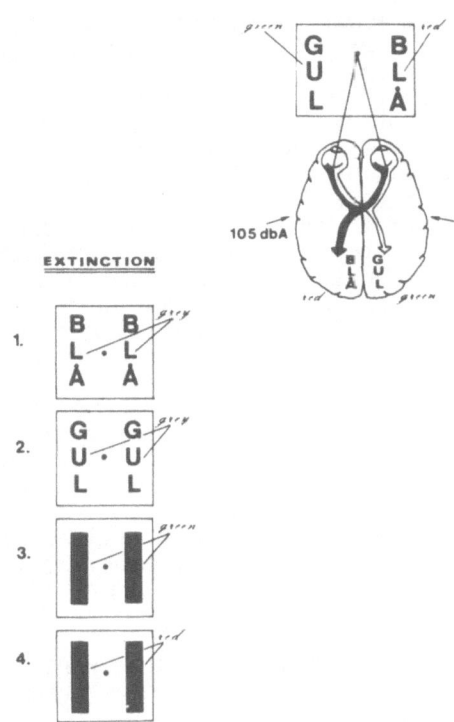

A description of the basic outline of the dichoptic acquisition paradigm. In the example in the Figure, two Norwegian color-words "GUL" and BLÅ" (yellow and blue) written in two conflicting colors (green and red) were simultaneously presented in the left and right visual half-fields as a compound CS during the acquisition phase. The stimuli were presented against a black background. A 105 db noise was used as the UCS, and presented contingent upon the display of the CS compound. During the extinction phase, each of the four CS cues (BLÅ, GUL, green, and red) were presented separately. The color-word cues (1-2) were written in grey against a black background. The color cues (3-4) were presented as green and red color-bars against a black background. Note the omission of the habituation phase in the example.

The data presented with the use of the dichoptic acquistion paradigm are from a study performed together with G. Kvale, H. Nordby at the University of Bergen and J.B. Overmier at the University of Minnesota. The basic set-up of the study was a modified CS compound paradigm (Kamin, 1969), in which the different cues in the compound are initially fed only to the left or to the right cerebral hemispheres. During acquisition, a compound of four different CS cues (two verbal and two non-verbal) were simultaneously presented to the subject followed by an aversive noise as the UCS. The four CS cues were two different color-words each written in two other different and conflicting colors (like the word "BLUE" written in red, and the word "YELLOW" written in green). The CS cues in the compound were tachistoscopically presented during acquisition with the help of the visual half-field technique. Thus, during this phase one color-word written in a conflicting color was presented in the right half-field (i.e. with initial left hemisphere input) with another color-word written in another conflicting color simultaneously presented in the left half-field (i.e. with initial right hemisphere input). The UCS-noise was delivered contingent on each CS compound presentation during acquisition, but not during extinction. Both the word and color cues were counterbalanced across half-fields and subjects.

During the extinction phase, each CS cue of the compound was tested separately for conditioned carry-over effects from the acquisition phase. This was done by presenting each of the four cues separately in a bilateral display (see Figure 4 for an explanation). The paradigm is thus called a dichoptic conditioning paradigm as a visual analogue to the dichotic conditioning paradigm previously described. The following predictions were made:

First of all, it was predicted that the word CS element of the compound presented to the left hemisphere during acquisition would show greater conditioned responding during the extinction phase than the word CS element presented to the right hemisphere. Following the same logic, it was predicted that the color, i.e. the non-verbal CS element of the compound presented to the right hemisphere during acquisition would show greater conditioned responding than the color CS element presented to the left hemisphere. This prediction was based on the findings by Pennal (1977) of a right hemisphere superiority for pure colors presented in the left half-field, although this effect has not always been replicated.

The basic findings during the extinction test-phase of the experiment are seen in Figure 5. It shows the trial 1 of extinction. From the figure it is evident that differential conditioned responding was demonstrated depending on which hemisphere the word and color CS elements were presented to during acquisition. There is further a difference between the hands, with overall larger skin conductance response amplitudes in the left hand.

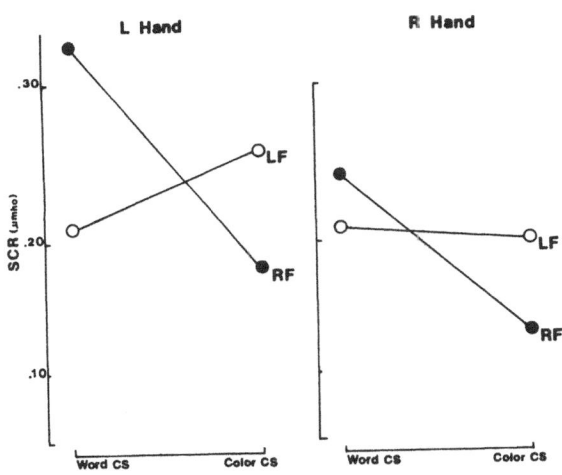

Mean skin conductance responses (SCRs) to the first extinction trial. LF=left field presentations during acquisition. RF=right field presentations during acquisition.

When analyzed across the entire extinction period, the left hemisphere effect for the word CSs still remained whereas the right hemisphere effect for the color-CSs faded away. Thus, it seems that the right hemisphere asymmetry of conditioning to a color-CS is a more transient phenomenon than is the left hemisphere asymmetry of conditioning to a word-CS.

It should however be kept in mind that differential effects of lateralization of CS presentations during acquisition were observed for both types of CS elements during a subsequent first trial extinction test. This finding thus further argues for a functional relationship between hemispheric asymmetry and associative learning in terms of classical conditioning.

However, although the present results for both the auditory and visual CSs have revealed effects of stimulus lateralization in conditioning paradigms, precautions should be taken before it is therefore concluded that lateralization of associative processes have been demonstrated. This is especially important concerning the interpretation of the visual CS data.

First of all, it is still possible that the effect to the visual CSs may have been caused by differential perceptual sensitivity observed between the hemispheres, and not by differential associative sensitivity. This possibility cannot be ruled out solely on the basis of the present results since a non-UCS control was not included. Some preliminary findings from our laboratory have however shown that perception is probably not the main cause for the observed extinction-effect.

This possible criticism is however not applicable to the results for the auditory CSs since a non-UCS control group was included. However, the different response outcomes for the CS+ Right ear and CS+ Left ear groups may have been influenced by biased attention, rather than by differential hemispheric associations. The CS+ Right ear effect may have been caused by the subjects selectively attending to the right ear (where the CS+ was presented during the dichotic extinction phase). One way to test this would be to have half of the subjects in each of the CS+ Right ear and CS+ Left ear groups respectively to deliberately attend only to the right ear, and the other half to deliberately attend only to the left ear. In other words, each of the CS+ Right and Left ear groups is split into a right ear attend group, and a left ear attend group. If superior resistance to extinction is still observed in the CS+ Right ear group despite that the subjects are instructed to actively attend only to the left ear, then an explanation of the observed phenomenon in non-associative terms is less likely. This experiment is presently under way in our laboratory, and some very preliminary results indicate that selective attention to the right ear is not the main cause of the phenomenon.

CONCLUSION

In the present paper I have presented data relevant for an interpretation of effects of lateralization on associative processes in the intact brain. Based on a selective review of the literature I have further argued that conditioning to semantically meaningful CSs should be primarily regulated by the left hemisphere, whereas conditioning to emotionally relevant CSs should primarily be regulated by the right hemisphere.

REFERENCES

Archer, T. (1982). DSP4, a new noradrenaline neurotoxin, and the stimulus conditions affecting acquisition of two-way active avoidance. J. C. Physiol. Psyhol., 96, 476-490.

Beaumont, J.G. (ed.) (1982). Divided visual field studies of cerebral organization. Academic Press, London.

Bradshaw, J.L., Burden, V., and Nettleton, N.C. (1986). Dichotic and dichaptic techniques. Neuropsychologia, 24, 79-91.

Bradshaw, J.L. and Nettleton, N.C. (eds.) (1983). Human cerebral asymmetry. Prentice-Hall, Englewood Cliffs, N,J.

Brodal, A. (1981). Neurological anatomy 3rd Edition. Oxford University Press, New York.

Bryden, M.P. (1982). Laterality: Functional asymmetry in the intact brain. Academic Press, New York.

Bryden, M.P., Munhall, K., and Allard, F. (1983). Attentional biases and the right-ear effect in dichotic listening. Brain Lang., 18, 236-248.

Dawson, M.E., Schell, A.M., Beers, J.R., and Kelly, A. (1982). Allocation of processing capacity during human autonomic classical conditioning. J. Exp. Psychol.: Gen., 111, 273-295.

Garcia, J., McGowan, B.K., and Green, K.F. (1972). Biological constraints on conditioning. In Biological boundaries of learning. (eds. M.E.P. Seligman and J.L. Hager). Appelton Century Crofts, New York.

Hellige, J.B. (1975). Hemispheric processing differences revealed by differential conditioning and reaction time performance. J. Exp. Psychol.: Gen., 104, 309-326.

Hugdahl, K. (1984). Hemispheric asymmetry and bilateral electrodermal recordings: A review of the evidence. Psychophysiology, 21, 371-393.

Hugdahl, K. (in press). Pavlovian conditioning and hemispheric asymmetry: A perspective . In Conditioning in humans. (ed. G. Davey). Wiley, Chichester.

Hugdahl, K., Broman, J.E., and Franzon, M. (1983). Effects of stimulus content and brain lateralization on the habituation of the electrodermal orienting reaction (OR). Biol. Psychol., 17, 153-168.

Hugdahl, K. and Brobeck, C.G. (in press). Hemispheric asymmetry and human electrodermal conditioning: The dichotic extinction paradigm. Psychophysiology.

Hugdahl, K. and Andersson, L. (in press). The "forced-attention paradigm" in dichotic listening to CV-syllables: A comparison between adults and children. Cortex.

234

Kamin, L. (1969). Predictability, surprise, attention, and conditioning. In Punishment and aversive behavior. (eds. B.A. Campbell and R.M. Church). Appelton Century Crofts, New York.

Kimble, D.P., Bagshaw, M.H. and Pribram, K.H. (1965). The GSR of monkeys during orienting and habituation after selective partial ablations of the cingulate and frontal cortex. Neuropsychologia, 3, 121-128.

Lacroix, J.M. and Comper, P. (1979). Lateralization in the electrodermal system as a function of cognitive/hemispheric manipulation. Psychophysiology, 16, 116-129.

Moscovitch, M. (1979). Information processing and the cerebral hemispheres. In Handbook of neurobiology. (ed. M.S. Gazzaniga). Plenum Press, New York.

Oke, A., Keller, R., Mefford, I., and Adams, R.N. (1978), Science, 200, 1411-1413.

Rescorla, R.A. (1972). Information variables in Pavlovian conditioning. In Learning and motivation Vol. VI. (ed. G.H. Bower). Academic Press, New York.

Rogers, L.J. (1986). Lateralization of learning in chickens. In Advances in the study of behavior, Vol. 16. (eds. I.S. Rosenblatt, C. Beer, M.C. Busnel, and P.I.B Slater). Academic Press, New York.

Seligman, M.E.P. (1970). On the generality of the laws of learning. Psychol. Rev., 77, 400-418.

Seligman, M.E.P., and Hager, J.C. (1972). Biological boundaries of learning. Appelton Century Crofts, New York.

Sergent, J. (1985). Role of input and task factors in asymmetric hemispheric processing of faces. J. Clin. Exp. Neuropsychol., 7, 158-159.

16
Laterality and Unconscious Processes

Nathan Brody

In recent years psychologists have studied the effects of
subliminal stimuli on cognitive processes. In this paper I shall
briefly outline three paradigms that have been used to study the
influence of subliminal stimuli in laboratory settings and I shall
present data on the effects of such stimuli on psychophysiological
indices.   I shall then discuss the use of these paradigms in
experiments employing lateralized stimulus presentations.   The
paper will conclude with the presentation of data obtained by my
colleagues and me in a recently completed set of experiments.

STUDIES OF UNCONSCIOUS PROCESSES

Studies of subliminal stimuli in laboratory settings enjoyed a
brief popularity in the 1950's starting with research on perceptual
defense (Brody, 1983).   The early research on this topic was
subject to both methodological and conceptual criticisms.
Analyses based on "response indicator methodology" and signal
detectability theory suggested that stimuli that were alleged to be
subliminal were not in point of fact subliminal using more refined
indices of discriminative capacity.   In addition, the research was
subject to a logical objection.   How is it possible to defend
against unconscious material unless one first is aware of the
nature of the stimulus that is being presented?

Research on the effects of subliminal stimuli acquired renewed
vigor in the 1970's with the development of paradigms that appeared
to offer more refined demonstrations that stimuli that were below a
threshold of discrimination could nevertheless exert influence on
cognitive processes. (For a critique of the adequacy of these
demonstrations see Holender (1986). In addition, the development of
models of cognition that emphasized the existence of preconscious
information processing relying on relatively automatic processes
requiring minimal attentional responses provided a theoretical

235

response to the logical objection raised to earlier research on
perceptual defense (see Posner, 1978; Shiffrin & Schneider, 1977)

In what follows I shall discuss results obtained in three
different experimental paradigms used to study the influence of
subliminal processes.   Corteen and Wood (1972) used a dichotic-
shadowing procedure in which previously shocked words were
presented to an unattended ear while Ss were required to "shadow",
i.e. verbally repeat, material presented to the attended ear.
They found that words previously associated with shock elicited
electrodermal responses although Ss were alledgedly unaware of the
emotional stimuli being presented to their unattended ear.

Kunst-Wilson and Zajonc (1980) have studied the development of
preferences for previously presented stimuli that individuals are
not aware of having encountered.   In their study Ss were presented
with a series of irregularly shaped polygons.   The stimuli were
presented tachistoscopically for brief durations.   Ss were then
asked to discriminate in a forced choice paradigm between the
previously presented stimuli and other irregular polygons that they
had not seen.   Other Ss exposed to the same stimuli were presented
with the same stimulus pairs, but they were asked which of the two
stimuli they preferred. Kunst-wilson and Zajonc found that Ss
developed a preference for those stimuli that they had previously
encountered even though their ability to distinguish between
previously presented stimuli and new stimuli did not exceed chance.
My colleague John Seamon and I have replicated this phenomenon in a
series of studies and we have shown that the effect occurs reliably
even when Ss are required to judge stimuli one week after initial
stimulus presentations. (Seamon, et al., 1983a; b; Seamon, et al.,
1984).

Marcel (1983) has used a lexical decision task to demonstrate
the influence of subliminal stimuli on cognitive processes.   The
lexical decision task requires Ss to judge whether a series of
letters presented to them is either a word or a meaningless string
of letters.   The time to reach this decision is noted.   If a word
is preceded by a priming stimulus that is related in meaning to the
word to be judged, the reaction time required to reach the decision
is decreased.   Marcel has used backward masking procedures to
prevent Ss from being aware of the priming stimulus.   In his
experiments Ss are presented with a prime stimulus followed by a
pattern mask consisting of a dense field of letters in different
orientations.   The backward masking effects of the pattern mask
are assumed to prevent the conscious recognition of the prime
stimulus. Marcel reports that reaction times for lexical decisions
for words primed by masked semantically related words are lower
than for words primed by semantically unrelated masked prime
stimuli.   (See also Fowler et al., 1981.)

The use of pattern masks to demonstrate the influence of
subliminal stimuli has been extended to studies in which

psychophysiological dependent variables are used to index the influence of the masked stimulus. Ohman (1986) has demonstrated that masked stimuli that had been previously associated with shock are capable of eliciting skin conductance responses even though Ss are not able to discriminate with greater than chance accuracy the CS+ and CS- stimulus presentations. Brondeis (1984) has demonstrated differences in event-related scalp potentials for backwardly masked familiar and non-familiar words.

LATERALIZED STIMULUS PRESENTATIONS

Relationships between research on laterality and research on unconscious processes exist on both a conceptual and empirical level. Research on laterality provides an anatomical spatially distinct metaphor for the assumption of independent processing of information that is one of the theoretical foundations of the contemporary study of unconscious processes. In addition, the discovery of a relatively mute hemisphere capable of perceiving the emotional content of stimuli lends credence to the dissociation of verbal reports and affects that is the basis for the postulation of unconscious influences on behavior in clinical settings. Gazzaniga (1985) and Davidson (1984) have discussed the relationship between research on laterality and unconscious processes.

Davidson has suggested that individuals who are prone to use repression as a mechanism of defense may have difficulty in transferring information from the right to the left hemisphere. Thus they are prone to experience negative affects in the right hemisphere that are inadequately described by the left verbal hemisphere. Gazzaniga has suggested that many actions and affects may occur under the control of separate modular structures that are without linguistic representation. Subsequent to the occurrence of the action or affect the left hemisphere may construct a rationale or theoretical explanation of an action that has been triggered by a process without conscious or verbal representation.

Rather than pursue the speculative conceptual connections between unconscious processes and laterality, I propose to describe the results of studies in which lateralized stimulus presentations are used in each of the three paradigms used to study unconscious processes described above. Dawson and Schell (1982) used the dichotic shadowing technique previously used by Corteen and Wood to study the effects of variations in the unattended ear which received the stimuli previously associated with shock. They found that evidence for the occurrence of electrodermal responses to target words that were unattended occurred only when the shadowed material was presented to the right ear and presumably to the left hemisphere and the previously conditioned words were presented to the left ear. If the opposite ear arrangement was used there was

no evidence of conditioned electrodermal responses in the absence
of stimulus awareness. They explained their results by appeal to
the differential capacities of the right and left hemisphere. They
assumed that right ear presentations of material to be verbally
shadowed occupied the left hemisphere and left the right hemisphere
free to mediate the emotional response. If the left ear received
the shadowed material it would be necessary for the right
hemisphere to transfer information about the material to the left
hemisphere in order to verbally shadow the left ear input. The
left hemisphere's preoccupation with the task of verbal shadowing
would interfere with its capacity to mediate the emotional response
to the conditioned stimuli presented to the right ear. In the
analysis of their results Dawson and Schell believed that the right
hemisphere, by virtue of its limited linguistic capacities, would
be centrally implicated in demonstrations of subliminal influences.
In a subsequent response to a critique of their study they note
that their results might also be attributable to a predominant role
of the right hemisphere in processing emotional stimuli (Dawson and
Schell, 1983; Walker and Ceci, 1983.)

    Although Dawson and Schell's hypothesis implicating the right
hemisphere in the mediation of unconscious processes may be valid
for their experimental situation it is not the case that evidence
on unconscious processing invariably implicates the right
hemisphere. Hines et al. (1984) have used lateralized
tachistoscopic presentations of backwardly masked prime stimuli in
a lexical decision task. They found that evidence for semantic
priming for stimuli that could not be discriminated occurred only
for stimuli presented to the right visual field (left hemisphere)
and not for prime stimuli presented to the left visual field.
They attributed their results to the superior verbal processing
capacities of the left hemisphere.

    Brandeis and Lehmann (1986) used backwardly masked verbal
stimuli to study the effects of subliminal stimulus presentations
on event-related potentials recorded from different scalp
locations. In one of their tasks, Ss were required to judge
whether a stimulus was a word or not. Under conditions of
backward masking Ss were not able to make this discrimination with
greater than chance accuracy. However, the scalp location of
maximal response values to late components of the event-related
potential occurring between 296 and 416 msec after stimulus onset
were obtained from electrodes that were lateralized more to the
left when the stimulus was a word than when it was a non-word. The
results of this study support the findings of Hines et al.
suggesting greater left hemisphere involvement in the processing of
subliminal verbal stimuli.

    It is not necessary to use verbal materials to obtain evidence
of left hemisphere involvement in the response to stimuli that
cannot be discriminated with greater than chance accuracy. In two
of our studies on the development of preferences for stimuli that

Table 1

Mean Percentage of Target Stimulus Selection in Experiment 3

	Judgment					
	Affect			Recognition		
Condition	LVF	RVF	Total	LVF	RVF	Total
Nonshadowed	52.0	65.0	58.5	57.0	47.0	52.0
Shadowed	59.0	66.0	62.5	47.0	47.0	47.0

Note: LVF = left visual field; RVF = right visual field

Ss are not aware of having encountered, John Seamon and I used lateralized tachistoscopic stimulus presentations (Seamon et al., 1983, Experiments 3 and 4). In one of these experiments we had a shadowing condition which required Ss to orally repeat words they heard while viewing tachistoscopically presented polygons. Table 1 presents the percentage of times in a forced choice post exposure viewing in which Ss preferred a previously presented polygon to one not presented and the percentage of times Ss selected a previously presented polygon as one they had seen as opposed to a polygon they had not seen for lateralized stimulus presentations. An examination of the data in Table 1 indicates that Ss were not able to discriminate with greater than chance accuracy which of the polygons they had previously seen. However, under both conditions of shadowing, Ss exhibited a statistically significant preference for polygons that had been presented to the right visual field (left hemisphere) and not for polygons presented to the left visual field (right hemisphere).

In another experiment in this series we used lateralized stimulus presentations followed by pattern and energy masks that are assumed to have an effect on the continued cognitive processing of the stimuli. Marcel (1983) has suggested that pattern masks prevent the conscious recognition of stimuli without necessarily impairing subsequent cognitive processing. Energy masks (in our case a flash of light over the stimulus field) are assumed to prevent all subsequent cognitive processing. Table 2 presents the results of this experiment. Under pattern masking conditions Ss are able to discriminate previously presented polygons from polygons not presented with greater than chance accuracy provided the stimuli are presented to the left visual field (right hemisphere). Preferences for previously presented polygons as in the previous experiment occur only for polygons presented to the right visual field (left hemisphere). Under energy masking conditions both kinds of judgments do not exceed chance under all conditions of stimulus presentation.

Table 2

Mean Percentages of Target Stimulus Selection in Experiment 4

| Masking condition | Judgment | | | | | |
| | Affect | | | Recognition | | |
	LVF	RVF	Total	LVF	RVF	Total
Pattern	52.0	74.0	63.0	64.0	52.0	58.0
Energy	58.0	56.0	57.0	48.0	56.0	52.0

Note: LVF = left visual field; RVF = right visual field

The results for recognition judgments obtained in this experiment may be explained by reference to the usual right hemisphere advantage for processing nonverbal visual stimuli. It is somewhat more difficult to explain the development of preferences for stimuli presented to the right visual field (left hemisphere). These results may be explained by reference to the hypothesis of left hemisphere specialization for the mediation of positive affective states (Davidson, 1984).

This brief review of studies involving the use of lateralized stimulus presentation in conjunction with experimental paradigms used for the study of unconscious processes suggests that the separate processing capacities and specializations of the hemispheres may, under specialized and possibly non-representative laboratory conditions, result in a dissociation of affective experiences and conscious awareness and knowledge of the presence of a stimulus. Thus in Dawson and Schell's experiment the right hemisphere mediated the occurrence of an emotional response to a stimulus that was not capable of being discriminated or verbally described. In the Seamon et al. experiments the left hemisphere apparently mediated the development of a preference for a stimulus that Ss were not aware of having previously encountered. The right hemisphere was capable of mediating the recognition of previously presented polygons but not of a positive preference for such polygons.

EXPERIMENTS ON AFFECTIVE PRIMING

In this last section of this paper I shall present results from a series of studies done at Wesleyan in collaboration with my colleagues and students (Brody et al., 1986). In these experiments Ss were required to judge the affective content of a lateralized target stimulus preceded by lateralized backwardly masked prime stimuli that were either positive, negative, or

neutral words or faces. The target stimuli were words with strong
positive or negative affective meanings. In several experimental
conditions the prime stimuli were subliminal. That is, Ss were
not able to discriminate with better than chance accuracy whether
the prime stimuli were positive or negative under the conditions of
viewing used in the experiment proper. We found that the effects
of subliminal prime stimuli were identical to supraliminal prime
stimuli. Thus claims that I shall make based on these studies for
subliminal priming influences do not extend to the view that
subliminal influences are distinct and different from the
influences of supraliminal stimuli.

In two of the experiments in this series (one in which we used
schematic positive, negative and neutral faces as primes and one in
which we used positive, negative and neutral words as primes) we
presented lateralized prime stimuli to the same and to the opposite
visual field as the target stimuli. These experiments permit us to
study inter-hemispheric transfer effects. We found that the way in
which prime stimuli presented to the left hemisphere affected right
hemisphere targets was different from the effects of prime stimuli
presented to the right hemisphere on left hemisphere targets.
These asymmetric inter-hemispheric transfer effects are present in
the data reported in Table 3. Table 3 presents semantic prime
scores for positive prime stimuli indicating the improvement in
acuracy of judgment of the target words following the presentation
of a positive prime stimulus* (either a face or a word) relative to
a neutral prime minus the decrease in accuracy of judgment when a
positive prime stimulus precedes a negative target. Positive
affective prime scores may be taken as evidence for the benefits
obtained in processing affective target stimuli when they are
preceded by stimuli with compatible affective meanings.

There are several significant relationships that may be
observed in these data. Note that affective priming occurs most
dramatically for positive prime stimuli presented to the left
visual field (right hemisphere) when the prime is followed by a
target presented to the right hemisphere. The benefits of
affective priming obtained here may reflect the usual right
hemisphere advantage for processing emotional stimuli. However,
the benefits of right hemisphere affective meaning occur solely for
right hemisphere targets. Note that right hemisphere positive
primes do not exhibit any evidence of an inter-hemispheric transfer
effect. Such primes actually have slightly negative affective
priming scores for left hemisphere targets.

---

* Table 3 presents data solely for positive prime stimuli. The
results for negative primes were for the most part weak,
inconsistent and not clearly interpretable.

Table 3

Affective Prime Scores for Positive Primes

	Target Location	
Prime Location	LVF(RH)	RVF(LH)
LVF(RH	26.8	-2.9
RVF(LH)	7.2	12.9

Left hemisphere positive primes do have a smaller affective priming effect on left hemisphere targets than right hemisphere positive primes have on right hemisphere targets. However, in contrast to the effects of positive right hemisphere primes, left hemisphere primes do exhibit inter-hemispheric transfer effects and lead to statistically significant affective priming effects on right hemisphere targets.

Although there is no evidence that right hemisphere positive primes result in the specific transfer of information about the affective value of the prime stimulus to the left hemisphere there is evidence for inter-hemispheric generalized affective priming of the left hemisphere following the presentation of right hemisphere primes. We calculated the accuracy of judgement of left hemisphere target stimuli following the presentation of either a positive or negative prime stimulus to the left hemisphere and the accuracy of judgment of left hemisphere targets following the presentation of neutral prime stimuli to the left hemisphere. We found that the former accuracy score was 3.2 percent higher than the latter accuracy score - a small but statistically significant ($p < .01$) effect. This result suggests that right hemisphere affective primes do prepare the left hemisphere to process affective targets. It is also interesting to note that left hemisphere affective primes do not result in an increase in the accuracy of judgment of left hemisphere targets relative to left hemisphere neutral primes. Affective left hemisphere primes result in accuracy scores for left hemisphere targets that are 2.3 percent lower than those obtained for left hemisphere neutral primes of left hemisphere targets.

The results of these experiments, particularly the data on inter-hemispheric transfer effects, provide evidence for parallel and partially independent processing of information at different loci. These parallel and partially independent processes may in part occur as a result of distinctive processing capacities and roles of the two hemispheres. It is possible to assign a speculative and perhaps premature interpretation of these rsults by appeal to the results of studies of brain damaged people (see Heilman et al., 1985). The left hemisphere may signal the presence of positive emotional events. Hence left hemisphere damage may result in a catastrophic or permanent emotionally negative state and an inability to experience positive affect. The right hemisphere may have a central role in processing affective stimuli and in signalling the presence of an emotionally relevant event. Hence right hemisphere damage may result in an indifference reaction and an inability to respond affectively. More generally, these results suggest that information processed by the right hemisphere about the emotional content of a stimulus may not invariably be transferred without significant distortion to the left hemisphere. The existence of separate and partially independent cognitive processes associated with hemispheric specialization may be a mechanism that contributes to the dissociation of affect and awareness that is a focal feature of unconscious processes.

References

Brandeis, D.V. (1984). An Electrocortical Investigation of Word Recognition in a Backward Masking Paradigm. Biol. Psychol., 18, 294-295.

Brandeis, D.V. and Lehman, D. (1986). Event-related potentials. Neuropsychologia, in press.

Brody, N. (1983). Human Motivation: Commentary on Goal-Directed Action. Academic Press, New York.

Brody, N., Goodman, S.E., Halm, E., Krinzman, S. and Sebrechts, M.M. (1986). Lateralized Affective Priming of Lateralized Affectively Valued Target Words. Submitted for publication.

Corteen, R.S. and Wood, B. (1972). Autonomic Responses to Shock-Associated Words in an Unattended Channel. J. exp. Psychol., 94, 308-313.

Davison, R.J. (1984). Affect, Cognition and Hemispheric Specialization. In Emotion, Cognition and Behavior (eds. C.E. Izard, J. Kagan and R. Zajonc). Cambridge University Press, New York.

Dawson, M.E. and Schell, A.M. (1982). Electrodermal Responses to Attended and Nonattended Significant Stimuli During Dichotic Listening. J. exp. Psychol., Human Percept. Perf., 8, 315-324.

Dawson, M.E. and Schell, A.M. (1983). Lateral Asymmetrics in Electrodermal Responses to Nonattended Stimuli: A Reply to Walker and Ceci. J. exp. Psychol., Human Percept. Perf., 9, 148-150.

Fowler, C.A., Wolford, G., Slade, R. and Tassinary, L. (1981). Lexical Access with and without Awareness. J. exp. Psychol., Gen., 110, 341-363.

Gazzaniga, M.S. (1985). The Social Brain: Discovering the Networks of the Mind. Basic Books, New York.

Heilman, K.M., Bowers, D. and Valenstein, E. (1985). Emotional

244

Disorders Associated with Neurological Diseases.   In Clinical
Neuropsychology.   Second Edition (eds. K.M.Heilman and E.
Valenstein). Oxford University Press, New York.

Hines,, D., Sawyer, P.A., Dura, J., Gilchrist, J. and Czerwinski,
M. (1984).   Hemispheric asymmetry in use of Semantic Category
Information. Neuropsychologia, 22, 427-433.

Holender, D. (1986). Semantic Activation without Conscious
Identification in Dichotic Listening, Parafoveal Vision and
Visual Masking: A Survey and Appraisal.   Behav. Brain Sci.
In press.

Kunst-Wilson, W.R. and Zajonc, R.B. (1980). Affective
Discrimination of Stimuli that cannot be Recognized. Science,
207, 557-558.

Marcel, A. (1983).   Conscious and Unconscious Perception: An
approach to the Relations between Phenomenological Experience
and Perceptual Processes. Cogn. Psychol., 15, 238-300.

Öhman, A. (1986). Face the Beast and Fear the Face: Animal and
Social Fears as Prototypes for Evolutionary Analyses of
Emotion.   Psychophysiol., 23. 123-145.

Posner, M.I. (1978). Chronometric Explorations of the Mind.
Erlbaum, Hillsdale, N.J.

Shiffrin, R.M. and Schneider, W. (1977).   Controlled and Automatic
Human Information Processing: II. Perceptual Learning,
Automatic Attending, and a General Theory. Psychol. Rev., 84,
127-190.

Seamon, J., Brody, N. and Kauff, D.M. (1983a). Affective
Discrimination of Stimuli that are not Recognized: Effects of
Shadowing, Masking and Cerebral Laterality.   J. exp.
Psychol., Learn. mem. Cogn., 9, 544-555.

Seamon, J.G., Brody, N. and Kauff, D.M. (1983b). Affective
Discrimination of Stimuli that are not Recognized: II. Effect
of Delay between Study and Test. Bull. Psychonom. Soc., 3,
187-189.

Seamon, J.G., Marsh, R.L., and Brody, N. (1984). Critical
Importance of Exposure Duration for Affective Discrimination
of Stimuli that are not Recognized. J. exp. Psychol., Learn.
mem. Cogn., 10, 465-469.

Walker, E. and Ceci, S.J. (1983). Lateral Asymmetries in
Electrodermal Responses to Nonattended Stimuli: A Response to
Dawson and Schell.   J. exp. Psychol., Human Percept. Perf., 8,
145-147.

Session IV
Language and Hemispheric Specialization

Chairman: C. von Euler

17
Hemispheric Monitoring*

Eran Zaidel

FIRST OF ALL

In this paper I extend the concept of hemispheric independence to include cognitive control. Each hemisphere has a separate cognitive system, together with its own executive control that enables it to operate in some independence of the other. A central element of control is a self-monitor that plans and corrects behavior through feedback. I argue that each hemisphere incorporates a self-monitor. These two separate hemispheric self-monitors differ in their competencies and are "modular", i.e., they maintain relative functional independence, with respect to the central processors in each hemisphere. Occasionally, cerebral resources can be optimized by assigning processing to one hemisphere and monitoring to the other. This confers an adaptive evolutionary advantage to hemispheric independence.

Self-monitoring as an element of control. An important part of almost any complex system is dedicated to control, i.e., to processes or rules that determine when to start and stop computations, that effect planning, monitor performance and initiate error correction, and that implement strategy shifts. Self-monitoring, in turn, requires an internal model of the self as part of the environment, and access to that model is probably necessary for consciousness in any system.

*Supported by NIMH RSDA MH 00179, by NIH grant NS 20187, by a Biomedical Research Support Grant to UCLA, and by the Institute for Advanced Studies at the Hebrew University in Jerusalem. Thanks to Asa Kasher, Joseph Bogen, Avraham Schweigher, and Guido Gainotti for helpful discussions, to Risa Hiller for assistance, and to Rose Williams, Cynthia Smith, Diane Brodahl and Toney Dixon for Word Processing.

The normal hemispheres as independent parallel systems.

Research with complete commissurotomy patients suggests that each disconnected hemisphere is a complex cognitive system sustaining a separate mind. Each has its own sensory, perceptual, mnestic, linguistic and cognitive repertoires. Each cognitive repertoire has some necessary features including a sense of self and environment, past and future, plan and error. For example, each disconnected hemisphere can arrange a coherent sequence of pictures in response to the question "What would you like to be demonstrating a sense of future?" Each, moreover, has a capacity for self-monitoring. Each hemisphere can process much of the same everyday information. However, since different information processing strategies are often employed by the two hemispheres, they sometimes compete for subcerebral resources, interfere with each other, or obtain conflicting results. Conversely, some problems seem to benefit from interhemispheric interaction. In either case, a supra-hemispheric control system needs to coordinate hemispheric activities, maintain independence when necessary, effect cooperation when needed, and resolve conflict when it arises.

The corpus callosum as separate channels of communication. We know that the corpus callosum is partly organized by sensory modality: The caudal part, the splenium, mediates homotopic transfer of vertical midline visual information between areas 18 and 19; fibers rostral to the splenium are auditory and still more rostral fibers through the body of the corpus callosum are probably tactile. There is reason to believe that more anterior fibers mediate the transfer of more abstract information. In the case of printed words, for example, orthographic information may be transferred through visual fibers, phonetic information through auditory fibers, and semantic information through anterior routes.

Evidence is accumulating that there are individual differences in regional patterns of callosal size, connectivity and flexibility, and that those are associated with differences in cognitive repertoires. Moreover, there seem to be both internal and environmental conditions that modulate regional callosal connectivity and often effect inhibition of transfer so as to maintain hemispheric independence in the intact brain. Since different cognitive tasks may require various degrees of independence and interaction, the modulation of callosal traffic can be regarded as an aspect of supra-hemispheric cognitive control.

Error detection and error correction. In order to illustrate some of the foregoing claims, I will discuss evidence for the existence of different error monitors in the two hemispheres. In the context of the following discussion, an error is defined as the result of a computation classified as an "acceptable attempt" yet yielding the wrong solution to a problem relative to some norm. The error may result, among others, from noisy or misperceived input, from a misassembled routine, or from random inaccuracies in intermediate computations. Errors due to random inaccuracies can be detected through sequential computation by the same system or module. (By a "module" I mean a functionally separate system. Modularity admits of degree so that two systems are more modular relative to each other if their neural-functional independence is greater.) A more effective means of detecting errors, including those due to misperceived input or misassembled routines, could be through parallel computation by a different module with different and equally reliable computational characteristics. A discrepancy between the original answer and the recomputation would signal a possible error and result in backtracking.

How is a discrepency between two computations resolved? One heuristic is to initiate a third computation. Another way is to make use of built-in hierarchies of criteria for correctness. For example, in a lexical decision task it may be possible to apply either a phonological or an orthographic code to achieve lexical access. But an orthographic code may be preferable since it is unique: Some letter strings (e.g., "phail") are orthographically wrong yet generate a valid phonological code (pseudohomophones). Thus, in cases of discrepency, decision goes with the strategy that generates a better match with the input or that is known from experience to yield more accurate results.

Degrees of awareness of errors. How does the brain generate deterministically a graded "feeling of certainty" or confidence (certitude) in the accuracy of a response? A logical possibility is that certitude is the result of sampling multiple computations that operate at different speeds (Rabbitt and Vyas, 1981), but this implies considerable redundancy and too redundant a system seems incompatible with the cerebral localization of specialized processes that require specific monitors. Alternatively, computations may be undertaken on a selective basis or be only occasionally successful so that certitude may be proportional to the relative size of the monitored portion of the problem. Yet another interpretation of certitude of response is in terms of the cumulative effect of progressing heuristics. Responses, especially in speeded tasks, are often made on the basis of partial computations that provide guesses but occasionally fail (cf. Rabbitt and Rodgers, 1977). Even after a speeded response is produced, a more complete computation may proceed and lead to error correction. Here, since the scope of the heuristic employed is known, the degree of certitude in initial response is inversely proportional to the amount of available evidence not used by the heuristic. Microgenetically, as responses become more accurate they are accompanied by more confidence, require less monitoring and become more automatic.

In principle, if a new computation is done in a way different from the original computation then it should be possible to find some variable that affects differently the two computations (cf.

Rabbitt, 1967). Such a variable can be discovered if we can behaviorally isolate the two computations, for example, if they are at least partly sequential.

Monitoring within and across the hemispheres. Two separate principles underly hemispheric functions: hemispheric specialization and hemispheric independence (Bogen, 1985, 1987; E. Zaidel, 1985). Hemispheric specialization spans a wide range of tasks: At one extreme are tasks that can be performed equally well by either hemisphere although the computations used in each are different. At the other extreme are tasks that can only be performed by one, specialized hemisphere. The latter are rare. An example of a common psycholinguistic task that is exclusively specialized in the left hemisphere is phonological analysis of printed words.

Whether hemispheric specialization is conceived of in terms of tasks (the left is linguistic, analytic, sequential and digital; the right is nonverbal, synthetic, parallel and analog) or in terms of information processing stages (the left is particularly important during late-stage processing of familiar stimuli; the right is particularly important during early processing of novel information) each hemisphere needs its own self-monitor. In addition, each hemisphere can also monitor the other. Consider first those tasks that can be performed by either hemisphere. A useful division of labor would be to solve the problem simultaneously in each hemisphere, preferably using different kinds of computations in order to detect possible errors. If both computations reach the same result then chances are good that the solution was correct, assuming that the computation in each hemisphere is reasonably effective. This scheme suggests an adaptive evolutionary advantage for hemispheric specialization superimposed on hemispheric independence: The combination of these two principles provides for more efficient error detection, again assuming good bilateral competence.

Consider next those tasks that can only be performed in one hemisphere. Then one might suppose that only that hemisphere can monitor its own performance, say by resorting to a sequential recomputation. However, it may be possible for one hemisphere to monitor the results of processing in the other without a new computation of the original input. Consider, for example, a person who writes out a complex word in order to check its spelling. Only the left hemisphere is probably capable of writing but the right hemisphere is able to read and can monitor the correctness of the spelling by appeal to orthographic rules or templates.

A task may be assigned for exclusive processing in one hemisphere even if it can be performed by the other hemisphere as well. In that case the other hemisphere could provide parallel computation for error detection. Indeed, one reason for such a division of labor in the first place, even without specialization, would be precisely to provide for efficient error correction (assuming bilateral competence). This is because in this view each hemisphere contains specialized processing and dedicated monitoring machinery.

An evolutionary account of hemispheric specialization could take various forms. For example, an alternative conceptualization regards the right hemisphere as a set of independent special-purpose information processing modules that only have in common a shared dependence on certain parallel hardware architectures that are unique to the right hemisphere (Umilta, personal communication). In turn, the left hemisphere may be regarded as a general problem solver that can apply a universal arsenal to virtually any problem, with more or less success. On this view, the left hemisphere would contain a universal monitor whereas the right hemisphere would lack a general-purpose monitor and at best contain simple module-specific monitors that operate by sequential recomputation.

Error detection and error correction are modular. The possibility that error detection is performed in the opposite hemisphere than the one performing the initial task means that error detection has a degree of modularity with respect to initial processing. Error correction, in turn, can involve a new computation in the same or the other hemisphere so that it, too, is potentially independent of the other two modules. How can we check for independence among (i) initial performance, (ii) error detection, (iii) attempted error correction, and (iv) successful error correction? A sufficient (though not necessary) condition for independence would be a 3-way interaction between some experimental stimulus parameter (say, words and nonwords in a lexical decision task), visual field of stimulus (left and right), and control function (initial performance, error detection and error correction). But which performance measures can index these three control functions? Latency, accuracy and especially the signal detection measures of sensitivity and bias are useful indices.

SOME TOOLS

Models of behavioral laterality effects. Consider a half-field tachistoscopic choice reaction time experiment using linguistic stimuli with normal subjects and where the distal phalanges of different hands are used to respond in different blocks of trials. The task may consist of a sequence of component processes, each of which can either be done in each hemisphere or else needs to be processed in one hemisphere which is exclusively specialized for the component process. Let us focus on one such component. If it can be processed by either hemisphere then the hemisphere which receives the visual input first can process the information independently of the other ("direct access" model). In that case, it should respond faster with the contralateral hand, whose motor centers are localized in the same hemisphere. In other words, the ipsilateral hand-field combination should show a latency and accuracy advantage, assuming that callosal transfer costs time and accuracy. Thus, in direct access processes, there may be an overall visual field-advantage signalling hemispheric specialization, but there must be a visual field x response hand interaction signalling hemispheric independence (Figure 1).

If available resources within each hemisphere are insufficient, so that response programming with the contralateral hand interferes with central processing of the input from the contralateral visual hemifield (i.e., when the same hemisphere receives the stimuli and generates the responses), then there may be an interference pattern which is minimized when the visual field is paired with the contralateral hand, i.e., when central processing and response programming are performed in opposite hemispheres. In that case, we should still observe a visual field x response hand interaction, but with a contralateral rather than an ipsilateral advantage (Figure 1). Either pattern, however, is diagnostic of direct access processing.

If, however, the process in question can only be performed in one, say the left, hemisphere, then input to the left hemifield will have to be relayed through the corpus callosum to the left hemisphere prior to processing, at some cost in time and stimulus quality and, consequently, at some increase in response latency and some decrease in accuracy ("callosal relay" model). In this case, responses with the right hand should always be faster than responses with the left hand. Thus we should observe both a right hemifield advantage and a right hand advantage (Figure 1).

A combination of two component processes, at least one of which involves callosal relay, will result in an overall callosal relay pattern with the hemifield advantage determined by the first callosal-relay component, and the hand advantage determined by the last callosal-relay component. Thus, a task showing a callosal relay pattern can mask a direct access component, but a task showing a direct access pattern must contain components all of which are direct access. Of course, a task that can be performed by either hemisphere (direct access) can be assigned to one hemisphere and show a callosal relay pattern in particular circumstances of resource allocation. Thus, again, a callosal relay pattern may mask a potentially direct access task.

More generally, it seems conceptually easier to prove the existence of hemispheric independence (direct access) than of hemispheric specialization (callosal relay). This is because evidence for independence consists in positive demonstrations of parallel competence in the two hemispheres, whereas evidence for specialization consists of showing that one hemisphere cannot perform some task as well as the other. Negative evidence of this sort is always susceptible to the possibility that future data, with newer probing techniques, will reveal hitherto hidden abilities in the inferior hemisphere.

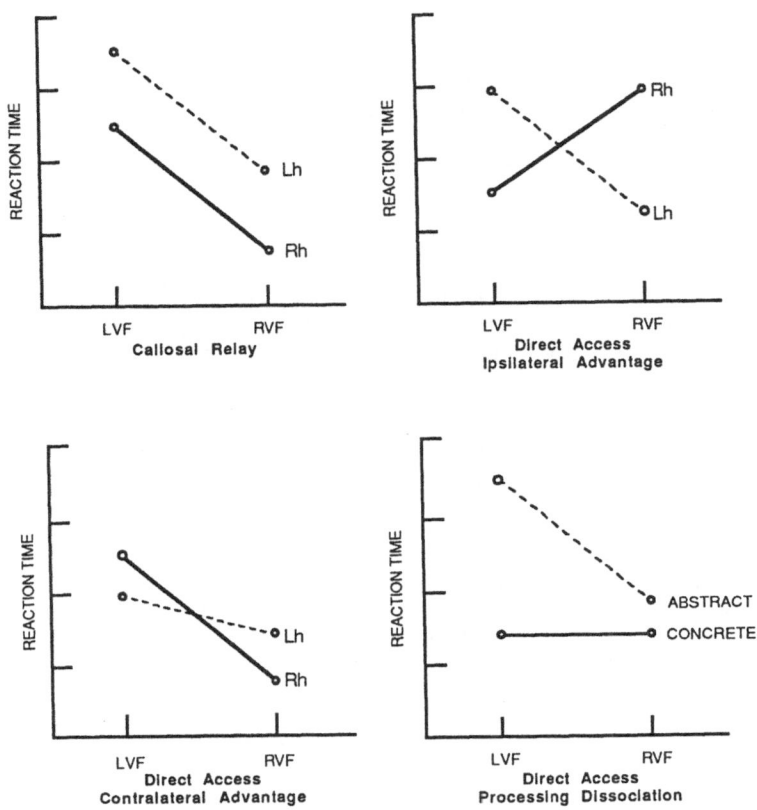

Figure 1. Models of laterality effects in the normal brain. LVF = left visual hemifield, RVF = right visual hemifield, Lh = left hand, Rh = right hand.

Another test for direct access is an interaction between hemifield of presentation and some stimulus dimension which should not affect callosal transfer ("processing dissociation" criterion). For example, in a lateralized lexical decision task we may obtain no visual hemifield advantage for concrete nouns but a right hemifield advantage for abstract nouns, where the concrete and abstract nouns are equally frequent, long, visually complex, etc. (Figure 1). Such an interaction must mean that each hemisphere uses a different strategy for processing the same stimuli, assuming that both values of the stimulus dimensions (e.g., concrete and abstract nouns) are handled the same in a given visual hemifield, i.e., both are handled in a direct access manner or else both are callosally relayed to the other side. If a dissociation can occur in the processes that apply to each value in the same field (say the left hemifield), i.e., if one value (say concrete nouns) is processed direct access and the other (say abstract nouns) is relayed callosally (to the left hemisphere) then we can no longer rely on "processing dissociation" to guarantee "direct access" and need to check the response hand x visual field pattern for each stimulus value in each field in order to determine which model applies.

Hemispheric differences in strategy may also be observed in visual field differences in the signal detection measure of bias, or in a different relationship between accuracy and latency in the two hemispheres. For example, one hemisphere may show a positive and the other a negative correlation between latency and error rate.

Measures of error detection and correction. Consider a choice reaction time experiment with self corrections. Initial performance can be measured by the usual indices, such as reaction time, accuracy, or sensitivity and bias in a signal detection paradigm. Error correction can again be measured by its latency and accuracy. The signal detection measures of sensitivity and bias for error correction include the proportion of successful and unsuccessful corrections. Thus, initial performance can be indexed simply by accuracy of first responses and correction accuracy can be indexed by accuracy of self-corrections. Error detection reflects the subject's awareness of his errors, but not necessarily his ability to correct them. We can distinguish local or tactical error detection from global or strategic error detection.

Local error detection refers to the subject's ability to correct real errors and to his tendency to correct non-errors. These are expressible, again, in the signal detection measures of changes of mind (in a two-choice experiment a change of mind following an error is a "hit", a change of mind following a non-error is a "false alarm", etc.). In reaction time experiments with more than two choices, the detection of wrong initial responses may be good, leading reliably to changes of mind, but may not be accurate, resulting in wrong self-corrections. (For example, a semantic category decision task may require the subject to decide whether a target word is inanimate, a plant or an animal. The subject may mistakenly classify "beaver" is inanimate then change his mind and classify it as a plant.) In two-choice reaction time experiments, signal detection measures of local area detectability are the same as of correction accuracy.

Global error detection refers to a more statistical, average self-confidence rating leading to a strategy of using a certain overall percentage of error correction (whether successful or not), rather than, or in addition to, a trial-by-trial evaluation. For example, responses may be accompanied by feelings of confidence so that low confidence may lead to a high number of self-corrections. Global error detection is measured by the proportion of self-corrections to total number of errors in initial performance.

SOME DATA

Hemispheric damage

The effect of localized brain-damage on general error detection and correction has not been studied systematically for at least two separate reasons. The first is that clinical neurological syndromes have traditionally been defined in terms of either content or input domains (e.g., agnosia) or in terms of typical effects or outputs (e.g., amnesia) but rarely in terms of control functions. A notable exception is the frontal lobe syndrome, which is considered a disorder of planning or strategy-shifting. The second reason is that all cognitive deficits may be said to include deficient central processing and deficient control since defective behavior is not detected and inhibited. This is not surprising considering that the most important element of control is self-monitoring and that the simplest kind of error detection and correction mechanism is the one involving recomputation. When central processing is disabled, this sort of error monitoring is also impaired.

The kind of monitoring posited hereby is hierarchic: each modular cognitive system has its own local monitor and higher levels of integration have theirs. Similarly, each level of cerebral organization has control structures. In particular, each hemisphere has its own monitors and the left may contain as well the monitor that oversees interhemispheric interaction.

The frontal-parietal balance. Many clinical observations may be summarized by noting that localized lesions differ systematically in the kind of control deficits that accompany the main symptoms, depending on how anterior or posterior the damage is (Denny Brown, cited in Mesulam, 1986). Anterior or frontal syndromes are said to be accompanied by relatively intact error detection but poor error correction. Posterior, or parietal syndromes, on the other hand, are accompanied by poor error detection ability. Frontal perseverations imply a loss of self-monitoring (start-stop rules). Indeed, an important element of many theories of frontal lobe functions is the construction of internal models of solution processes, which are periodically compared to intermediate steps in actual performance so as to regulate action through error-correction (cf. Luria, 1980 and see

Fuster, 1980, Stuss and Benson, 1986). Perseverations can be domain-specific, consistent with an hierarchic concept of control (Goldberg and Bilder, 1987). Frontal perseverations, for example, may represent selective disorders of motor control.

Consider the following four examples of the anterior-posterior dissociation in control functions:

1. ANTERIOR VERSUS POSTERIOR APHASICS. There is considerable evidence that anterior, nonfluent or Broca's aphasics maintain relatively intact error detection but relatively ineffective error correction. Broca's aphasics are painfully aware of their deficits and often get depressed over their inability to improve their performance. They do respond to cues but their spontaneous error correction is poor because it is too "costly" or difficult. Conduction aphasics show similarly intact error detection but improved error correction. Their production errors often show successive self-correction through progressive phonological approximations to the target ("conduit d'approche"). By contrast, posterior, or Wernicke's aphasics can show dramatic anosognosia (Lecours, Lhermitte and Bryans 1983). They may be surprised at their own output when it is presented to them later, out of the conversational context. It has also been observed that there is greater depression in anterior aphasias and that the more posterior the aphasic lesion, the more indifferent or euphoric the reaction of the patient to his own deficit (Robinson and Benson, 1981).

2. ANOMIA WITH AND WITHOUT COMPREHENSION DEFICITS. Gainotti et al (1986) distinguish the error correction symptomatology of anomic patients with and without comprehension or conceptual disorders of posterior etiologies. Those without comprehension disorders will often move on from the wrong choice, and detect wrong phonemic cues or options. By contrast, those with conceptual deficits will occasionally accept the wrong cue or option.

3. NEGLECT IN REMISSION. Often right parietal patients starting with severe neglect and anosognosia will show a transitional stage in recovery where the line bisection task shows over-correction rather than neglect. That is, the patients bisect the lines closer to their left rather than right edges. This can be interpreted as a symptom of awareness of deficit leading to a global error-correction strategy (compensation), but without local error-detection or error-correction ability to monitor performance and execute the line bisection task correctly.

4. USE OF PARTIAL MODELS. Partial models of the answer to contructional tasks refer to intermediate steps in the solution process. Anterior patients, unable to plan an

overall solution, can benefit from partial models whereas parietal patients with perceptual deficits do not benefit from such models (Hecaen and Albert, 1978). There is a related distinction in the ability to imitate a model in apraxia. Ideomotor apraxia, said to occur with left supra-marginal gyrus lesions, is said to include poor imitation, whereas ideational apraxia, with diffuse, bilateral localization, is said to include preserved ability to imitate. Heilman and Rothi (1985) distinguish two forms of ideomotor apraxia. One is induced by lesions of the left supra-marginal or angular gyrus. These patients not only perform poorly to command and imitation, but also cannot discriminate poorly performed from well performed acts. The other form is induced by lesions anterior to the supra- marginal gyrus. These patients can also not perform well to command or imitation, but they can discriminate between well-performed and poorly-performed acts. Thus, monitoring of own performance, as determined by imitation, may be impaired when monitoring of others' performance remains functional.

In the case of constructioral apraxia, the deficit due to right brain damage is said to be unhelped by a model, whereas left brain-damaged patients are helped by a model of the solution. The ability to use a model implies preserved monitoring function but poor representation of the goal. More generally, there has been an attempt to associate frontal symptomatology with left hemisphere disfunction and parietal symptomatology with right hemisphere dysfunction, as in the case of Luria's, Pribram's and Jakobson's sequential versus simultaneous synthesis disorders. Such an association would assign a special role to the right hemisphere in monitoring cognitive and linguistic functions.

Can there exist cases with preserved error correction but without error detection? Surely one cannot correct an error he cannot detect. But suppose his errors are pointed out to him first? This should only help if there is residual error detection ability which is enhanced by external corrections and can then facilitate error correction, for which it is a prerequisite.

Disconnection Syndrome

1. Error detection with error correction: Trial and error with the Raven Matrices

The Raven Progressive Matrices is a nonverbal and allegedly culture-free test of intelligence. It requires no complex perception or manual manipulation, it is largely self-explanatory and is therefore thought to be a suitable candidate for assessing hemispheric contributions to "general intelligence" or "g". The test is a figure analogy task which requires the subject to complete a pattern by pointing to or placing one of six or eight alternatives which best fits the missing part. The Raven Colored Progressive Matrices is the children's version and comes in two forms: In the book form the subject points to the correct choice on a printed page. In the board form the missing part is empty. The wooden choices can be removed physically, puzzle-fashion, and the subject has to physically place the correct answer in the missing part.

The book and board versions of the Colored Matrices were administered three or four years apart to two complete commissurotomy patients (LB and NG). Both versions were presented in one visual hemifield at a time, using a contact lens method which permits free ocular scanning of the stimulus page in the subject's lab, first by the right hemisphere, then, a week later, by the left hemisphere. Responses were made with the hand ipsilateral to the scanning visual hemifield and visually guided movements were possible. With the board version, the subject was encouraged to use a trial-and-error approach until s/he was satisfied with the answer. We scored separately the subjects' first attempted solution and their last, final one in order to determine whether they used a different strategy from the one they used in the book form (E. Zaidel, 1978a; E. Zaidel, D.W. Zaidel and Sperry, 1981).

The results for the book version show a substantial left visual hemifield advantage for patient NG and a ceiling effect with a small but consistent left hemifield advantage for patient LB. However, the results for the board version show that the two hemispheres are not equally able to benefit from the opportunity for self-correction through overt trial-and-error: The final score of the right hemisphere on the board version was equal to that obtained on the book version several years previously. The left hemisphere, on the other hand, had benefitted substantially from error correction so that its final score on the board version surpassed its own as well as the right hemisphere score on the book version. Moreover, the first responses by both hemispheres on the board version are inferior both to their own book scores as well as to their final board solutions, showing that both hemispheres used a new strategy and did attempt to take advantage of trial-and-error (Figure 2).

Thus, the right hemisphere, which was attracted to close alternatives of the correct choices, was unable to use the discrepancy in order to improve its characterization of the desired answer. It is as if the right hemisphere does not have access to an internal model of its own solution process which it can interrogate to update intermediate steps. The conclusion that the book versions of the Colored Matrices favors the right hemisphere, whereas the board version favors the left hemisphere is supported by the finding that verbalization of strategy during and after solution of the problem by children improved scored on the board version of the test but actually decreased scores on the standard book version.

The right hemisphere of patient NG often used a "gestalt strategy" of rotating a chip which she had placed in the empty aspace on the board in an attempt to get a better "perceptual" fit. And yet, even when the misaligned pattern was the correct choice, the right hemisphere tended to choose another final pattern. This suggests that the right hemisphere is unable to use spontaneous image rotation as part of an error correction strategy.

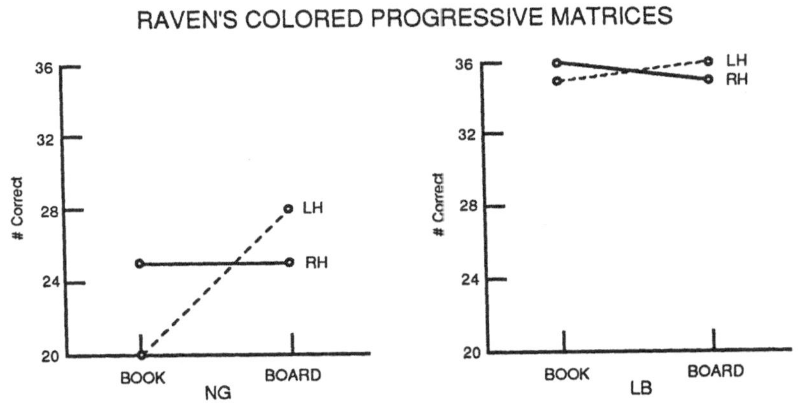

RAVEN'S COLORED PROGRESSIVE MATRICES

Figure 2. Comparison of performance (Number correct out of 36) by commissurotomy patients LB and NG on the book and board (trial and error) versions of Raven's Colored Progressive Matrices. LH = left hemisphere scores, RH = right hemisphere scores using the contact lens method for free hemispheric scanning.

2. Error correction without error detection: getting a second chance

Suppose one disconnected hemisphere engages in a multiple-choice decision. Suppose further that a response was wrong and the examiner corrected it. Will the two hemispheres benefit equally from the second chance? Of course, each could benefit from the consequent restriction in search space. Indeed, there is some evidence that the right hemisphere is more sensitive than the left to the number of response choices in a task. The right, but not the left hemisphere performs better relative to chance on three-than on four-choice tasks.

In fact, lateralized tactual tests of the right hemispheres of commissurotomy patients have often employed only three choices, since careful preliminary observations revealed a reduced ability by the disconnected right but not left hemisphere to deal effectively with more than that number of choice alternatives (Nebes, 1971; D. Zaidel and Sperry, 1973). But beyond such gains, is there any hemispheric difference in error correction when detection is provided by the examiner? I will use two examples, each involving linguistic functions: The first is the ability to comprehend spoken sentences involving various grammatical structures by pointing to a picture corresponding to the target sentence among several choices. The second is the ability to comprehend various conceptual semantic relations represented pictorially. Responses are signalled by pointing to a quartet of pictorial choices. The right hemisphere is relatively poor on the first type of task and relatively good on the second type.

Syntactic comprehension. Four tests for evaluating comprehension of syntactic structures were administered to NG and LB in lateralized presentations on different occasions to the right and left hemispheres. The tests were designed to assess grammar comprehension in children and in aphasics. They included Frazer, Bellugi and Brown's 1963 test, Lea's 1971 Northwestern Syntax Screening Test, Carrow's 1969 Auditory Test for Language Comprehension, and Shewan and Canter's 1971 Sentence Comprehension Test (see E. Zaidel, 1973, 1978, 1978a for details). The first three were designed for use with children, and the last two were designed to be used with aphasics. In all four tests, the three or four multiple-choice pictures differ from each other in precisely one dimension involved in the target grammatical contrast. Thus, eliminating one wrong choice helps to focus on the correct solution by removing possible confusions.

For the purpose of this study, the multiple-choice array was exposed for scanning in one hemifield and the examiner read the target word, phrase or sentence aloud. The patient was required to point to the correct picture with the hand on the side of the scanning field. Whenever the patient made an error the examiner indicated this to her/him verbally and asked the patient to try again. For comparison, initial and corrected performance were expressed as z-scores relative to chance, computed by the normal approximation to the binomial guessing distribution, $z = (x-c)/(npq)^{\frac{1}{2}}$, p = correct guessing probability, $q = 1 - p$, n = number of items, c = chance correct score. For a four-choice test the first attempt has p = 1/4 and the first and second attempt together have p = 1/2. The question is whether a second chance improves the z-score for one hemisphere more than for the other.

As Table 1 shows, in general second chances following incorrect first choices result in bilaterally higher raw scores but statistically less significant differences from chance (z-scores) on both sides. The only exceptions are the

Table 1. Unilateral raw and z-scores (relative to chance) on four syntex test for first and second chance.

			Shewan and Canter		Fraser, Bellugi and Brown		Lea		Carrow	
			N = 42		N = 52		N = 38		102(112)	
			4 choices		4 choices [1]		4 choices		3 choices	
			raw score	z-score	raw score	z-score	raw score	z-score	raw score	z-score
NG	RH	1st Choice	13	.8909↑	33	6.4051↓	17	2.8097↓	87	12.8748↓
		2nd Chance	24	.9258	46	5.5469	25	1.9467	101	8.5042
	LH	1st Choice	16	1.9599↑	48	11.2090↓	28	6.9307↓	95	14.6205↓
		2nd Chance	34	4.0119	49	6.3790	34	4.8667	103	8.8822
LB	RH	1st Choice	24	4.8108↑	37	7.6861↓	18	3.1844↓	86	13.3465↓
		2nd Chance	38	5.2463	45	5.2696	22	.9733	90	7.1875
	LH	1st Choice	31	7.3053↓	48	11.2090↓	36	9.9277↓	108	17.4573↓
		2nd Chance	41	6.1721	51	6.9337	38	6.1644	110	10.2050

[1] modified version of the original 2-choice version

↑ = improvement in 2nd chance relative to 1st choice

↓ = 2nd chance is worse than 1st choice

performances of NG's left and right hemispheres and of LB's right hemisphere on Shewan and Canter's test. Furthermore, the differences in z-scores between the left and right hemispheres are invariably smaller for first choices than for second chances. The left hemispheres' actual gain in raw scores is smaller than the gains by the right hemispheres though this reflects a ceiling effect which is exacerbated by loss in significance relative to guessing in second chances. Thus, the indirect beneficiary from error correction in those syntax comprehension tests, if any, seems to be the right hemisphere. Even a definitive selective right hemisphere advantage from second chances need not signal an advantage in error correction. It could have reflected release from initial left hemisphere interference which is commonly observed in testing the right hemisphere of commissurotomy patients when part of the stimulus is available to the left hemisphere. Moreover, as already mentioned, there is evidence that the right hemisphere would benefit more than the left simply from the reduction in the number of multiple choices, independently of error correction. Consequently, the data do not provide strong evidence for a hemispheric difference in benefit from error correction without detection on tasks that show left hemisphere superiority in initial performance.

Visual concepts. The Visual Reception and Visual Association subtests of the Illinois Test of Psycholinguistic Abilities were administered to complete commissurotomy patients LB and NG in lateralized presentation using the contact lens for free hemispheric ocular scanning. The Visual Reception subtest requires the subject to match by pointing a stimulus photograph (e.g., a fan) to another photographic exemplar (e.g., curtains blowing in the wind) of the same concept, such as "windy", even when they are perceptually dissimilar and when some of the decoys are perceptually quite similar to the stimulus. This then is a test of class inclusion. The Visual Association subtest has two parts. The first requires the subject to select by pointing one of four line drawings that most meaningfully relates to a stimulus drawing. In this case, a variety of semantic relations are sampled. The second part of the test consists of conceptual visual analogies (E. Zaidel, 1979).

As Table 2 shows, although the number of correct identifications in both tests improves considerably after error correction in both right and left visual field presentations, second chance scores are again generally less significantly different from chance since the probability of guessing increases from .25 to .5. This time, the z-scores show an increase from first choices to second chances only in left hemisphere performance, and only in patient NG. It is noteworthy that NG's right hemisphere performance on both tests is superior to her left hemisphere (and identical to her free field score, which is presumably controlled by the right side).

Matched t-test statistical comparisons between number of correct responses in the left and right hemispheres show that this time first choices are as likely to show differences (significant t values are actually higher) as do second chance scores. In this case, then, the left hemisphere has a tendency to benefit more from correction (larger increase in z-score) but this does not translate into a greater hemispheric difference.

Taken together, the data from second chances in multiple choice tasks in the split brain do not provide a sensitive index of hemispheric error correction without detection. z-scores relative to chance tend to show a general decline with second chances in spite of increases in accuracy (because the guessing rate increases also from .25 to .5). When the index does indicate hemispheric asymmetries, these differ as a function of task. This would mean that neither hemisphere has a better overall error-correction-without-detection ability. Conceivably, nonverbal indication of

Table 2. Number correct and z-scores (relative to chance) on the Visual Reception and Visual Association subtests of the Illinois Test of Psycholinguistic Abilities.

Association			Visual Preception N = 40 4 Choices		Visual N = 42 4 Choices	
			# correct	z-score	# correct	z-score
NG	RH	1st Choice	21	4.02↓	20	3.39↓
		2nd Chance	26	1.09	30	2.70
	LH	1st Choice	15	1.83↑	17	2.32↑
		2nd Chance	27	2.22	31	3.09
LB	RH	1st Choice	19	3.29↓	24	4.81↓
		2nd Chance	27	2.21	32	3.39
	LH	1st Choice	35	9.13↓	29	6.59↓
		2nd Chance	39	6.01	39	5.56

↑ = improvement in 2nd chance relative to 1st choice

↓ = 2nd chance performance is worse than 1st choices

error by the examiner would benefit the right hemisphere more than a verbal signal (D.F. Benson, personal communication), although the right hemisphere does seem to comprehend the signal and shows an improvement in accuracy in second chances. Finally, even when the z-score shows a selective increase in one hemisphere, this does not necessarily translate into a greater hemispheric difference in accuracy as determined by a matched t-test.

The Normal Brain

No interhemispheric monitoring?

Golding, Reich and Wason (1974). For some time now the idea has been around that there is poor interhemispheric monitoring even with intact neocortical commissures and that this can explain some discrepancies between direct verbal reports and other means of interrogating the solution process. But without an explicit account of hemispheric independence, these occasional reports remained speculative interpretations of results that have yet to be explained definitively. As early as 1974, Golding, Reich and Wason presented normal subjects with a difficult deductive problem (Wason's "selection problem"). Subjects not only frequently fail to solve the problem but also tend to maintain their incorrect solutions even when shown the correct solution and confronted by inconsistencies in their thinking (cf. Wason and Johnson-Laird, 1972). In this case, errors reflect the application of inadequate heuristics, when logically provable solutions are not available. The authors speculated that many of the errors (especially of one kind called "perceptual matching responses") originate in the right hemisphere, whereas the error-correction procedure (i.e., making explicit the logical inconsistency between the subject's initial choice and his acceptance of the examiner's arguments) only affects the left hemisphere. On this analysis, error correction involves a reformulation of the solution process but only the results rather than the processes themselves are available for interhemispheric comparison. Moreover, Golding et al conclude that the right hemisphere inhibits performance whereas the left hemisphere facilitates it (cf. also Golding, 1981 and Gellatly, 1985). This then is an extreme view of hemispheric independence of computational machinery together with shared resources, in complex problem solving with discrete responses.

Golding et al proceeded to test the hypothesis that the right hemisphere was responsible both for the errors and for the resistance to the correction by lateralizing the test and correction items to unimanual palpation by one hand or the other. The results did not show worse initial performance with the left hand but did yield more improvement after error correction with the right hand. But the authors' interpretation is not compelling. In so far as subjects verbally insist that their original solutions are correct, their "left hemispheres" do have access, i.e., can monitor, their right hemispheres. The failure of the error correction procedure to resolve the contradiction may simply reflect the failure of the right hemisphere to benefit from error correction. Thus, this result, though consistent with hemispheric independence does not suggest poor interhemispheric monitoring. The result can, however, be naturally interpreted in terms of two weakly interfaced solution models with poor intertranslation rules between them.

Landis, Graves and Goodglass (1981). More recently, Landis et al (1981) administered lateralized object and facial expression recognition tasks (chance = 50%), keeping error rates constant (about 25%) and encouraged subjects to tell the experimenter whenever they thought they had made a mistake. Previous studies showed a left hemifield advantage for the facial expressions task and a right hemifield advantage for the object recognition task. Results showed no difference in total number of "verbal corrections" (whether right or wrong) between the two tasks, but a much higher percentage of correct verbal corrections (93%) in the object recognition task than in the facial expressions task (58% which is not different from chance). The authors interpreted this to reflect a "disconnection affect" between the verbal monitor in the left hemisphere and right hemisphere judgment of emotional facial expressions.

The analysis presented in this remarkable paper is incomplete. The authors do not report the accuracy of verbal corrections of left hemifield and right hemifield presentations separately, although the data (floor for faces, ceiling for objects) and absence of statistical effects suggest no hemifield difference in accuracy of verbal corrections for either task. The disconnection interpretation can only be correct if right hemifield presentations of facial expressions results in shuttling of the stimuli for processing to the right hemisphere via the corpus callosum, i.e., if the task is what we called "callosal relay" specialized in the right hemisphere. Conversely, given the authors' interpretation, good verbal corrections of objects exposed in the left hemifield must mean that object stimuli are always shuttled for processing to the left hemisphere where they are available to verbal monitoring, i.e., the object recognition task is also callosal relay, specialized in the left hemisphere. As described above, there now exist techniques for determining whether a task is direct access or callosal relay. If either the object recognition or the facial expressions task turns out not be callosal relay, then Landis et al's disconnection analysis must be wrong since it would mean either that the verbal monitor in the left hemisphere has access to object recognition in the left hemifield or that it has no access to facial expressions in the right hemifield.

It is interesting that Landis et al believed that the left hemisphere monitor may not have access to right hemisphere results, whereas Golding et al believed that the left hemisphere monitor has access to right hemisphere results, though not to details of right hemisphere processing.

Hemispheric error detection and error correction in lexical decision are modular

In order to study the ability of each normal hemisphere to spontaneously correct its own errors, we administered a hemifield tachistoscopic lexical decision task which is already known to be "direct access," i.e., to be processed independently by each hemisphere of the normal brain (E. Zaidel, 1983). Subjects were induced to correct their own errors by appeal to a special reward system that penalized heavily all uncorrected errors but rewarded moderately all self-corrected errors. We compared initial performance to the number of attempted corrections (as a measure of global error detectability) and to the accuracy of those correction (as a measure of error correction ability). Since the task was a two-choice identification (word versus nonword), the measure of local error detection discussed above is the same as error correction in this case.

The experiment was carried out by Mr. Richard Stein, a UCLA Psychology senior as part of an honors thesis and an independent research project (Stein, 1986; Stein and E. Zaidel, 1987). Stimuli consisted of 64 four to six letter, concrete English nouns (high in frequency and imageability) and 64 orthographically regular nonwords. Trials were arranged so that the first and second halves of the test each consisted of 64 trials, half words and half nonwords, in order to allow for an analysis of practice effects. Stimuli were presented in a random order to left and right of fixation for 40 msec. Word/nonword responses were signalled on a two-way toggle switch (direction counter-balanced). Subjects were instructed to be fast and accurate and to correct themselves by pushing the switch in the opposite direction to the initial response. Thirty-two right-handed subjects, half males and half females, participated.

Accuracy of initial decisions showed a significant right hemifield advantage and a significant right hand advantage but also a visual hemifield x response hand interaction (interference pattern) suggesting direct access (Figure 3). Words were recognized more accurately than nonwords and this difference was greater in the right hemifield. The latter word x visual field interaction was due to a selective right hemifield advantage for words and again suggests a direct access pattern by the processing dissociation criterion discussed above. Performance was more accurate in combinations pairing opposite hand and hemifields, presumably because there was a shortage of resources and the opposite pairing minimized interference between central processing and response programming.

By contrast, accuracy of error corrections showed no hemifield advantage and a massive advantage for nonwords (Figure 3).

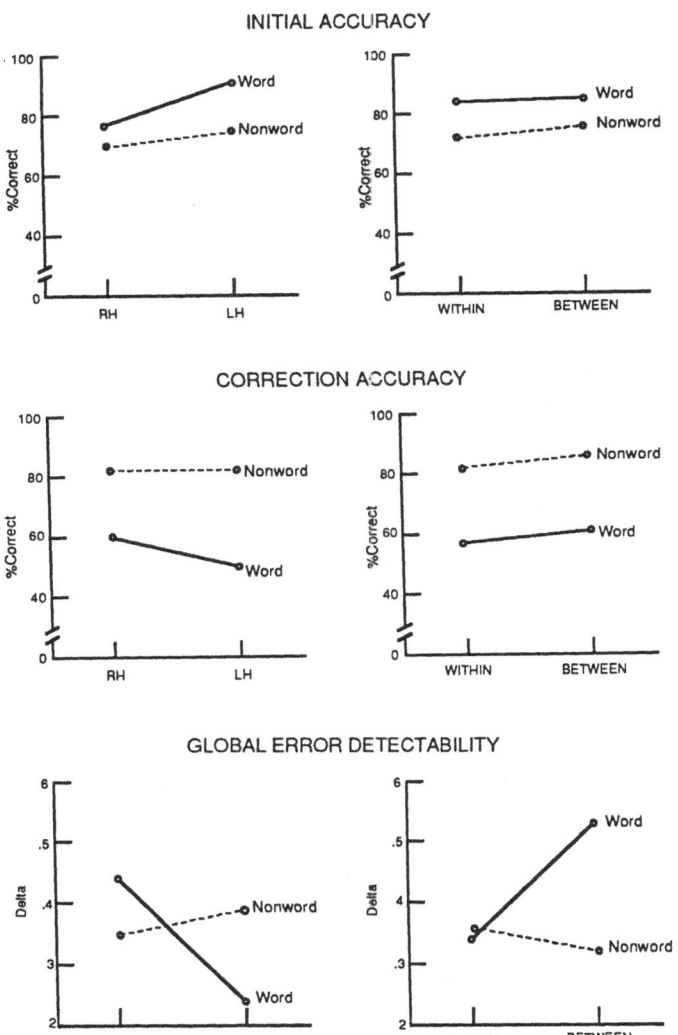

Figure 3. Lexical decision with self-corrections in normal subjects (Stein, 1986). Initial performance, error correction accuracy and global error detection ability. LH = left hemisphere, i.e. left visual hemifield-left hand performance. RH =
right hemisphere, i.e. right hemifield-right hand performance.
Within = mean of RH and LH. Between = mean of left hemifield-
right hand and right hemifield-left hand.

Again there was a significant word x visual hemifield interaction signalling direct access, and a significant visual hemifield x response hand interaction (ipsilateral advantage) confirming it. The fact that initial decisions showed an advantage for words, whereas self corrections showed an advantage for nonwords, suggests that error correction is not done by recomputation and that there are separate modules for error correction. (Since the task

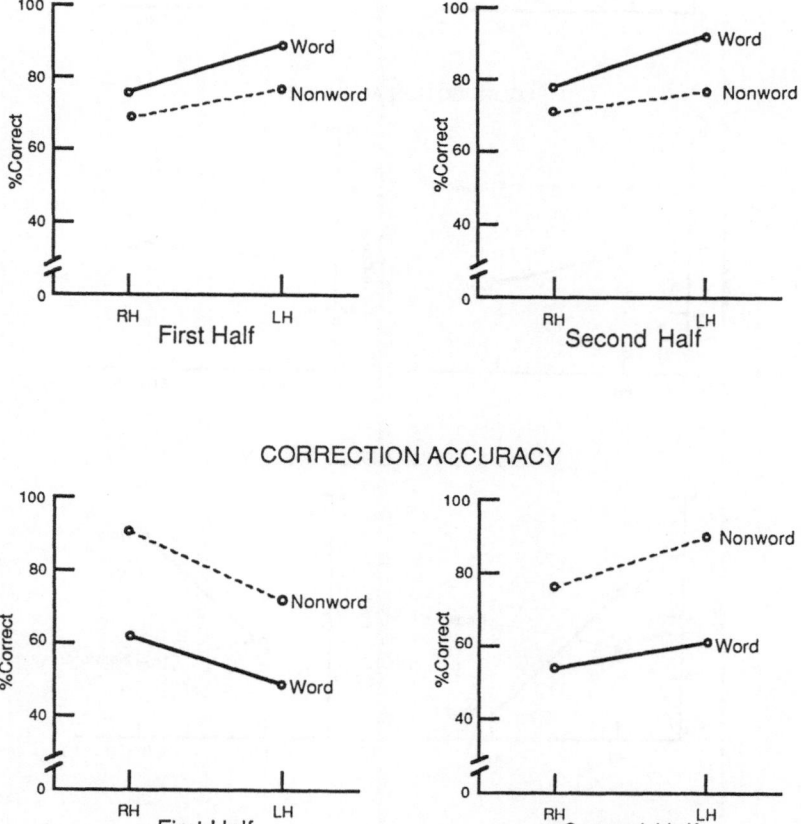

Figure 4. Changes in performance (initial and correction accuracies) of normal subjects in lexical decision from first to second half of the test. See figure 3 for legends.

involves word and nonword responses, there is no necessary tradeoff between initial decisions and self-corrections. Thus, the fact that initially words are decided more accurately than nonwords does not mean that nonwords should necessarily be corrected more accurately.)

Instructive sex differences emerged. Women showed a direct access pattern in initial decisions. The visual hemifield x response hand pattern for initial decisions by males did not show an interaction, but rather a right hand advantage together with a right hemifield advantage signalling callosal relay with left hemisphere specialization. And yet, for males too, error correction showed a direct access pattern. This suggests that left hemisphere performance of the tasks was done for "convenience" rather than by "necessity" and that even while the left hemisphere performed the initial decisions for left hemifield stimuli, the right hemisphere corrected them! This means that error monitoring can occur just as effectively across the commissures as within a hemisphere.

For global error detectability, the pattern of visual hemifield x response hand, though not statistically significant, is consistent with direct access. Glcbal detectability shows no hemispheric difference in number of self corrections applied to nonwords, but a large right hemisphere superiority in number of self corrections applied to words (Figure 3). Moreover, error detectability for words is better across than within the hemispheres! Thus, certain types of monitoring are actually better performed by the non-acting hemisphere.

What is the effect of experience or practice on initial accuracy, on glcbal error detection and on error correction in each hemisphere? Stein compared responses on the first and second halves of the trials administered in the test. He found that there was a mild across-the-board improvement in initial accuracy. Surprisingly, correction accuracy showed a right hemisphere decrease and a left hemisphere increase with experience (Figure 4). The observed shift in relative hemisphere superiority from a right to a left hemisphere advantage in correction accuracy is novel. Others have reported a similar shift in hemispheric performance level as a function of novelty and experience, but in this case there is no comparable shift in initial accuracy, the variable equivalent to the usual performance measures.

Global error detectability shows only an overall net increase for nonwords in the left hemisphere with experience, but there are large differences in size and direction of changes in global detectability as a function of wordness, hemisphere and sex.

In sum, all these measures, i.e., initial accuracy, correction accuracy, and global error detectability show very different patterns of responses to experience, verifying yet again their independence or relative modularity of each other. In particular, there is no overall hemispheric advantage in the ability to spontaneously correct errors on this task: the right hemisphere has an initial advantage which shifts to the left hemisphere with experience, all while initial decisions maintain the same mild left hemisphere advantage.

SOME THOUGHTS

Monitoring through callosal transfer? Experiments on hemispheric specialization in the normal brain often reveal interhemispheric transfer times of 50 msec and more, whereas the callosal transmission of simple motor commands takes around 2 msec. (E. Zaidel, 1985). One could speculate that a plausible reason for the difference is that interhemispheric transfer of complex information involves "translation" from the code or language of one hemisphere into that of the other. The two languages are not likely to be co-extensive. Relations in one may not be expressible in the other or, conversely, they may have much simpler structural descriptions in the other. In the latter case mere translation, i.e., the act of interhemispheric transfer itself, may discover errors that could not be easily detected in the more complex description, but are transparent in the simpler description (A. Ruskin, personal communication).

Speech errors. Speech errors are notoriously common: they variously involve disordering and substitution of linguistic units and misapplication of linguistic rules. But why do errors occur? Freud (1958) believed that (semantic) errors reveal repressed thoughts and arise from parallel competing actions, one reflecting the conscious message the other reflecting the latent content. Baars, Motley and MacKay (1975) adapted this "competing plans hypothesis" and argued for the existence of an internal monitor (editor) which usually detects and blocks errors but sometimes "slips" itself, causing an error to surface. This explains lexical errors in the form of some word blends, such as "instantaneous/momentary" yielding "momentaneous" (Fromkin, 1987). Of course, now it is necessary to explain why and how the monitor malfunctions. But the concept of hemispheric independence provides a natural candidate for the origin of competing lexical plans because the lexicon in each hemisphere has a different organization.

MacKay (1987) contrasts his own Node Structure theory (of perception and action) with a class of models denoted "editor theories." For MacKay, the defining characteristic of an editor theory is a mechanism that "listens to the visual, auditory, kinesthetic or other feedback which results from output, compares this feedback with the intended output, identifies errors, and computes corrections using a duplicate copy of the information originally available to the motor control system." Recent variants have added a feedforward or efference-copy component, i.e., a preliminary representation of an about-to-be-produced action which enables editing before the actual occurrence of the output.

But how does the editor know the intended output in order to compare it with the feedback? It must have performed its own computation in parallel so that the interhemispheric monitoring model is a special case. (If this computation itself has an error then the correction it will initiate will not be effective.) But then monitoring need not apply only to actual output and can in principle apply to intermediate internal results as well. MacKay does not discuss the possibility of error detection following the response by computing back from the output, top-down fashion, to obtain the original initial conditions. This would have to be a sequential computation, i.e., following the original computation, and therefore could in principle (though not necessarily) be carried out by the same module that performed the original computation.

Feedback is required during the early stages of skill acquisition but is unnecessary in later stages of acquisition and for skillful executions. Our data from self-corrected lexical decision indicate that the right hemisphere is important for initial error correction, suggesting that it is perhaps selectively involved in processing external feedback.

Semantic activation without conscious identification: Processing without monitoring? There are several dramatic examples, both experimental and clinical, for lexical semantic access, demonstrated indirectly, say through its effects on subsequent processing, without conscious identification, demonstrated directly by failure to name, discriminate, categorize or classify the word, etc. The alleged experimental examples include (1) the effect of semantic information in the unattended ear during dichotic listening, (2) the semantic effect of irrelevant primes in parafoveal vision, and (3) the semantic effect of backward visually masked words that could neither be identified nor even detected (see review in Holender, 1986). Assuming that these demonstrations are valid, do they necessarily constitute examples of semantic processing without monitoring? Our analysis suggests that the answer is no. In this view, monitoring can be both conscious and unconscious. Thus, consciousness is not necessary for monitoring and the failure of semantic information to reach consciousness does not preclude adequate monitoring.

The clinical examples include indirect evidence for semantic processing together with verbal denial of recognition or even detection in (1) blind sight, i.e. for information flashed in the hemianopic field of neurological patients with visual field defects (cf. Campion et al, 1983) and (2) in pure alexia (Landis et al, 1980). These clinical examples are even weaker than the experimental ones because failure of conscious identification rests mostly on verbal reports. It is possible that other, nonverbal, probe techniques will demonstrate conscious awareness.

This is best illustrated by the disconnection interpretation of pure alexia. Verbal denial may reflect the ignorance of the aphasic but still dominant left hemisphere whereas semantic processing may reflect the residual linguistic competence of the "released" right hemisphere. Though mute, the (disconnected) right hemisphere is certainly conscious and can demonstrate competent lexical semantic analysis without naming. In addition, the clinical examples are subject to the same criticism of confounding monitoring with consciousness that were leveled at the experimental examples above.

SOME PHILOSOPHY

In this paper I have been concerned not with the occurrence of errors but with their recognition and correction. The concept of hemispheric specialization leads naturally to the realization that the two hemispheres can produce different types of errors. But the realization that these errors can be monitored separately and simultaneously in the two hemispheres can only be inferred from the concept of hemispheric independence. Does the realization that two potentially conflicting evaluations of the same act can occur simultaneously have any interesting philosophical implications?

Types of errors. Goldman (1986) mentions three types of errors as prominent in the skeptical challenge to knowledge: (1) fallibility of the cognitive system, (2) problems in the relation between the mind and the objects of cognition, and (3) problems in the logical relation between hypothesis and evidence. It seems plausible that each type has its psychologically equivalent error or noise and is resolved in the organism by different cognitive operations. At the same time, it would seem that the psychological mechanisms for reducing these errors share the general strategy of seeking convergent data from systematic sampling of a whole range of parameters.

"Noisy perception" (the fallibility of the senses) is a common source of errors and is often resolvable by seeking more information. These are errors that can be corrected if the cognitive system has access to the complete information or adequate resources. Here error correction is effective if it provides more information or economizes in resources by restricting the search-space, effecting a better strategy, etc. "Errors in reasoning," however, may be systematic and resistant to error correction, as in some abstract logical problems. Here the ineffectiveness of error correction may often stem from the fact that the error was reached through the operation of one cognitive subsystem, typically in the form of heuristics, whereas error detection is expressed in the vocabulary of a separate cognitive subsystem, often deductive, and the two subsystems have a weak interface. Before error correction becomes effective, the interface needs to be enriched. What makes error detection possible in many such cases is the emergence of contradictions.

Errors inherent in the relation between the mind and the objects of cognition may be ultimately explained by a thorough-going "neuro-psychologism" where objects, cognitions and the possible relations between them are expressible as brain states that can be studied empirically.

Error, refutation and falsity. Systematic errors or contradictions can serve as primitive concepts in a cognitive reconstruction of a philosophy of science. In such a reconstruction, error detection is the awareness of a discrepancy between an abstract theory, internally represented plan, on the one hand, and models of action (which are cognitively more stable and more empirical than the plan) or actual behavior, on the other. Error detection is a result of empirical evaluation of behavior after it (or an internal representation of it) has occurred. Thus, error detection is a mechanism of cognitive control and it leads to a modification of internal representation and of ensuing behavior. The evaluation of a thought, strategy or action implies conceptualization of its denial or failure, hence reference to and consciousness of it. Indeed, for C.S. Pierce, cognition by the self of ignorance and error makes possible self awareness (Hartshone and Weiss, Vol. 5, p. 236). For Donald Davidson, in turn, the argument that language is necessary for thought rests on the possibility of belief, which may be wrong and thus presupposes the possibility of error (1985).

Unlike falsity, error presupposes commitment or purposefulness and suggests that there may be the possibility of correction. Thus, errors may be classified by the degree of "criticism" they evoke: slips are slight, inadvertant errors, usually in speaking and writing, and they usually carry no criticism; "mistakes" result from carelessness, inattention, misunderstanding, etc. and do not carry strong criticism; "blunders" imply stupidity, clumsiness, inefficiency, etc. and carry more criticism. Thus, error is a psychological concept and belongs in the "context of discovery" rather than the "context of justification" of scientific research program. In fact, error correction may be interpreted as the process of testing a scientific idea against logical theoretical adequacy criteria, or testing a theory against reality.

Popper's concept of refutation is closer to the concept of falsity than to our psychological concept of error. According to Popper, scientific progress consists in modifying theories in response to error detection. But he denies that there is a systematic way of correcting scientific theories (Popper, 1968). Thus, Popper may be said to believe that science has a methodology for error detection but not one for error correction and that scientific progress consists of error detection without error correction (A. Kasher, personal communication). On the contrary, I believe that science has heuristics (if not formal methodologies) for both error detection and error correction and that scientific progress consists of iterative cycles of error detection followed by responsive error correction. Popper's insistence in specifying precisely the conditions under which one is willing to give up one's position, as opposed to trying to entrench one's position by proving it, ignores the psychological commitment behind the idea found to be in error. In I. Lakatos' words (1970), "Belief may be a regretably unavoidable biological weakness to be kept under control of criticism, but commitment is for Popper an outright crime." The proper logic of discovery does not engage in either avoiding or in ruthlessly eliminating errors, but rather in understanding how, why and when they occur. A refuted conjecture carries with it the idea of its own fallibility in focal attention whereas error detection is always a surprise, requires a focusing of attention through reevaluation and is usually not predictable. That is why the process of making errors is necessary in the first place. And that is why there may be a bias in the scientific literature towards citing experimental evidence that confirms a theory in vogue, while systematically ignoring negative evidence consisting of failures to support theoretical predictions (Greenwald et al, 1986).

Piaget, too, believed that error is necessary for cognitive development through "equilibration." Here error is a logical contradiction between opposing internal representations, plans and reality, and working out the contradiction develops a more sophisticated understanding and representation. According to Piaget, then, the mind is built upon logic and by eliminating contradictions, the mind achieves logical equilibrium and discovers truth (Piaget, 1986). In this way, Piaget has transformed error from a microgenetic concept into an ontogenetic one. But he too overemphasizes the role of logical contradiction as the paradigm case for error, as if logic is the operating principle of the human mind, rather than merely one of its products.

Double simultaneous conciousness? Hemispheric independence means not only that each hemisphere is a complex and separate cognitive system, not only that it incorporates control mechanisms for evaluating its own performance, but also that each can thus reach decisions that apply to the organism as a whole. These evaluations and decisions dictate action. What happens when the evaluations in the two hemispheres differ and the chosen actions conflict? Suppose one hemisphere interprets the person's immediately preceding act as right and the other interprets it as wrong. Suppose, in other words, that each has its own rules of justification. Can the person then be said to be rational, to possess unified consciousness, moral agency?

The fact that each hemisphere has an independent self-monitor can be said to mean that each is independently self-conscious.

This can indeed lead to internal conflicts and occasionally to indecison and even contradictory behavior, as we all well know from our own experiences. But we all nonetheless usually function as unified, consistent, and responsible persons because we have at our disposal interhemispheric integrative control mechanisms, both callosal and sub callosal, that permit monitoring across the hemispheres and maintain unified, conflict-free behavior. The "interhemispheric monitor" may focus on the evaluative system of one hemisphere in particular circumstances but it will generally construct a mutually consistent local interface and endow it with priviledged access to focal attention and observable behavior.

FINALLY

I have argued that error is an important element of cognitive control and necessary for cognitive development. Error detection presupposes the ability to self-monitor and error correction presupposes access to an internal model of the solution process, hence also to an internal model of the cognizing self. Error detection seems, in turn, necessary but not sufficient for error correction. The clinical neurological evidence suggests that self-monitoring can be dissociated from problem solving per se and that self-correction is, in turn, dissociable from self-monitoring. Thus it is necessary but not sufficient to know what is wrong in order to know what is right. The commissurotomy data show that each disconnected hemisphere has the capacity for self-monitoring and self-correction but that these capacities are not equally effective in the two hemispheres and vary in relative hemispheric effectiveness as a function of task. This supports the claim for hemispheric independence but not for hemispheric dominance in general self-monitoring control operations. It remains to be found whether the anterior-posterior dissociation between error detection and correction applies to both hemispheres equally. But the data already support the claim that each hemisphere is independently conscious in the sense of possessing a complex enough cognitive system that includes internal models of self and environment, plans, a concept of goal and of error, and the capacity to change a strategy adaptively in response to external measures of success.

The data I presented on error correction without detection showed different patterns of hemispheric superiority with different tasks. But the index we used -- z-score relative to chance -- turned out to be relatively insensitive and the paradigm itself was neither representative nor comprehensive. First of all, multiple choice responses restrict error correction to the elimination of alternatives prescribed by the test designer. This is a very specific form of correction and possibly biased in favor of one hemisphere. Second, there is evidence that error correction can interact with error detection and that trial-by-trial feedback actually inhibits effective error correction, presumably because it suppresses self-monitoring

(Salmoni et al, 1984). Sometimes, an effective feedback is a summary-form knowledge-of-results for K trials back, corresponding perhaps to our global error detection index. Such knowledge of results improves retention though not acquisition of motor tasks and one wonders whether there are hemispheric differences in improvement as a function of K. Third, there are alternative interpretations of the data on second chances, that do not involve error correction. Responses on second choices may simply reflect partial knowledge that is available to the subject from the start and may be elicited directly by asking subjects not only to indicate their choice for the right answer but also to rank the wrong choices (Coombs et al, 1956).

Results from the normal brain support the conclusion that each hemisphere has independent self-monitoring and error-correction control mechanisms. But monitoring can be just as effective or even more effective across than within the hemispheres. In fact, I have argued that hemispheric resource assignment may be used to optimize error detection and correction by restricting the initial computation to one hemisphere and monitoring to the other. This is a sense in which the integrated unified sense of cerebral consciousness transcends the independent, parallel and simultaneous consciousnesses of the two hemispheres. This analysis predicts that unilateral brain damage should sometimes result not only in loss of function controlled by the damaged tissue but also in loss of monitoring of function controlled by the intact side. Thus, hemispheric damage, certainly hemispherectomy and perhaps also commissurotomy, can lead to a selective loss of error detection and a failure of error correction of preserved functions.

Conversely, unilateral brain damage may impair a localized hemispheric processor leaving intact the appropriate monitor on the other side. This is possible if the monitor contains a procedure for a partial check or a heuristic for detecting errors, but not a separate computation that could substitute for the original process. For example, the right hemisphere which can comprehend and read but cannot speak or write may be able to detect a spelling error in the written output of the left hemisphere or to detect a speech error in the spoken output of the left hemisphere. The resulting syndrome will then include symptoms of an impaired function, such as agrammatism, together with a paradoxical preservation of the corresponding metalinguistic function, such as grammaticality judgement. In that case, preserved grammaticality judgement (cf. Linebarger et al, 1983) need not deny the existence of agrammatism; it can simply signal a preserved right hemisphere (or even left hemisphere) ability to perform some grammaticality judgement (monitoring) but to support little grammatical comprehension and no grammatical production.

Moreover, lesions to either hemisphere can result in maladaptive overall control: right parietal lesions can result in left-sided neglect and denial that are not seen with the disconnected left hemisphere. Similarly, left parietal lesions can result in global reading deficits that are not seen with either the disconnected or the normal right hemisphere. Surprisingly, even twenty years after the surgery, patients with complete cerebral commissurotomy persist in neglecting the competent right hemisphere and denying its purposeful, successful performance when the speaking left hemisphere remains ignorant. These are striking examples of a need for integrated, purposeful, unified interhemispheric behavior which transcends hemispheric independence.

REFERENCES

Baars, B.J., Motley, M.T. and MacKay, D.G. (1975). Output editing for lexical status in artificially elicited slips of the tongue. Journal of Verbal Learning and Verbal Behavior, 14, 382-391.

Bogen, J.E. (1985). One brain, two brains, or both? Two Hemispheres - One Brain: Functions of the Corpus Callosum. Neurology and Neurobiology, 17, 21-34. (eds. F. Lepore, M. Ptito and H.H. Jasper). Alan R. Liss, New York.

Bogen, J.E. (1987). Partial hemispheric independence with the neocommissures intact. Brain Circuits and Functions of the Mind. (ed. C. Trevarthen). Cambridge University Press, London. In press.

Campion, J., Latto, R. and Smith, Y.M. (1983). Is blindsight an effect of scattered light, spared corted, and near-threshold vision? The Behavioral and Brain Sciences, 6, 423-486.

Coombs, C.H., Milholland, J.E. and Womer, F.B. (1956). The assessment of partial knowledge. Educational and Psychological Measurement, 16, 13-37.

Davidson, D. (1985). Thought and Talk. In Inquiries into Truth and Interpretation. Oxford University Press, London, 155-170. (Originally published in 1975.)

Freud, S. (1958). Psychopathology of Everyday Life. Translated by A.A. Brill. Mentor, New York. (Originally published in 1901.)

Fromkin, V.A. (1987). Grammatical aspects of speech errors. Linguistics: The Cambridge Survey. (ed. F.J. Newmeyer). Cambridge University Press, New York. In press.

Fuster, J.M. (1980). The Prefrontal Cortex: Anatomy, Physiology, and Neuropsychology of the Frontal Lobe. Raven Press, New York.

Gainotti, G., Silveri, M.C., Villa, G. and Miceli, G. (1986). Anomia with and without lexical comprehension disorders. Brain and Language, 29, 18-33.

Gellatly, A.R.H. (1985). Hemisphere-specific effects in reasoning performance. Psychological Research, 47, 217-221.

Goldberg, E. and Bilder, R.M., Jr. (1987) The frontal lobes and hierarchical organization of cognitive control. Frontal Lobes Revisited. (ed. E. Perecman). JRBN Press, New York.

Golding, E., Reich, S.S. and Wason, P.C. (1974). Interhemispheric differences in problem solving. Perception, 3, 231-235.

Golding, E. (1981). The effect of unilateral brain lesion on reasoning. Cortex, 17, 31-40.

Goldman, A.I. (1986). Epistemology and Cognition. Harvard University Press, Cambridge, MA, 30-32.

Greenwald, A.G., Pratakanis, A.R., Lieppe, M.R. and Baumgardner, M.H. (1986). Under what conditions does theory obstruct research progress? Psychological Review, 93, 216-229.

Hartshone, C. and Weiss, P. (eds.) (1931-1935). The Collected Papers of Charles Sanders Pierce. 6 Volumes. Harvard University Press, Cambridge, MA.

Hecaen, H. and Albert, M.L. (1978). Human Neuropsychology. Wiley Press, New York.

Heilman, K.M. and Rothi, L.J.G. (1985). Apraxia. In Clincial Neuropsychology, 2nd ed. (eds. K.M. Heilman and E. Valenstein). Oxford University Press, New York, 131-150.

Holender, D. (1986). Semantic activation without conscious identification. The Behavioral and Brain Sciences, 9, 1-66.

Lakatos, I. (1970). Falsification and the methodology of scientific research programs. In Criticisms and the Growth of Knowledge. (eds. A. Musgrave and I. Lakatos). Cambridge University Press, Cambridge.

Landis, T., Graves, R. and Goodglass, H. (1981). Dissociated awareness of manual performance on two different visual associative tasks: A "split-brain" phenomenon in normal subjects? Cortex, 17, 435-440.

Landis, T., Regard, M. and Serrat, A. (1980). Iconic reading in a case of alexia without agraphia caused by brain tumor: A tachistoscopic study. Brain and Language, 11, 45-53.

Lecours, A.R., Lhermitte, F. and Bryans, B. (eds.) (1983). Aphasiology. Bailliere Tindall, London.

Linebarger, M., Schwartz, M. and Saffran, E. (1983). Sensitivity to grammatical structure in so-called agrammatic aphasics. Cognition, 13, 361-392.

Luria, A.R. (1980). Higher Cortical Functions in Man, 2nd rev. ed. Basic Books, New York.

MacKay, D.G. (1987). The organization of Perception and Action: A Theory for Language and Other Cognitive Skills. MIT Press, Cambridge, MA. Provisionally accepted for publication.

Mesulam, M.-Marsel (1986). Frontal cortex and behavior. Editorial, Annals of Neurology, 19, 320-324.

Nebes, R.D. (1971). Investigations on lateralization of function in the disconnected hemispheres of man. Unpublished Doctoral Dissertation. Division of Biology, California Institute of Technology, Pasadena, California.

Piaget, J. (1986). The Central Problem of Intellectual Development. The University of Chicago Press, Chicago.

Popper, K.R. (1968). Conjectures and Refutations: The Growth of Scientific Knowledge. Harper Torchbooks, New York. (Originally published in 1963.)

Rabbitt, P.M.A. (1967). Time to detect errors as a function of factors affecting choice-response time. Acta Psychologica, 27, 131-142.

Rabbitt, P. and Rodgers, B. (1977). What does a man do after he makes an error? An analysis of response programming. Quarterly Journal of Experimental Psychology, 29, 727-743.

Rabbitt, P. and Vyas, S. (1981). Processing a display error after you make a response to it. How perceptual errors can be corrected. Quarterly Journal of Experimental Psychology, 33A, 223-239.

Robinson, R.G. and Benson, D.F. (1981). Depression in aphasic patients: Frequency, severity, and clinical pathological correlations. Brain and Language, 14, 282-291.

Salmoni, A.W., Schmidt, R.A. and Walter, C.B. (1984). Knowledge of results and motor learning: A review and critical reappraisal. Psychological Bulletin, 95, 335-386.

Stein, R. (1986). Hemispheric monitoring in a lexical decision task. Undergraduate Honors Thesis, Department of Psychology, UCLA.

Stein, R. and Zaidel, E. (1987). Hemispheric monitoring in lexical decision. In preparation.

Stuss, D.T. and Benson, D.F. (1986). The Frontal Lobes. Raven Press, New York.

Wason, P.C. and Johnson-Laird, P.N. (1972). Psychology of Reasoning: Structure and Content. Batsford, London.

Zaidel, D. and Sperry, R.W. (1973). Performance on the Raven's Colored Progressive Matrices test by subjects with cerebral commissurotomy. Cortex, 9, 34-39.

Zaidel, E. (1973). Linguistic competence and related functions in the right cerebral hemisphere of man following commissurotomy and hemispherectomy. (Doctoral Dissertation, California Institute of Technology). Dissertation Abstracts International 34, 2350B (University Microfilm No. 73-26, 481).

Zaidel, E. (1978). Auditory language comprehension in the right hemisphere following cerebral commissurotomy and hemispherectomy: A comparison with child language and aphasia. In Language Acquisition and Language Breakdown: Parallels and Divergencies. (eds. A. Caramazza and E.B. Zurif). Johns Hopkins University Press, Baltimore.

Zaidel, E. (1978a). Concepts of cerebral dominance in the split brain. Cerebral Correlates of Conscious Experience. (eds. P. Buser and A. Rougeul-Buser). Elsevier, Amsterdam, 177-197.

Zaidel, E. (1979). Performance on the ITPA following cerebral commissurotomy and hemispherectomy. Neuropsychologia, 17, 259-280.

Zaidel, E., Zaidel, D.W. and Sperry, R.W. (1981). Left and right intelligence: case studies of Raven's Progressive Matrices following brain bisection and hemidecortication. Cortex, 17, 167-186.

Zaidel, E. (1983). Disconnection syndrome as a model for laterality effects in the normal brain. In Cerebral Hemisphere Asymmetry; Method, Theory and Application. (ed. J. Hellige). Praeger Press, New York, 95-151.

Zaidel, E. (1985). Callosal dynamics and right hemisphere language. Neurology and Neurobiology, 17, 435-459. In Two Hemispheres - one Brain: Functions of the Corpus Callosum. (eds. F. Lepore, M. Ptito and H. H. Jasper). Alan R. Liss, New York.

18
The Intrahemispheric Localization and Some Electrophysiological Correlates of Language and Memory

George A. Ojemann

Issues in the neurobiology of human language involves both lateralization and intrahemispheric organization. Lateralization has been addressed by many of the papers in this volume. Here, we turn to the organization of language within the "dominant" hemisphere. The localization of different language processes there and some electrophysiologic correlates of those processes in the electrocorticogram (ECoG) and single neuronal recordings are discussed.

The intrahemispheric localization of language functions has traditionally been derived from the study of language changes after spontaneous brain damage, most commonly strokes. Observations derived from that case material provided the basis for the traditional model of a posterior inferior frontal area involved predominantly in speech production, and a larger posterior temporal parietal area associated with language understanding.

The models of the intrahemispheric localization of language considered in this paper were derived from a different data base, the localization of language changes associated with electrical stimulation mapping during neurosurgical operations under local anesthesia. This technique was first developed by Penfield and his associates (Penfield and Roberts, 1959). It depends on a somewhat different effect of the application of an electric current to cortex than that usually considered with "stimulation". An electric current applied to the cortical surface excites neural elements, cells and "en passage" fibers of both excitatory and inhibitory varieties, but also, through depolarization blockade, blocks function of these same neural elements (Ranck, 1975).

Supported by National Institutes of Health Grants NS21724, NS1711 NS20482.

Thus, physiologically a surface electric current might either evoke an effect or block a function, the latter by either depolarization blockade or excitation of inhibitory elements. Which effect predominates becomes an empirical question.

Stimulation of the lateral cortex in the quiet awake patient evokes excitatory effects from motor, somatosensory and visual cortex. No effect is generally evoked from the remainder of the cortex. However, if the patient engages in an ongoing language task such as naming simple object pictures, applying the current to some of these "silent" sites will disrupt performance on the language task while stimulation of other "silent" sites will not. Penfield and Roberts (1959) reported that the sites where stimulation disturbed the naming of pictures, particularly when speech was otherwise intact, were in the same general location as the traditional areas considered to be essential for language based on lesion data. Subsequently, Ojemann and Dodrill (Ojemann, 1983a; Ojemann and Dodrill, 1985) demonstrated that when an anterior temporal resection encroached on a site where stimulation had repeatedly disturbed a language function, a postoperative deficit in that function was likely, whereas when the resection did not encroach on such sites, a deficit was unlikely. Ojemann and Dodrill were able to demonstrate that effect for both a measure of general language function when the resection had encroached on a site where stimulation had disturbed naming, and also for a specific measure of recent verbal memory when the resection encroached on sites related to recent verbal memory by stimulation mapping. These effects could not be explained by differences in the size of the resections or the characteristics of the patients with or without the postoperative language or memory deficits. Thus, both lesions and stimulation mapping seemed to identify the same areas of cortex as essential to a language function. This stimulation effect, then, seems to be analagous to a temporary lesion.

Cortical Localization of One Language
Function, Object Naming

The stimulation mapping technique has been used by Ojemann and his associates to investigate the detailed localization of one language process, naming of simple objects (Ojemann and Whitaker, 1978a; Ojemann, 1979, 1983a). They found that in perisylvian lateral cortex of the dominant hemisphere of an individual patient, the areas that seemed to be essential for naming, based on repeated disruption with stimulation, were localized to several quite discrete areas, each of about 1 sq cm in extent. An individual patient generally showed at least one such area in the posterior inferior frontal region and one or two additional areas in the temporal or parietal lobes. These areas had very sharp boundaries, and the surrounding cortex did not seem to be essential for naming, even when wider electrode spacings were used or more subtle effects on the language process, such as changes in latency, were sought. From the perspective of stimulation mapping, it appears that the essential areas for a given language task in an individual patient are quite localized, to several discrete mosaics. For naming (and many other language functions) there is at least one such mosaic in inferior frontal cortex and one or more in temporal and parietal cortex.

On the few occasions when stimulation mapping has been repeated after intervals of a few months without an intervening resection, the patterns of localization have generally been similar (Ojemann, 1983). Perisylvian sites essential for naming in a four-year old who was just acquiring vocabulary also showed localization to discrete mosaics when identified with stimulation mapping (Ojemann, unpublished observation). Both these findings suggest that the highly localized "mosaic" pattern of sites essential for a language function in association cortex is a relatively fixed, structural, "wired-in" characteristic.

The variability in the location of these mosaics essential for naming has been examined in several populations of left brain dominant patients. Penfield and Roberts (1959) noticed that sites where naming was disturbed sometimes extended beyond the traditional areas related to language based on lesion studies, in particular involving the supplementary motor area. An even wider distribution of these sites into mid frontal cortex and quite far posteriorly near the parietal occipital junction, was reported by VanBuren, et al (1978). Ojemann and his associates (Ojemann, 1979, 1983a; Ojemann and Whitaker, 1978a) found that the sites essential for naming showed a high degree of variability when assessed with the stimulation mapping technique in perisylvian cortex of left language dominant hemisphere of 21 patients. They found that these sites were sometimes localized well beyond traditional language cortex. Within the traditional language cortex only the portion of posterior inferior frontal gyrus immediately in front of face motor cortex demonstrated sites where stimulation disturbed naming in almost all of these patients. Elsewhere in traditional language cortex, particularly in temporal and parietal lobes, there was no area where more than two-thirds of the patients demonstrated sites essential for language. In some patients, sites essential for language in superior and occasionally even in middle temporal gyrus were found to be rather far forward, in front of the plane of motor cortex. In the Ojemann series, differences in these patterns of localization were found between males and females (Ojemann, 1983a). Larger frontal areas related to naming were present in males compared to females. In addition, different patterns were observed in the more or less verbally bright patients of that series, as measured by preoperative verbal IQ. Patients with lower verbal IQs demonstrated naming changes in parietal operculum; this was not evident in the patients with the higher IQs. There was no difference in the total extent of area related to naming in these two populations, but rather the specific localization to parietal operculum seemed to be the concomitant of the poorer overall verbal function (Ojemann, 1983a).

Electrophysiological Correlates of Naming

Electrophysiologic correlates of these areas apparently essential for naming were sought in the ECoG by Fried and Ojemann (Fried, et al 1981; Ojemann and Fried, 1982). The ECoG was recorded from a series of lateral temporal sites whose relationship to naming had been previously established by electrical stimulation mapping. Two behavioral tasks were used during the ECoG recording, a naming task and a spatial matching task using the same visual inputs. Changes in the ECoG during the naming task were sought that were localized to the sites that had been independently related to naming by stimulation mapping and were less evident in the surrounding areas of cortex that had not been related to naming. At the naming site, those changes were then assessed during the spatial task to test the hypothesis that they were unique to naming and not to a spatial matching task using the same visual inputs.

Two changes were observed in ECoG responses averaged to the presentation of material to be named, that seemed to be localized to naming sites only during the naming task: a slow potential at frontal naming sites and local desynchronization, loss of activity in the 7-12Hz range, at temporoparietal naming sites. These two changes appeared simultaneously, within 200 msec of the presentation of the material to be named, and lasted over 1 sec. This suggested some event occuring in parallel at frontal and temporoparietal naming sites during that language process. No ECoG events related to naming suggestive of serial processing between temporal and frontal areas were identified.

Local desynchronization lasting up to 1 second seemed to differentiate temporoparietal sites essential for naming from the surrounding cortex during the process of naming. The relationship of this desynchronization to naming sites during the naming task was confirmed by statistical analysis of quantitative spectral density measures (Ojemann and Lettich, 1985). A mechanism for local desynchronization is activity of the thalamocortical activating circuit (Jasper, 1960; Skinner and Yingling, 1977). Activity of that circuit, then, seems to be one mechanism operative during the generation of the language process, naming, at the sites essential for naming. That mechanism is operative at or very shortly after initiation of the naming process at both frontal and temporoparietal sites essential for naming and persists for the duration of the naming process.

Thalamic Language Mechanisms

Additional evidence of the importance of thalamic mechanisms in the generation of human language has come from observations of alterations in language function during electrical stimulation of the thalamus. These investigations were made during stereotaxic thalamotomies for the treatment of dyskinesias (Ojemann and Ward, 1982). Stimulation parameters included current levels below the threshold for reported sensation. Disruption of naming was observed during stimulation of localized portions of left lateral thalamus, including the lateral pole, medial portions of ventral lateral (VL) nucleus (Ojemann and Ward, 1971; Ojemann, 1975), and the anterior-superior-lateral pulvinar (Ojemann, et al 1968). The nature of the naming errors evoked with stimulation of the pulvinar and posterior part of VL thalamus was similar to those evoked by cortical stimulation, but the naming errors evoked from stimulation of medial central and anterior portions of VL thalamus were of a kind not seen with cortical stimulation. In the medial central portion of VL, patients perseverated on the first syllable of correct or incorrect object names for the duration of stimulation. This was interpreted as indicating an area integrating motor and language systems. Further evidence that this area of medial central VL integrates motoric mechanisms with language was the finding that this same area was one of the locations where respiratory inhibition could be evoked with thalamic stimulation, with a significantly lower threshold in left than right thalamus (Ojemann and VanBuren, 1967). The thalamic respiratory change that was evoked there was prolonged expiration, the appropriate respiratory substrate for speech.

In anterior VL, the object naming errors represented production of a particular wrong word each time naming was required during stimulation above a threshold current. This wrong word was often the last object correctly named at a subthreshold current. Language function in anterior VL were found to share a common mechanism with recent verbal memory mechanisms. The nature of that mechanism was best defined from changes in recent verbal memory measures evoked by stimulation of anterior VL (Ojemann et al 1971; Ojemann, 1975, 1977). Stimulation of left anterior VL electrodes at the time of input of information to memory increased the probability of subsequent correct recall of that material compared to nonstimulation control conditions. Stimulation of the same electrodes at the same currents but at the time of retrieval, accelerated the retrieval process initially and then at somewhat larger currents, the process failed.

This stimulation effect was modeled as a "specific alerting response". The change in function in thalamic neurons or fibers evoked by the stimulating current directed attention to incoming verbal information, enhancing the probability of subsequent recall of that information, while simultaneously blocking retrieval of already internalized information. There was evidence of some anatomic dissociation of these stimulation effects within lateral thalamus. The effect of directing attention to incoming information was particularly evident with anterior VL stimulation, while blockage of retrieval was more evident with posterior lateral thalamic and anterior superior pulvinar stimulation (Ojemann and Fedio, 1968). These stimulation effects on language and memory were limited to left thalamic electrodes. Stimulation of right thalamic electrodes showed analagous effects on measures of visuospatial memory (Ojemann, 1977).

These thalamic stimulation effects on language and memory were related to activity of thalamocortical activating circuits. Those circuits within left brain apparently play an important role in the generation of both language and recent verbal memory. As demonstrated in the investigations of cortical ECoG correlates of naming, some of this effect seems to be mediated by the selection of the essential cortical mosaics for the language function.

<u>Cortical Localization of Other Language Functions: Naming in Second Languages, Reading, Recent Verbal Memory, Speech Sound Identification and Orofacial Motor Control</u>

Localization of language functions in addition to naming in the patient's primary language has also been investigated in dominant hemisphere perisylvian cortex using electrical stimulation mapping techniques. Stimulation effects on the ability to name the same set of object pictures in two different languages have been examined in a small number of bilingual patients (Ojemann and Whitaker, 1978b; Ojemann, 1983a; Rapport et al 1983). In each of these patients at least one area was identified where stimulation disturbed naming in the first language and not the second, and another area where the pattern was reversed, with the naming in the second language disturbed and not the first. These findings seem to indicate that subtly different areas of cortex are, in part, utilized for the same linguistic process in two different languages.

The localization of naming, reading, short-term verbal memory, orofacial motor control for single or sequential speech gestures and speech sound identification was assessed with the stimulation mapping technique in a series of 14 patients at 118 sites in the perisylvian cortex of the dominant hemisphere (Ojemann, 1983b). None of these language functions were altered at 23% of the sites, while only a single language function was altered at 44% of the sites. The most common function to be altered alone when the other functions were intact, was recent verbal memory. However, sites were also identified where stimulation altered only naming or only reading, of the five language-related functions tested. Thus, this stimulation mapping study suggested a high degree of differential localization of language functions within perisylvian cortex of the language dominant hemisphere.

From the patterns of localization of sites where stimulation altered these different language related functions, a model of the organization of language within perisylvian cortex of the dominant hemisphere was derived (Ojemann, 1983 a and b). Immediately in front of face motor cortex, in inferior frontal gyrus was an area which acted as a final motor pathway for speech. Any type of speech, no matter how generated, was blocked by stimulation of this area. Stimulation there also interfered with mimicry for even single orofacial movements, although this area was distinguished from face motor cortex by the absence of spontaneous evoked face movements.

Inferior frontal gyrus anterior to this area and also the mid portion of superior temporal gyrus and parietal operculum were also involved in motor speech function. Stimulation there disturbed the ability to mimic sequences of orofacial movements, but not single movements. Naming and reading, but not memory were also usually altered by stimulation of these sites. This is further evidence of the interrelationship between areas of the brain related to motor function in the left hemisphere and language. In lateral cortex it appears that there are areas of the brain related to control of sequential orofacial motor abilities that are also important to language. At a lateral thalamic level, attentional mechanisms which may be important to motor function also seem to be involved in language and verbal memory function. Thus, language functions in the left hemisphere may have developed by incorporating specialized motoric mechanisms that were already present.

At 82% of the sites where stimulation altered the ability to mimic orofacial movements either singly or in sequence, the ability to detect speech sounds was also altered. The speech sound test, which involved the identification of stop consonants embedded in a carrier phrase, was designed so that stimulation only occurred at the time the item was heard, with a response period without stimulation. Thus, this was a measure of the effects of stimulation on speech sound perception. Nevertheless, there was a very substantial overlap in the area where speech sound perception was altered by stimulation and the area where the ability to mimic orofacial movements was altered by stimulation. This area of overlap identified a common cortex essential for the production of speech sounds and production of speech gestures that could represent the anatomic substrate of the motor theory of speech perception (Lieberman et al, 1967), or it might represent a cortex with some other common function important to both speech decoding and production such as precise timing.

Surrounding this area related to motor speech function anteriorly in the frontal lobe, posteriorly in parietal lobe, and inferiorly in temporal lobe were sites related to short-term verbal memory by stimulation mapping . Sites related to only naming or only reading were located between those sites related to motor function and memory function. Sites related to only one of each of these language functions, naming, reading or short-term verbal memory, were present frontally as well as temporoparietally. The sites related only to reading showed a unique type of reading error that involved only the syntactic aspects of the short sentences that were used as a measure of reading performance during stimulation. This suggested that those sites might have a special relationship to syntactic aspects of language.

Lateral cortical sites related by stimulation mapping to short-term verbal memory generally did not overlap with other functions. The measure of short-term verbal memory used for stimulation mapping separated the effects of stimulation on the input to memory from those during the time the memory had to be stored or at retrieval. Stimulation of temporoparietal sites most often altered memory when applied during the input or storage phases, rather than at retrieval, while stimulation of frontal sites most often altered memory when applied at the time of retrieval (Ojemann, 1978, 1983a. Ojemann and Dodrill (1985) demonstrated that damage to these lateral cortical sites related to short-term memory accounted for some of the memory deficit after left temporal lobectomy where hippocampus was left intact. They demonstrated that even when memory was totally dependent on the left hemisphere based on preoperative intracarotid amytal assessment, if the lateral cortical memory component were spared in a left temporal resection along with hippocampus, no memory deficit followed. Thus, there seems to be a lateral cortical component related to short-term verbal memory that functions in addition to the mesial temporal and peri third ventrical components previously described.

Further evidence for a role for lateral temporal cortex in recent verbal memory has come from recording of single neuronal activity in that cortex during measures of naming, reading and recent verbal memory (Ojemann et al 1985). In that study, single neuronal activity was recorded in the margins of the cortex that was subsequently resected for the treatment of medically intractable epilepsy. Recordings were made during a series of behaviors that included a measure of short-term verbal memory, in which naming was the input to memory and a simple reading task part of the distraction during which the memory had to be stored. Performance during that memory measure was compared to naming and reading tasks that did not involve a memory component, and to spatial matching tasks that used the same visual inputs as the naming and reading tasks. Neurons were identified in several patients that increased firing during the input and/or storage phases of the memory task compared to naming or reading without a memory component or the spatial matching task. These neurons demonstrated rather prolonged periods of firing at the time of input to memory and again at the time of retrieval: to date, no neuronal populations have been identified that sustained increased firing during the entire time the memory was stored. It has been suggested that the neural substrate for the recent memory "engram" includes ongoing neuronal events. The findings to date from the single neuronal recordings suggest that these events are a period of neuronal firing at input to memory and again at each time of retrieval.

Other neuronal populations in lateral temporal cortex of the dominant hemisphere were related to both language and memory function. Neurons there often fired relatively briefly during the language task without a memory component and for a more prolonged period when the same language task was part of the memory measure. A few neuronal populations were identified where increased activity was related primarily to naming or reading. Since these recordings were made in areas that were not considered to be essential for those language functions, based on stimulation mapping, this finding suggests that there are neuronal aggregates participating in language functions in areas of cortex that are not essential for those functions.

Localization of Language Functions in the Nondominant Hemisphere

Electrical stimulation mapping of lateral perisylvian cortex of the nondominant hemisphere during measures of naming, reading, short-term verbal memory, mimicry for orofacial movement or speech sound identification was generally not associated with any disturbance of these functions. Thus, stimulation mapping, like lesions and amytal studies, suggests that the nondominant hemisphere does not play any essential role in the generation of language. On the other hand, single neuronal recordings from the nondominant hemisphere during a number of auditory language tasks (Cruetzfeldt et al 1985) have demonstrated a variety of changes. These included neuronal aggregates that increased activity when hearing the second or subsequent syllables of multisyllable words, with no firing on single syllable words, and neuronal aggregates that increased activity at the time of speech output. This evidence for neuronal activity during some language functions in a hemisphere that is not essential for language may explain some of the increased metabolic activity that has been recorded in the nondominant hemisphere during language tasks (Ingvar, 1983). Neuronal aggregates that are inhibited during speech output have also been recorded in the temporal lobe of the nondominant hemisphere. However, these neuronal aggregates inhibited during speech do not seem to be unique to the nondominant hemisphere, for they have been recorded at least as often in middle temporal gyrus of the dominant hemisphere.

Thus, under conditions of intact callosal connections, nondominant hemisphere neuronal aggregates may play a participatory role in language processes. However, this role is clearly not essential, since their removal and, indeed, inactivation of the anterior two-thirds of that entire hemisphere with intracarotid amytal does not seem to alter speech function. These findings highlight the difference between participatory and essential areas for language and may explain some of the differences in the role for the nondominant hemisphere that have been described with different patient populations.

References

1. Creutzfeldt, O.D., Ojemann, G., and Lettich, E. (1985). Single neuron activity in the human temporal lobe: I. Listening and speaking. Soc. Neurosci. Abst. 11:879.

2. Fried, I., Ojemann, G and Fetz, E. (1981). Language related potentials specific to human language cortex. Science 212:353-56.

3. Ingvar, D.H. (1983). Serial aspects of language and speech related to prefrontal cortical activity. Human Neurobiology 2:177-189.

4. Jasper, H. (1960). Unspecific thalamocortical relations. In Handbook of Phsyiology. (eds. J. Fields, H. Magoun and V. Hall). Williams and Wilkins.

5. Lieberman, A.M., Cooper, F.S., Shankweiler, D.P. and Studdert-Kennedy, M. (1967). Perception of the speech code. Psychol. Rev.74:431-61.

6. Ojemann, G. (1975). Language and the thalamus: Object naming and recall during and after thalamic stimulation. Brain and Language 2:101-20.

7. Ojemann, G. (1977). Asymmetric function of thalamus in man. Ann. N.Y. Acad. Sci. 299:380-96

8. Ojemann, G. (1978). Organization of short-term verbal memory in language areas of human cortex: Evidence from electrical stimulation. Brain and Language 5:331-48

9. Ojemann, G. (1979). Individual variability in cortical localization of language. J. Neurosurg. 50:164-69.

10. Ojemann, G. (1983a). Brain organization for language from the perspective of electrical stimulation mapping. Behvaioral and Brain Sciences 6:189-206.

11. Ojemann, G. (1983b). Electrical stimulation and the neurobiology of language. Behavioral and Brain Sciences 6:221-230.

12. Ojemann, G., Blick, K. and Ward, A.A. Jr., (1971). Improvement and disturbance of short-term verbal memory during human ventrolateral thalamic stimulation. Brain 94:225-40.

13. Ojemann, G., Creutzfeldt, O.D. and Lettich, E. (1985). Single neuron activity in the human temporal lobe: II. Naming, reading memory, face and figure matching. Soc. Neurosci. Abst. 11:879

14. Ojemann, G. and Dodrill, C. (1985). Verbal memory deficits after left temporal lobectomy for epilepsy. J. Neurosurg. 62:101-107

15. Ojemann, G. and Fedio, P. (1968). The effects of stimulation of human thalamus and parietal and temporal white matter on short-term memory. J. Neurosurg. 29:51-59.

16. Ojemann, G. and Fried, I. (1982). Event related potential correlates of human language cortex measured during cortical resections for epilepsy. Advances in Epileptology 13:3385-388.

17. Ojemann, G. and Lettich, E. (1985). Electrocorticographic correlates of naming and verbal memory. EEG Clin. Neurophysiol. 63:832

18. Ojemann, G., Fedio, P., and VanBuren, J. (1968). Anomia from pulvinar and subcortical parietal stimulation. Brain 91:99-116.

19. Ojemann, G. and VanBuren, J. (1967). Respiratory, heart rate and GSR responses from human diencephalon. Arch. Neurol. 16:74-88.

20. Ojemann, G. and Ward, A.A., Jr. (1971). Speech representation in ventro lateral thalamus. Brain 94:669-680

21. Ojemann, G. and Ward, A. A., Jr. (1982). Abnormal movement disorders. In Neurological Surgery. (eds. J. Youmans). Saunders

22. Ojemann, G. and Whitaker, H., (1978a). Language localization and variability. Brain and Language 6:239-60.

23. Ojemann, G. and Whitaker, H. (1978b). The bilingual brain. Arch. Neurol. 35:409-12.

24. Penfield, W. and Roberts, L. (1959).   Speech and brain mechanisms. Princeton University Press.

25. Ranck, J., Jr. (1975). Which elements are excited in electrical stimulation of mammalian central nervous system: a review. Brain Res. 98:417-40.

26. Rapport, R., Tan, C., and Whitaker, H. (1983). Language function and dysfunction among Chinese and English-speaking polyglots: Cortical stimulation, Wada testing and clinical studies. Brain and Language 18:342-66.

27 Skinner, J. and Yingling, C. (1977). Central gaiting mechanisms that regulate event-related potentials and behavior: a neural model for attention. Prog. Clin. Neurophysiol. 1:30-69.

28. VanBuren, J., Fedio, P. and Frederick, G. (1978). Mechanism and localization of speech in the parietotemporal cortex. Neurosurgery 2:233-39.

19

Single Neuron Activity in the Right and Left Human Temporal Lobe During Listening and Speaking

O. Creutzfeldt, G. Ojemann and E. Lettich

SUMMARY: Single unit activity in the temporal lobe was recorded from 23 consenting, conscious patients during epilepsy operations. Units in the superior and middle temporal gyrus responded to language signals with excitation or suppression in close temporal relation with the speech signal. No specific responses to phonetic or semantic contents of the words could be recognized, but pure tones and other noises were ineffective. A few units responded specifically to the second part of composed words. Language spoken by the patient himself could also lead to excitatory or inhibitory responses in the same or other units. Melodic, quiet music induced a suppression of spontaneous discharge rate in a majority of units. Responses to language were confined to the middle and posterior section of the superior and middle temporal gyrus. No responses were recorded from the temporal pole region and the inferior temporal gyrus. Suppression by music was found in the same regions which showed positive language responses, but other areas were not explored. As yet, we have not been able to find significant differences of single unit responses to these acoustical signals between the left and right temporal lobes.

INTRODUCTION

The temporal lobe is involved in various functions of the brain such as audition, visual cognition, smell, attention, memory, emotional responses and – in humans – language. However, it is not required for these functions as a whole, but can be divided in various compartements: The temporal operculum including part of the superior temporal gyrus is necessary for audition, the posterior perisylvian cortex for language, the

infero-temporal cortex (the precise delineation of
which in humans is not clear, see Discussion) contributes to visual cognition and visual memory, and the
temporo-basal cortex, which can be further subdivided
according to its anatomical and functional connections,
participates in smell, memory and emotional responses.
While for most of these functions both temporal lobes
appear to be nearly equivalent, so that after lesion of
one side the other can essentially sustain the function, language is clearly lateralized and depends on
the integrity of one, the speech-dominant side. In addition, for certain cognitive functions some asymmetric
contribution can also be assumed (for review see Creutzfeldt 1983).

Although linguistic functions clearly depend on
the language areas of the dominant hemisphere, there
are several indications that also the non-dominant
hemisphere may participate in passive and active
language functions. Thus, slow motor potentials preceding speech onset are recorded over both hemispheres
with maximal amplitude over the medial premotor regions
(supplementary motor area) (Grözinger et al. 1977, Desmedt 1977, Hillyard and Woods 1979), and the regional
cerebral blood flow (RCBF) maybe slightly increased
over both temporal lobes during listening and speaking
(Larsen et al. 1979, Roland et al. 1985, Ingvar et al.
1986).

In the course of temporal lobe surgery, we had the
opportunity to record the activity of single neurons or
groups of neurons in various locations of the temporal
lobe during various tasks including those involving
linguistic functions. These recordings have demonstrated, that neurons in both temporal lobes may participate in auditory analysis of language signals.

METHODS:

All patients suffered from pharmacologically
intractable temporal lobe seizures and were referred to
the Department of Neurological Surgery, University of
Washington, Seattle/WA for right or left temporal lobe
operation. Preoperative diagnosis was performed by the
Epilepsy Center of Haborview Hospital in Seattle/WA.
Craniotomy, electro-corticography and functional
mapping of the temporal lobe and the perisylvian cortex
were done under local anesthesia after sedative, preoperative premedication (Ojemann 1985, 1986). Microelectrode recording was done following these diagnostic
procedures and before the resection of the epileptic

region.
For microelectrode recording, lacquer insulated,
electrolytically sharpened tungsten wires with a tip
diameter of 1-5 µm and a tip resistance of 3-7 MΩ
were used. The microelectrodes were held by a hydraulic
micromanipulator which was fixed to the skull. In order
to reduce tissue movement due to pulsation and respira-
tion in the region of the microelectrode penetration, a
"pressure foot" of 1.5 cm diameter was attached to the
microelectrode holder and lowered onto the cortical
surface with slight pressure (Ward 1969). The electrode
was introduced through a small (1-2 mm) whole in the
center of the pressure foot. Microelectrode recordings
were only done from regions which had to be included in
the subsequent resection, and were done with written
consent of the patient after extensive information on
the purely scientific nature of the procedure. In 10
patients, we recorded simultaneously from two microelectrodes, which were mounted together with a vertical
tip distance of 2-4 mm.

TESTS: When a unit or a cluster of units was identified, a number of visual and auditory tests were performed. These were the following: 1) The patient should
listen to single words collected from a word list for
speech audiometry (2-3 syllable nouns), with a pause of
about 5 sec between the words. 500 ms before each word,
a 1000 Hz tone (duration 100 ms) was presented. 2) The
patient was asked to repeat each word. 3) Patients were
shown drawings of objects, animals or persons on a
TV-type projection screen and had to name them aloud or
silently. 4) Single or double words were shown and had
to be read silently or aloud. 5) The patient listened
to sentences, which were acoustically distorted by
various filter procedures, so that they could be understood only after the subject knew the sentence. 6) More
complex tasks involving attention and memory consisted
of a sequence of slides shown at 5 sec intervals. The
first slide showed an object which the patient had to
name to himself and remember. This was followed by a
slide with two words, which also had to be remembered.
The third slide showed a sentence with one word m-
issing. This missing word was one of the previously
shown words and the patient had to fill in the correct
one and pronounce it aloud. On the last slide, the word
RECALL appeared, and the patient had to say which
object he had seen on the first slide.

In addition to these formalized tests, unit activity was recorded during listening to conversations,
noises, tones and music as well as during conversational speech. All auditory stimuli were presented in

an open field situation. Language signals and music were played back from tape recordings.

Evaluation: Unit activity as well as ECoG-responses from several recording sites, speech and stimulus markers were recorded on magnetic tape and off-line photographically recorded and/or analysed with a computer.

RESULTS:

We have performed microelectrode recordings on 23 patients (10 right, 13 left hemispheres), from at least 2 recording sites in most patients. Although not in all patients a full series of tasks could be completed, limited observations on auditory responses were available from all patients. Recording sites included, on both hemispheres, the superior temporal gyrus lateral from the motor face area as well as slightly posterior and anterior to it, middle temporal gyrus at various rostro-caudal positions, inferior temporal gyrus and the temporal pole. Recordings were only done from regions which were later included in the resection. Therefore, no recordings are available from regions the electrical stimulation of which had interfered with language functions during the preceding electrical stimulation mapping, as these regions were left intact also in the subsequent resection. Mostly, the microelectrodes were introduced at the edge of a gyrus, so that they could penetrate horicontally or obliquely through the cortical wall of the sulcus. Thus, cellular unit activity could be recorded over a long penetration track. We were able in some instances to record from single units over considerable time periods (up to several minutes), but most recordings were from clusters comprising several units, some of which could be distinguished and followed over several minutes, however, by their amplitude.

Responses to word listening:
Excitatory and inhibitory single unit responses to words were found, but both were unspecific with respect to the phonetic and semantic content of a word. Usually, excitatory responses were largest immediately after word on-set with a latency of 80-200 ms, and then decreased during the word (Phasic-tonic responses). Within 200 ms after the end of the word, the discharge rate had usually returned to the spontaneous background level (Fig. 1 A1, A3 and Fig. 3). Responses of a single neuron mostly varied considerably from one word to the next.

8509, A.K., Right middle temporal gyrus

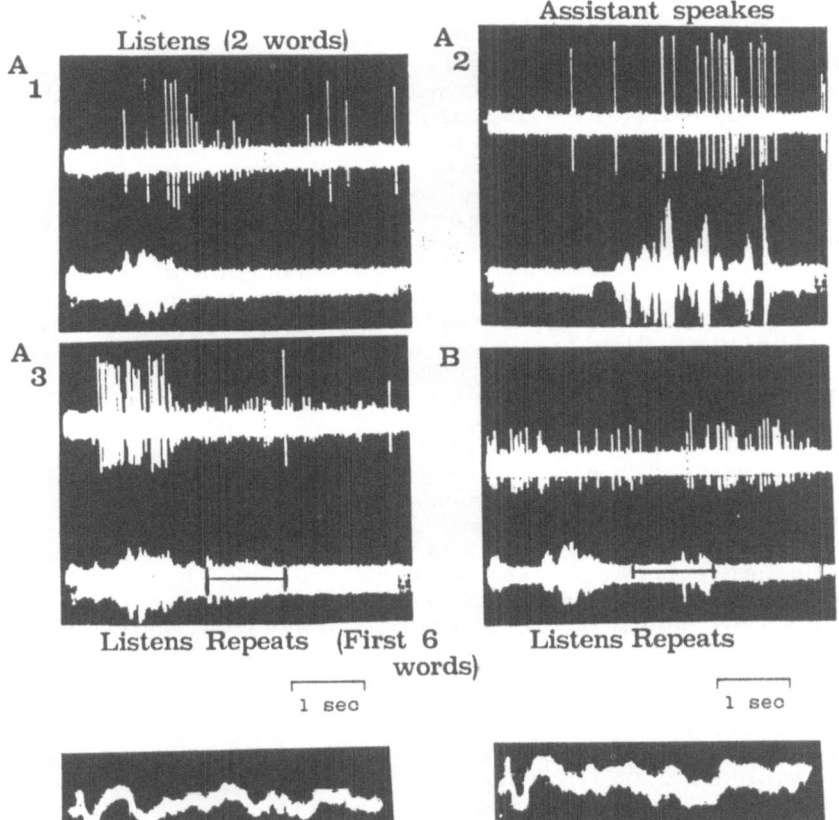

Fig. 1: Recordings from two different electrode positions (A,B) in the right middle temporal gyrus. Top
records: Microelectrode recordings. Bottom: Sound
track. A1-3: Recordings of single units distinguished
by their amplitude. A1: Responses to a word (see
oscillogramm on the sound track), preceded by 1000 Hz
tone (beginning of record). Two sweeps superimposed.
A2: The large unit responds vigorously during an information exchange between team members over the loud-
speaker. A3: Responses of the same neurons to single
words, which the patient is asked to repeat. No
response during repeat. Six sweeps superimposed. B:
Recording from another neuron at different depth. The
activity is slightly suppressed during listening and
during repeat of the word. Below: ECoG-responses from
an electrode in the pressure foot, just above the
microelectrode recording site.

Individual syllables (phonemes) of multisyllable words sometimes lead to temporally segregated activations. In a few units, only the later parts (syllables) of a word produced an activation. A most striking observation was from a unit, which was activated only by the second part of a composite word (e.g. Christmas-tree), but not by the third syllable of single 3-syllable words (e.g. Cadillac) (see Fig. 2). It is not clear yet, whether the slightly longer pause between the two parts of a composed word as compared to the shorter pause between the second and third syllable of a multisyllable word, or whether the semantic specification by the second part of the word caused this activation. The first interpretation (longer pause) may be suggested by the response to the word Heli-copter. Nevertheless, also in such a unit, the responses were not semantically specific.

The responses of all units were speech specific, however. Neither the initial alerting tone of 1000 Hz nor inadverted noises in the operation room had an effect. Also, occasional conversations at low voice not related to the experimental procedure or to the patient usually did not lead to responses. On the other hand, if the patient was directly addressed, or if a conversation between team members was specifically related to the patient and the experimental procedure, units could respond vigorously (Fig. 1 A2). Also, the excitation of units could vary dependant on whether the patient was asked to just listen or whether he was asked to repeat a word (see Fig. 3, left column). Thus, attention appeared to be an important variable which could modify the responses. On the other hand, it did apparently not matter much whether a word or sentence was understood or not. This was formally tested by comparing responses to words and sentences which were phonetically intact or were made ununderstandable by filtering out some spectral components or even reversing them. However, our data are still incomplete on this aspect.

Inhibitory responses typically consisted in a slight or, less frequently, a complete suppression of the spontaneous discharge activity (Fig. 1B). The time course of this suppression was about the same as that of excitatory responses. In some instances with two-microelectrode recordings, it could be observed that one group of units was activated, the other inhibited during the words. This suggests that suppression and activation of discharge rate by speech signals could take place within the same narrowly confined region.

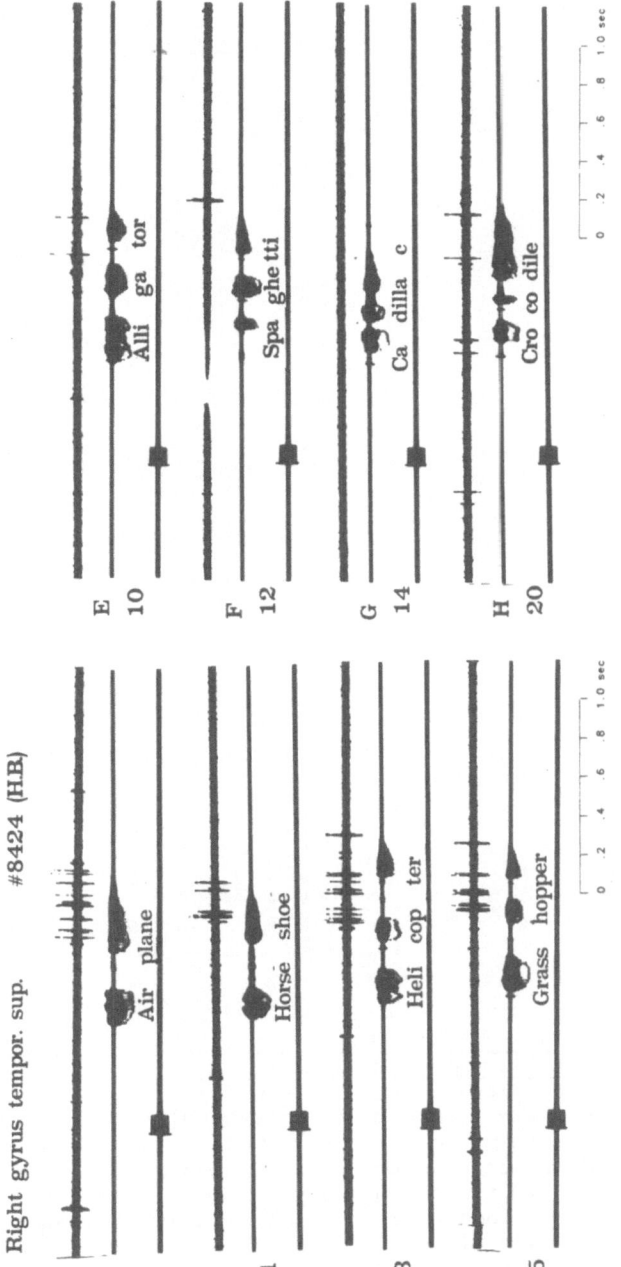

Fig. 2: Recording from a single neuron in the right superior temporal gyrus. Top records: Microelectrode recording. Second line: Sound track. Third line: Synchronization signal. A-H: Responses to different words as indicated. The numbers refer to the sequential appearance of the words during the test. A-D: Composite words. E-H: Threesyllable words. Note the strong activation of the neuron during the second half of the composite words in A-D, but only little activity during the last syllables of the threesyllable words in E-H.

302

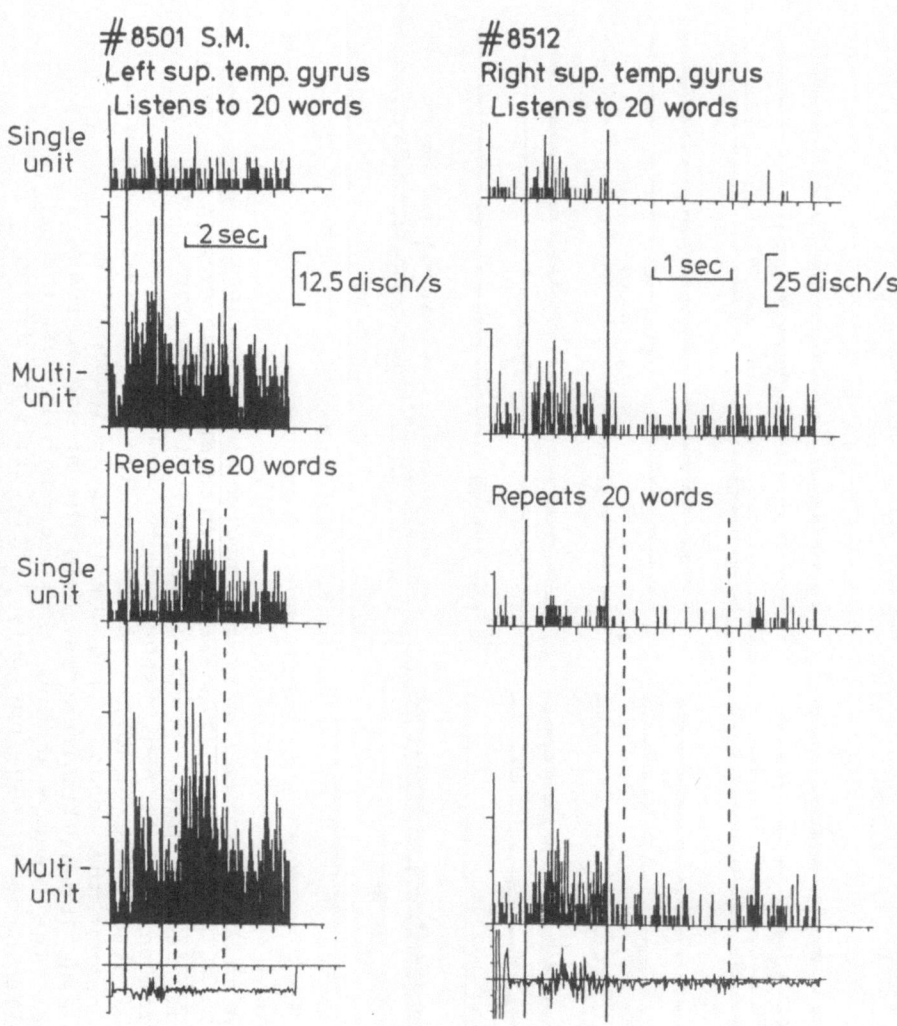

Fig. 3: Unit activity in the left and right superior temporal gyrus during listening and speaking.
Left column: Responses of a single unit (top histogram) and a multiunit cluster (second histogram)) recorded with the same microelectrode. Upper two records: Patient listens to 20 words (words appear between the two vertical lines, see also phonogramm at the bottom). Lower two records: Patient repeats each word (between the two broken vertical lines). Note the activation of both the single unit and the unit cluster during word repetition. Right column: Same formate as left column. At this recording site, units are not strongly affected by the patients voice.

Responses to speaking:

Some neurons which responded to a speech signal, also responded to the speech of the patient (see Fig. 3, left column). The latencies between sound onset and the excitatory responses were about the same as after external speech signals. Also with self-produced speech, the most typical responses were phasic-tonic. Responses were not related to phoneme content or semantic meaning of the word. There was also no difference whether the words had just to be repeated, whether it involved cognitive processes (naming, reading) or memory retrieval (names or birthdays of relatives, addresses etc.). Silent speech (e.g. silent naming and reading) did never produce any response. An exception were silent reading tasks during test 6. Here, the patient had to read on one slide silently two words and on the next slide a whole sentence in which one word was missing. He then had to fill in and say the missing word which was one of the words he had seen in the preceding slide. However, as these neurons were not active during a control task involving only silent reading of the same words without the task to remember them, it can be assumed that in these cases, the activation was related to the increased attention or other processes related to memorizing.
Some neurons were inhibited during self-produced speech. Although, typically restricted to the sound itself with a delay of 50 to up to 100 ms and returning to baseline discharge rate within 100-200 ms (see Fig. 4), the inhibition could start in a few neurons up to some hundred ms before voice onset. When only single words had to be spoken, this inhibition could last in such a neuron up to 1-2 sec following the sound. During word sequences or sentences, the activity could be suppressed for most of the speaking time.

In some recordings, neuronal activity changes during speech could be erroneously simulated by mechanical artefacts due to speech synchronous movements of the brain. Usually these artefacts were easily recognized because of concomitant amplitude changes of action potentials, because typical injury discharges developed eventually and/or the unit activity disappeared completely after a few repetitions of the same speech task. Movement artefacts could be easier recognized when two microelectrodes were used simultaneously.

Location of speech related responses:

Activation and inhibition could be recorded in the superior and middle temporal gyrus of both temporal

304

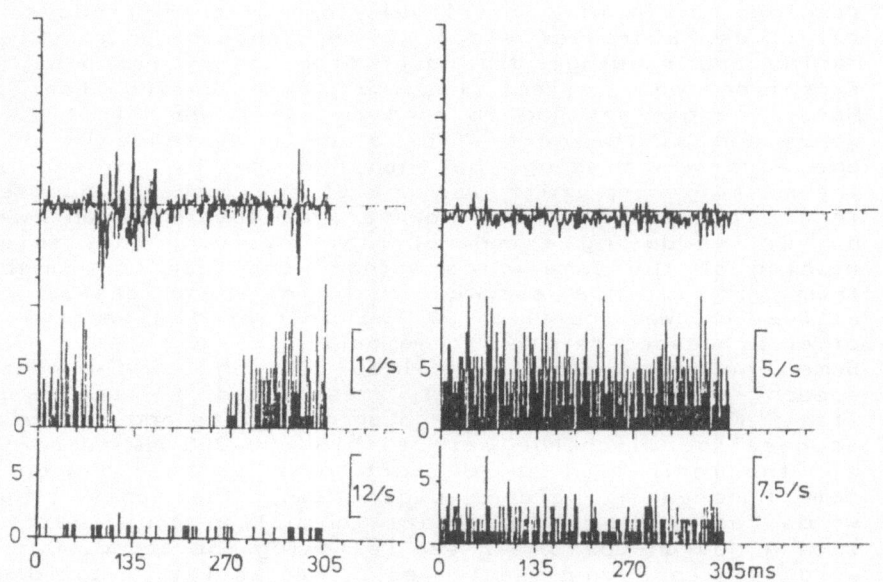

Fig. 4: Suppression of spontaneous discharge rate of a
single unit in the right middle temporal gyrus during
naming. At zero, a picture showing an object appears.
In the left panel, the patient is asked to name the
object (see voice recording, upper trace), in the right
panel he is asked to name the object to himself. Note
the suppression of discharge rate of the first unit
when the patient speaks (middle trace), but no significant change of activity in the second unit (bottom
trace). No significant change of discharge rate in
either unit during silent naming. Voice record and
histograms averaged from 6 presentations.

lobes lateral to the face representation of the sensory
and motor cortex. In the right (speech-nondominant)
hemisphere, recordings could also be performed more
posteriorly close to the posterior end of the sylvian
sulcus. Also in this location, activation and inhibition during listening to language could be recorded.
However, in none of these recording sites every neuron
or recording cluster was modulated by language.
Responses to active speech were found in the same
regions.

In the anterior third of the superior and middle
temporal gyrus, i.e. the temporal pole region, and in
the inferior temporal gyrus, speech related activity
neither during listening nor during speaking was observed.

Listening to music:
In eight patients (5 right, 3 left hemispheres),
temporal lobe unit activity was recorded from 18 sites
(including nearby recordings with two microelectrodes)
while the patient listened to music. Various pieces of
music were presented, some having a simple tune either
familiar to the patient (Twinkle - twinkle - little
star, in the piano version of W.A. Mozart " Ah -
dirai-je vous madame", two North-American folk-tunes in
orchestra arrangement by Steve Foster), or not known to
the patient (R. Schumann: Nachtlied, B.Galuppi: Piano-
Sonata No. 5, C-Major). While the piano pieces were of
about equal loudness, the orchestra arrangements varied
considerably in volume. In 9 out of the 18 recordings,
the unit activity was strongly suppressed during the
music (Fig. 5). This suppression was most clearly seen
during the piano music with simple and clearly recognizable tunes, but was also present during the more
complex Foster-arrangements. The suppression lasted
throughout the more than 20 sec-presentation of the
music with some tendency to become less conspicuous at
the end of the recording. In some neurones a loose
entrainment of the reduced unit activity with the individual tones could be observed. Some units showed a
slight rebound activation after termination of the
music, but no neurons showed an activation during the
music. In five patients, we also played to the patients
a contemporary unmelodic, noisy and essentially rhythmi-
cal piece (predominantly percussion and some electronic
effects, from the TV-series "Miami vice"). Here the
effects were less uniform and could vary during one
presentation with slight excitation, suppression or no
change of discharge rate. However, in three patients
the overall activity tended to be higher than during
the calme music, but was lower in one patient. Music

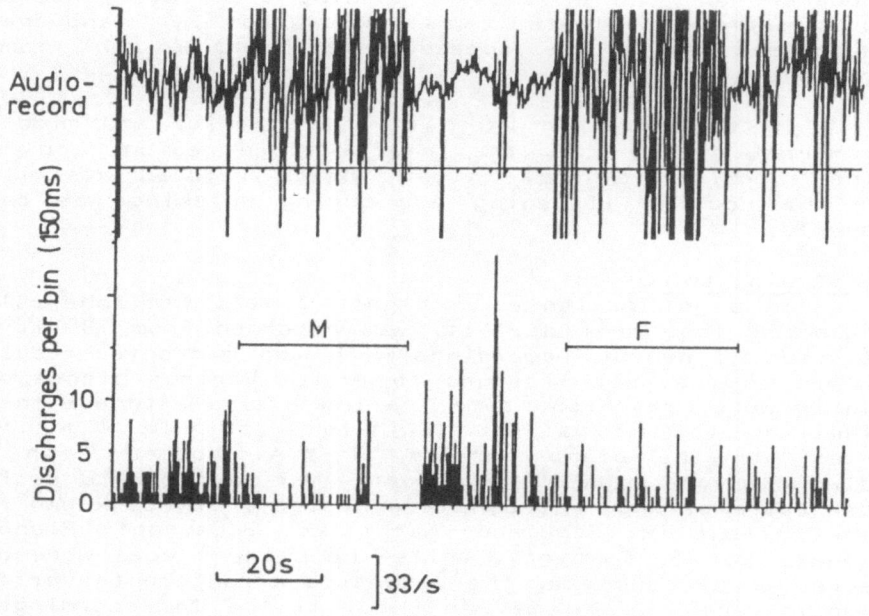

Fig. 5: Suppression of spontaneous activity of a single unit in the right superior temporal gyrus during music. M = A simple tune of W.A. Mozart is played on the piano. F = An American folk song arranged by Steve Foster is played by an orchestra.

effects were recorded in the right middle and superior temporal gyrus and in the left superior temporal gyrus.

DISCUSSION and CONCLUSIONS:

Our observations on single neuronal unit activity in the human temporal lobe justify, in the context of this symposion, the following conclusions:
1) Speech specific activations and inhibitions of neuronal activity can be found in both temporal lobes. These neuronal responses are related to the acoustic aspects of language, but are neither phonetically nor semantically specific.
2) Neuronal responses may differ depending on whether the patient listens to speech or whether he speaks himself. The responses to his own speech are most probably also auditory responses, but suppression of single unit activity may sometimes precede the phonation.
3) Responses may depend on the attention given to speech signals.
4) Unit activity in the temporal lobe is predominantly suppressed while listening to calm, melodic music.
Ad 1: An involvement of both temporal lobes in language perception was already earlier suggested by several lines of evidence: increased RCBF on both sides during conversation and fluent speech (Larsen et al. 1978, Roland et al. 1985, Ingvar et al. 1986); evoked potentials in the temporal lobe during listening and speaking (Desmedt 1971, Hillyard and Woods 1986); and last not least the fact, that left (speech-dominant) temporal lobe lesions or resections do not interfere seriously with language functions as long as the perisylvian speech areas are kept intact. On anatomical grounds, the areas involved in audition (primary and secondary auditory areas) were hitherto restricted to the superior temporal gyrus (Galaburda and Sanides 1980). Our findings suggest, however, that the intermediate segment of the middle temporal gyrus should also be included in the auditory cortex. The fact, that no neuronal responses to language signals were found in the anterior third of the temporal lobe (including temporal pole region) and the inferior temporal gyrus indicates that these regions do not participate in auditory signal analysis. These functional anatomical observations maybe relevant if homologies between the monkeys and the human temporal lobe are considered. Usually, the inferior and the middle temporal gyrus are thought to represent the human homologue of the monkey's "infero-temporal region", while our findings suggest that the middle temporal gyrus is still auditory association cortex.

Our findings stress the purely auditory nature of the neuronal responses. Even with this limitation, they appear to be relatively unspecific in that no specific phonetic elements or semantic structures could be related to the responses, so far. A more detailed analysis is in progress. On the other hand, the fact that language sounds are more effective than pure tones in affecting unit activity in the non-primary auditory cortex, is in agreement with experiments in cat and monkey which had demonstrated that cortical auditory neurons, even in the primary auditory cortex, may be more responsive to natural sounds than to pure spectral signals (Goldstein and Abeles 1975, Sovijärvi 1975). This maybe due to the continuous temporal modulation of frequency and amplitude in such signals (Creutzfeldt et al. 1980).

Ad 2: Also in agreement with the experience from animal experiments is the fact that neuronal responses may differ depending on whether language is just heard or whether it is spoken by the subject. Similar observations have been made in squirrel monkeys, in whom some auditory neurons responded differently if a species specific call was produced by the animal or if it was played back to him from a tape (Müller-Preuss 1979). We can as yet not safely decide whether differences of the acoustic composition of the air-conducted and the bone-conducted auditory signals or whether internal gating mechanisms were responsible for such different responses. The fact that at least in some neurons an activity depression - probably inhibition - could precede the phonation indicates that internal gating mechanisms may at least be partially responsible.

Ad 3: Gating mechanisms leading to response variations related to attention are suggested by some of our observations, such as the strong responses to low voice conversation, response differences depending on whether a word had to be repeated or not, and even the specific responses to a second part of a composed word, as this changed the semantic meaning of the first word segment. However, this aspect has to be further analysed and additional experiments are needed.

Ad 4: The widespread suppression of spontaneous discharge rate by music may be unexpected, if the yet ill defined contribution of temporal lobe functions to various aspects of musicality are considered (see in Critchley and Henson 1977). The type of music where we observed discharge suppression could be characterized as simple, familiar or easily remembered tunes arranged simply and easy to follow. Thus, what one might call soothing, quieting music. Such music apparently suppresses to some extent the spontaneous activity of neuronal systems in the temporal lobe which participate in the analysis and representation of linguistic acoustic material. It may be related to the quieting effect of such music and to the mutual interference between concentrated listening to music and speech. At this stage we cannot safely say whether the music effects were lateralized to some extent or not.

In conclusion, our recordings from single neurons in the human temporal lobe of awake patients have demonstrated, that language related acoustic signals are represented by neuronal activity modulations in both temporal lobes. Thus, it is reasonable to assume that the neuronal mechanisms for auditory processing of linguistic signals are bilaterally represented without obvious lateral specialization. One must further conclude, that these bilateral systems for auditory signal analysis each have direct access to the specific language systems in the speech-dominant hemisphere, where they are combined to semantically significant structures.
An extensive documentation and detailled discussion of our observations are in preparation.

Supported by N.I.H. Grants NS 21724, 17111 and 20482; and a Travel Grant to O.C. from Max-Planck-Institute, Göttingen.

REFERENCES:

Creutzfeldt, O.D. (1983). Cortex cerebri. Springer-Verlag, Berlin, Heidelberg, New York.

Creutzfeldt, O., F.-C. Hellweg and Chr. Schreiner (1980). Thalamo-cortical transformation of responses to complex auditory stimuli. Exp. Brain Res. 39, 87-104

Critchley, M. and R. A. Henson (Eds.) (1977). Music and the brain. Heinemann Medical Books, London.

Desmedt, J.E. (ed) (1977). Language and hemispheric specialization in man: Cerebral event related potentials. Progr. Clin. Neurophysiol., Vol.3. Karger, Basel.

Galaburda, A. and F. Sanides (1980). Cytoarchitectonic organization of the human auditory cortex. J. comp. Neurol. 190, 597-610.

Goldstein, M.H. and M. Abeles (1975). Single unit activity of the auditory cortex. pp. 119-219. In: Handbook of sensory physiology, Vol. V/2 (eds. W.D. Keidel and W.D. Neff). Springer-Verlag, Berlin, Heidelberg, New York.

310

Grözinger, B., H.H. Kornhuber and J. Kriebel (1977). Human cerebral potentials preceding speech production, phonation, and movements of the mouth and tongue, with reference to respiratory and extracerebral potentials. pp 87-103. In: Language and hemispheric specialization in man: Cerebral event related potentials (ed. J. Desmedt). Karger, Basel.

Hillyard, S.A. and D.L. Woods (1979). Electrophysiological analysis of human brain function. pp. 345-378 in: Handbook of Behavioral Neurobiology, Vol. 2 Neuropsychology (Ed. M.S. Gazzaniga) Plenum Press, New York.

Ingvar, D., B. Brädvik and E. Ryding (1986). Bilateral representation of speech functions. This volume.

Larsen, B., E. Skinhoj and N.A. Lassen (1978). Variations in regional cortical blood flow in the right and left hemispheres during automatic speech. Brain 101, 193-209.

Müller-Preuss, P. (1979). Processing of self-produced vocalization by single neurons in the auditory cortex of the squirrel monkey (Saimiri sciureus). pp. 146-151 in Hearing Mechanisms and Speech (eds. O. Creutzfeldt, H. Scheich, Chr. Schreiner). Exp. Brain Res. Suppl. 2.

Ojemann, G. (1985). Surgical treatment of epilepsy. Pp. 2517-2527 in Neurosurgery, Vol. 3 eds. Wikins and Rengarchary. McGraw Hill, London, New York.

Ojemann, G. and O. Creutzfeldt (1986). The investigation of human speech during neurosurgical procedures. In press. Handbook of Physiology, Vol. The American Physiol. Soc., Washington/DC.

Roland, P.E., L. Friberg, N.A. Lassen and T.S. Olsen (1985). Regional cortical blood flow changes during production of fluent speech and during conversation. J. Cereb. Blood Flow Metabol. Suppl. 1985

Sovijärvi, A.R.A. (1975). Detection of natural complex sounds by cells in the primary auditory cortex of the cat. Acta Physiol. Scand. 93, 318-335.

Ward, A.A. (1969). The epileptic neuron: Chronic foci in animals and man. pp. 263-288 in Basic Mechanisms of the Epilepsies (Eds. H.H. Jasper, A.A. Ward and A. Poppe). Little Brown and Comp., Boston

20
CT Scan Lesions and Language Behavior in Left-handed Aphasia Cases: Observation of Separate Hemispheric Dominances for Handedness, Speech Output and/or Comprehension

Margaret A. Naeser and Joan C. Borod

INTRODUCTION

Study of the relationship between handedness and aphasia dates back at least to the time of Broca (1865). Hécaen and Sauguet (1971) observed left-handers to recover more rapidly from aphasia than right-handers; Gloning and Quatember (1966) observed left-handers to have more transient aphasia. Brown and Hécaen (1976) reported left-handers fell into less clear-cut diagnostic categories of aphasia and Goodglass and Quadfasel (1954) reported less hemisphere specialization for language in left-handers than right-handers. Gloning et al. (1969), however, found no significant difference in language of left-handed (LH) or right-handed (RH) aphasics with similar left hemisphere lesions.

The relationship between morphological brain asymmetry and dominance for handedness has been studied on CT scan. The majority of RH people (aphasics and controls) show asymmetry in the occipital area on CT scan (LeMay, 1977; Chui and Damasio, 1980; Pieniadz et al., 1983; Koff et al, 1986). In approximately 60-70% of these cases, the left occipital length and width are greater (i.e., increased). The majority of LH controls also have greater (i.e., increased) left occipital length and width according to some studies (Chui and Damasio, 1980; Koff et al., 1986), but not according to others (LeMay, 1977).

Occipital CT scan asymmetry has been related to language in two ways, First, atypical increased right occipital asymmetry in right-handers has been related significantly to recovery from global aphasia in some language functions (Pieniadz et al., 1983). Second, occipital length on CT scan has been correlated significantly with length of the planum temporale at postmortem (Pieniadz and Naeser, 1984). The planum temporale is associated with cerebral dominance for language (Geschwind and Levitsky, 1968;

311

Galaburda et al., 1978). Atypical (increased right) occipital CT
scan asymmetry may have some relationship to anomalous dominance
for language. This chapter summarizes the results from a study
with 31 LH aphasia cases with left or right hemisphere lesions
where CT scan occipital asymmetry was also considered (Naeser and
Borod, 1986). For more detailed information on the cases presented
here, the reader is referred to that study.

LEFT-HANDED APHASIA CASES

    The 31 LH aphasia cases studied included 29 men and 2 women
who had ischemic infarcts (90%) or hemorrhages (10%).
Left-handedness was determined by either self-report or family
inquiry. Age ranged from 36 to 69, with a mean of 52.9 years (S.D.
= 12.0). Twenty-two cases had a single left hemisphere lesion, 5
had multiple left hemisphere lesions and 4 had a single right
hemisphere lesion. Lesion localization was determined by MN
without prior knowledge of language behavior. Most cases (26/31,
84%) could be classified into a diagnostic category of aphasia as
described in Goodglass and Kaplan (1972) or Naeser et al. (1982)
(e.g. subcortical aphasia cases with capsular/putaminal lesion
sites). For the 22 cases with single left hemisphere lesion, these
classifications included the following: transcortical motor (TCM),
5; mixed, 5; global, 4; conduction, 3; subcortical capsular/
putaminal, 2; aphemia, 1; Broca's, 1; unclassified, 1. For the 4
cases with single right hemisphere lesions, the classifications
included the following: global, 1; conduction, 1; Broca's, 1; and
unclassified, 1. Only five of the total of 31 cases (16%) were
unclassified and three of these had multiple left hemisphere
lesions.

    This was a retrospective study and only those LH aphasics who
were severe enough to present for clinical evaluation at one month
up to eight years post stroke onset were examined. Hence, the
results obtained from this study may not be applicable to the LH
population in general, as nothing is known about those LH cases who
may have had only transient aphasia or early recovery.

I. Distribution of CT Scan Asymmetries in Left-Handed Aphasia
   Cases

    The CT scan length asymmetries in the frontal and occipital
regions were measured using a previously published method (Pieniadz
et al., 1983; Pieniadz and Naeser, 1984). The length asymmetries,
only were chosen for measurement because CT scan occipital length,
not width, correlated with planum temporale length at postmortem
(Pieniadz and Naeser, 1984).

    The typically increased left occipital length was observed for
a significant majority (59%) of the LH aphasia cases (Chi Square =

8.56; df = 2; p<.02). The atypically increased right occipital length was observed for 17% and there was symmetry in 24%. For the frontal length measurement, asymmetries were randomly distributed across the three categories where 18% had left, 43% equal and 39% right (Chi Square = 3.07; df = 2; p<.30).

No significant differences were observed when the occipital and frontal distributions for CT scan asymmetries from the LH male aphasics were compared using Chi Square tests to those obtained from LH negative controls (Koff et al., 1986). When Chi Square tests were applied to the occipital length distributions, all distributions were significantly different from chance (p<.05) and, regardless of handedness or presence/absence of aphasia, increased left occipital length was the most common asymmetry within each group. No distinct frontal length asymmetry distribution was observed in any group. Thus, the CT frontal or occipital asymmetries do not appear to be useful in predicting handedness or which hemisphere, if damaged, will produce aphasia.

II. Comparison of Language Behavior between Left-Handed and Right-Handed Aphasics with Similar Left Hemisphere Lesion Sites

In 8 of the 22 LH aphasia cases with single left hemisphere lesion, the location and extent of the lesion was similar to eight RH aphasia cases whose CT scans and language behavior data were on file at the Aphasia Research Center. The Boston Diagnostic Aphasia Examination (BDAE) (Goodglass and Kaplan, 1972) and Token Test (Spreen and Benton, 1969) scores for the 16 matched cases, taken at approximately the same time after the stroke onset, were similar for the LH and RH aphasics with comparable left hemisphere lesion sites. T-tests for correlated samples were run for each of the 16 subtests; there were no significant differences (p<.05) between the LH and RH aphasia groups. The 16 matched cases included 2 aphemia cases; 2 Broca's; 4 globals; 2 TCM; 4 atypical TCM (some impaired repetition) and 2 subcortical capsular/putaminal cases. The CT scans for the LH and RH matched TCM, Broca's and global cases are shown in Figures 1, 2 and 3.

Among the 22 LH aphasia cases with left hemisphere lesion, there were four cases of global aphasia, a severe aphasia often seen in RH aphasia cases with left hemisphere lesion. Three of the LH cases had typical left occipital length asymmetry on CT scan. One of these cases who was matched to a RH global aphasia case with left hemisphere lesion is shown in figure 3. Both the LH case and the RH case had large left frontal, parietal and temporal lobe lesion; each case also had the typical left occipital length asymmetry on CT scan; each case had severe deficits. For example, at 11 months post onset (MPO), the LH case scored only -1.3 on the BDAE auditory comprehension z-score; at 15 MPO the RH case scored -0.8. One of the LH global aphasia cases with left hemisphere lesion and typical left occipital asymmetry was still globally

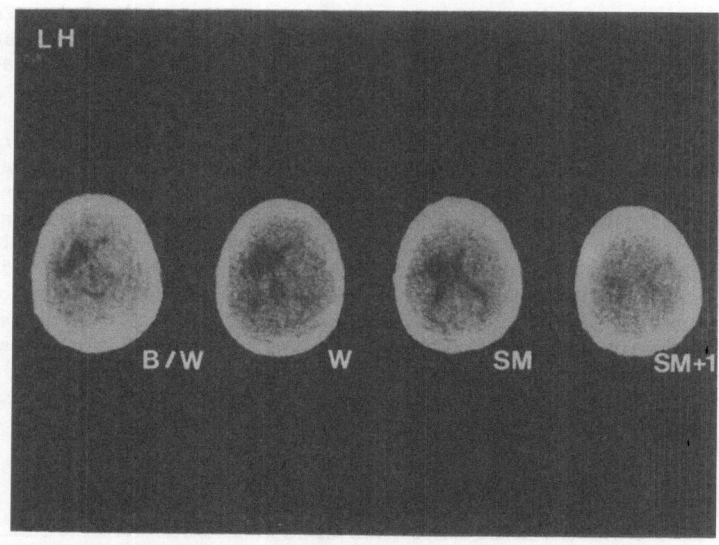

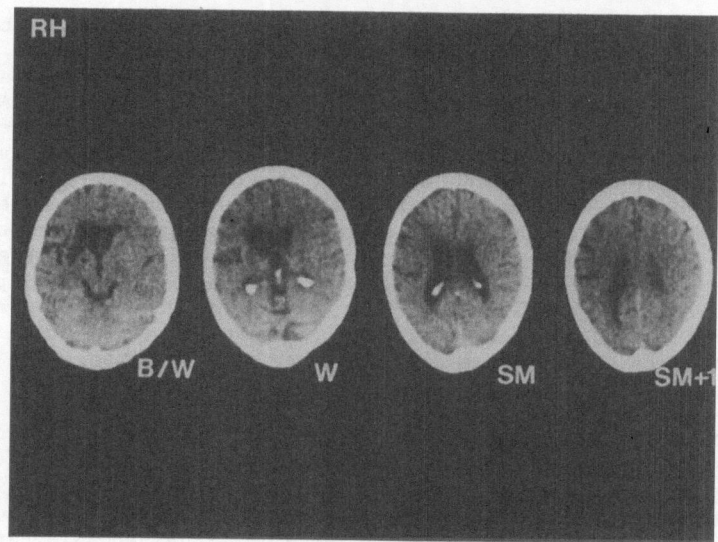

Figure 1. CT scans for two matched transcortical motor aphasia cases with small subcortical capsular/caudate and putamen left hemisphere lesion; top, LH case; bottom, RH case. Cases were tested at 7 MPO (LH case) and 13 MPO (RH case). The LH case scored 90% on the Token Test, 104 on Visual Confrontation Naming and 9 on Animal Naming in one minute. The RH case scored 94.5% on the Token Test, 104 on Visual Confrontation Naming and 9 on Animal Naming in one minute. Reproduced with permission from <u>Neurology</u>.

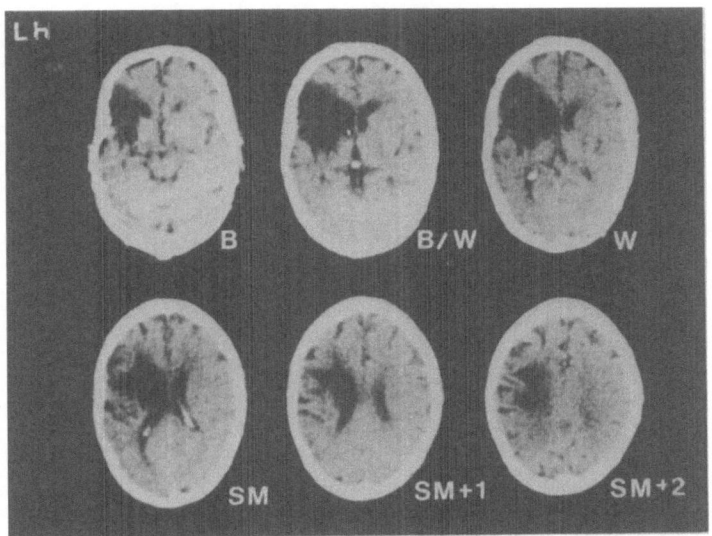

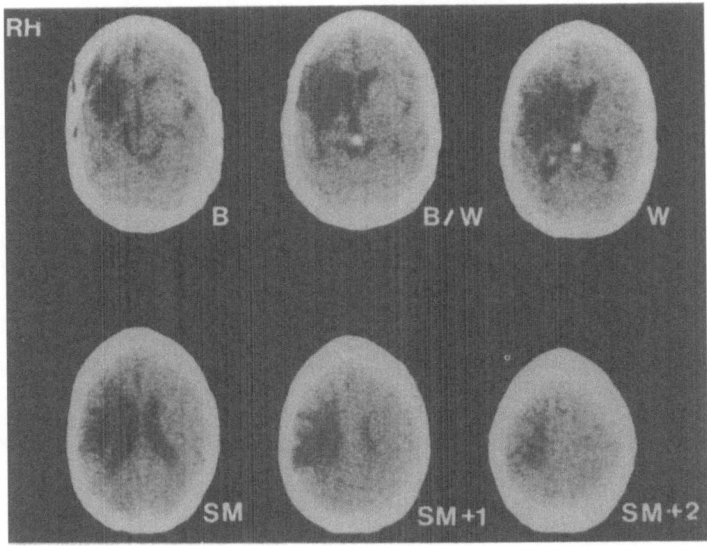

Figure 2. CT scans for two matched Broca's aphasia cases with large left frontal lobe lesion; top, LH case; bottom, RH case. Lesion distribution in each case is compatible with longer-lasting Broca's aphasia. Cases were tested at 11 MPO (LH case) and 53 MPO (RH case). Some of the LH case/RH case scores were as follows: BDAE Auditory Comprehension z-score, +0.8/+0.7; Phrase Length, 4/3; Artic. Agility, 2/2; Grammatical Form, 3/3. Reproduced with permission from Neurology.

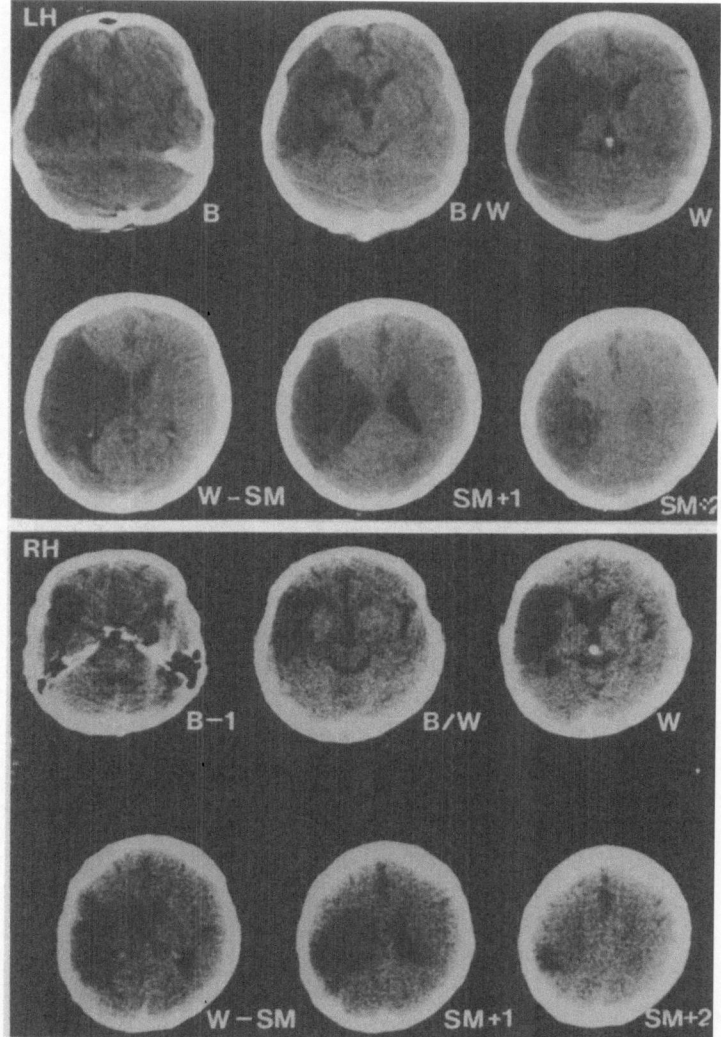

Figure 3. CT scans for two matched global aphasia cases with left hemisphere lesion; top, LH case; bottom, RH case. Each case had large frontal, parietal and temporal lobe lesion including Broca's area and lower motor cortex area, surface and deep, as well as Wernicke's area. Each case had typical increased left occipital length asymmetry (slices W and W-SM). Cases were tested at 11 MPO (LH case, -1.3 on the BDAE Auditory Comprehension z-score and 1 Word Phrase Length) and at 15 MPO (RH case, -0.8 on the BDAE Auditory Comprehension z-score and 0 word Phrase Length). Note occipital asymmetry was ipsilateral to the lesion side and there was no recovery of comprehension. Reproduced with permission from Neurology.

aphasic at 29 MPO. These severe LH global aphasia cases with large
left hemisphere lesion and typical increased left occipital length
asymmetry are similar to RH global aphasia cases with large left
hemisphere lesion and typical increased left occipital length
asymmetry who have poor recovery (Pieniadz et al., 1983).

There was one LH global aphasia case with large left
hemisphere lesion who had atypical right occipital length
asymmetry. This case recovered some single word comprehension on
Word Discrimination (BDAE), scoring 53/72 at 13 MPO. Thus, the
improved single word comprehension scores may have been associated
with some "anomalous dominance" for posterior language function.
This case is similar to RH global aphasics with improved
single-word comprehension after 8 MPO who have large left
hemisphere lesions and atypical right occipital length asymmetry
(Pieniadz et al., 1983: Galaburda et al., 1978).

III. Language Behavior in Left-Handed Aphasics with Single
     Right Hemisphere Lesions

Four LH aphasia cases had single right hemisphere lesions,
including 1 global, 1 conduction, 1 Broca's and 1 unclassified
aphasia, as mentioned above. None had a LH parent or sibling.
Three of the four had typical increased left occipital length
asymmetry.

The possibility of separate hemispheric loci for motor speech
output dominance and language comprehension dominance was observed
in two of these LH aphasia cases with right hemisphere lesion.
Both of these cases had large right frontal, parietal and
temporal (FPT) lobe lesion that damaged both the right "Broca's"
and right "Wernicke's" areas. The CT scan for one of these cases
is shown in Figure 4. Both had impaired speech output but good
comprehension. Both had left occipital length asymmetry. Thus,
the occipital length asymmetry was contralateral to the large right
temporal (frontal and parietal) lobe lesion, and there was recovery
or sparing of comprehension. One case scored +0.6 on the BDAE
Auditory Comprehension z-score at 72 MPO and the other case scored
at +0.8, at 4 MPO. Each probably had handedness dominance and
speech output dominance in the right hemisphere but comprehension
dominance in the left. Because the left occipital length asymmetry
was contralateral to the lesioned hemisphere producing the
aphasia, there may have been recovery or perhaps even sparing of
language comprehension.

One LH aphasia case had a large right FPT lobe lesion
including right "Broca's" and right "Wernicke's" areas, right
occipital length asymmetry, and lasting global aphasia. The CT
scan for this case is shown in Figure 5. This case is of
particular interest because he is the mirror image of the majority
of RH global aphasics who have left hemisphere dominance for

318

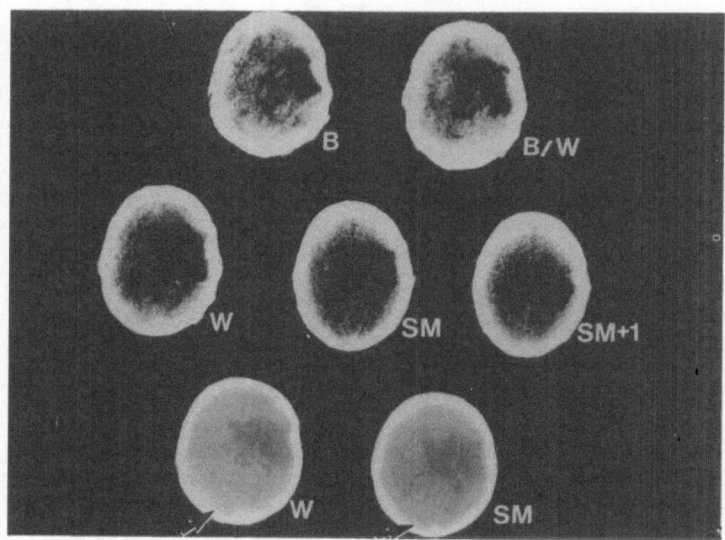

Figure 4. CT scan of LH case with right hemisphere lesion which
produced a nonfluent Broca's aphasia with good comprehension.
Large lesion is present in the right frontal, parietal and temporal
lobes including "right Broca's area" and lower motor cortex,
surface and deep, as well as "right Wernicke's area." There is
typical left occipital length asymmetry (see arrows on lowest two
slices, W and SM, made with bone windows). This case probably had
handedness dominance and speech output dominance ("Broca's area")
in the right hemisphere but comprehension dominance ("Wernicke's
area") in the left. Note occipital asymmetry was contralateral to
the lesion side and there was improved or spared comprehension
ability. Reproduced with permission from Neurology.

handedness, speech output, and comprehension, with large left FPT
lesion, left occipital length and lasting global aphasia. This
case was the only LH aphasia case examined who appeared to have all
three dominances in the right hemisphere - handedness, speech
output, and comprehension. This case did not show recovery in
language comprehension and the occipital asymmetry was
ipsilateral to the comprehension-deficit producing lesion (-1.5
on the BDAE Auditory Comprehension z-score at 24 months).
Left-handed aphasia cases with right hemisphere lesions, right
occipital length asymmetry, and all three dominances in the right
hemisphere, are probably very rare.

Levy and Reid (1976) have suggested that writing posture may be useful in predicting hemispheric dominance for language. The inverted or hooked writing posture is hypothesized to indicate language representation in the hemisphere ipsilateral to the writing hand. Although data on writing posture was available on only 5/31 cases examined in this study, this included 2/4 of the left-handers who were aphasic from right hemisphere lesions. Two cases who were aphasic from right hemisphere lesion (the lasting global aphasic in Figure 5, and a conduction aphasic) and two cases who were aphasic from left hemisphere lesion (the lasting global aphasic in Figure 3, top, and a conduction aphasic) all wrote with the inverted or hooked writing posture with the left hand. The Broca's aphasic in Figure 2, top, who was aphasic from a left hemisphere lesion, however, wrote with a regular writing posture. Therefore, in this study, writing posture (albeit available for only a small subsample) was not associated with hemispheric dominance for language. In fact, the left-hander who had a lasting

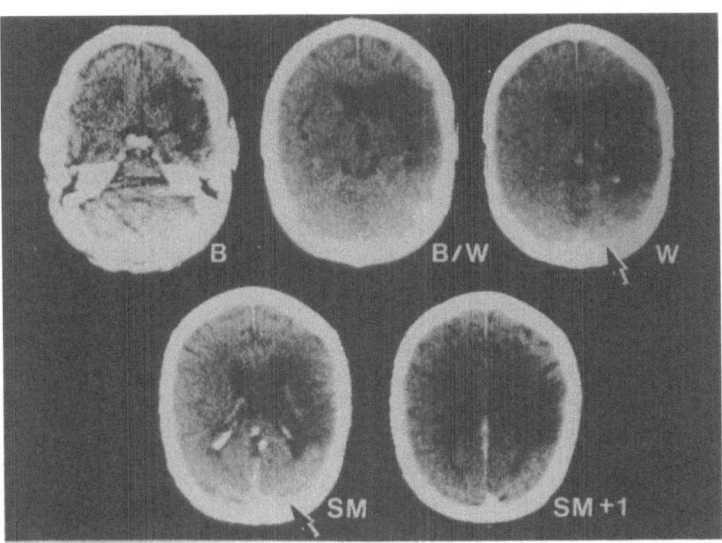

Figure 5. CT scan of LH case with right hemisphere lesion which produced a severe lasting global aphasia (BDAE Auditory Comprehension z-score was -1.5 at 24 MPO). Large lesion is present in the right frontal, parietal and temporal lobes including "right Broca's area" and lower motor cortex, surface and deep, as well as "right Wernicke's area." There is atypical right occipital length asymmetry (arrows, slices W and SM). This case probably had handedness, speech output and comprehension dominance all in the right hemisphere; a combination which is probably very rare. Note occipital asymmetry was ipsilateral to the lesion side and there was minimal recovery of comprehension. Reproduced with permission from Neurology.

global aphasia from a <u>right</u> hemisphere lesion (Figure 5), wrote with the inverted writing posture (suggesting left hemisphere language dominance). He was the only case in this study postulated to have both language dominances (speech output and comprehension) in the right hemisphere, a condition which the inverted writing posture would not have suggested. Additional case studies with writing posture information are necessary.

SUMMARY

The results from Part I are similar to those from studies with RH aphasia cases - i.e., occipital asymmetries cannot be used to predict handedness or which hemisphere, if damaged, will produce aphasia (Pieniadz et al., 1983; Henderson et al., 1984). In the Pieniadz et al. (1983) study with 89 RH aphasia patients who were aphasic from <u>left</u> hemisphere lesion, 33% had atypical <u>right</u> occipital length; these cases were, nevertheless, aphasic from <u>left</u> hemisphere lesion. Likewise, in the present study with 26 LH aphasia patients who were aphasic from <u>left</u> hemisphere lesion, 15% had atypical <u>right</u> occipital length. In a study of crossed aphasia cases (Henderson et al., 1984), 11 of 15 (73%) of the RH aphasia cases who were aphasic form <u>right</u> hemisphere lesion, had typical <u>left</u> occipital length asymmetry. Likewise, in the present study, 3 of the 4 (75%) of the LH aphasics with <u>right</u> hemisphere lesion had typical <u>left</u> occipital length asymmetry. Left occipital length asymmetry appears to be the most common CT scan asymmetry observed among both left-handed and right-handed aphasics with either left or right hemisphere lesions and thus appears not to be useful in predicting handedness or which hemisphere, if damaged, will produce aphasia.

In Part II, where left hemisphere lesion sites were matched between LH and RH aphasics, there were no significant group differences on any of the language measures. This finding is in agreement with the lesion-matched postmortem study of Gloning et al. (1969) where LH and RH aphasics demonstrated similar language behavior. Lesion sites and occipital length asymmetries, where appropriate (i.e. in cases with temporal lobe lesions), should be matched, in future studies regarding "severity of aphasia" or "recovery from aphasia" in left-handed vs. right-handed aphasia cases. For example, while no LH aphasics in our LH population had Wernicke's aphasia, no left-hander had a lesion site which was limited to the left temporo-parietal area which typically produces a Wernicke's aphasia.

In part III, among the four LH aphasics with right hemisphere lesion, <u>improved</u> recovery or perhaps even sparing of comprehension was observed when the posterior lesion was <u>contralateral</u> to the occipital length asymmetry. The results from this section also revealed that dominance for speech output and dominance for comprehension among this subsample may be located in separate hemispheres. Two LH aphasia cases with large right FPT lobe lesion and typical left occipital length asymmetry, who were nonfluent or severely limited in speech output but who had excellent comprehension, probably had handedness and speech output dominance in the right hemisphere, but comprehension dominance in the left hemisphere.

One case was globally aphasic from a large right FPT lobe lesion and this case had atypical right occipital length asymmetry. This aphasia case probably had dominance for handedness, speech output and comprehension all within the right hemisphere. This case is somewhat similar to a case published by Delis et al. (1983) where they described a LH man with Wernicke's aphasia after a right temporal lobe lesion. This case also had atypical right occipital length and width asymmetry on CT. Hence, their study demonstrated another LH aphasia case where dominance for handedness and dominance for language comprehension both appeared to be in the right hemisphere. There was minimal recovery of language comprehension and the CT scan occipital length asymmetry was ipsilateral to the temporal lobe lesion producing the comprehension deficit.

The most common locus-of-dominance pattern among right-handers is the left hemisphere for handedness, speech output and comprehension (as evidenced from Wada testing and most aphasia data). Left-handers may have a less common locus-of-dominance pattern and their locus for handedness dominance may be separate from that for speech output dominance and/or comprehension dominance. Separate hemispheric loci for handedness, speech output and/or language comprehension also have been suggested by Gur et al. (1984) (using handedness and Wada test results, which were controlled separately), Strauss et al. (in preparation) (using handedness, angiographic, Wada and dichotic listening data, which were not always highly intercorrelated) and Graves (1983) (using handedness, mouth asymmetry, dichotic listening, and tachistoscopic data, which also showed low intercorrelations). An anatomic study on the size of Broca's area on the right did not necessarily have a larger planum temporale on the right (Falzi, Perrone, Vignolo, 1982). Along these same lines, Eling (1984) recently published a translation of Broca's original statements regarding laterality for speech and handedness where Broca wrote "...it seems to me absolutely not necessary that the motoric part (handedness) and the intellectual part (speech) of each of the hemispheres are solidary to one another..." (Eling, 1984). The data from the left-handed aphasics examined in this study appear to be supportive of Broca's original statement (Broca, 1865).

Acknowledgements

The authors would like to thank Carole Palumbo and Kate Mora for assistance with the manuscript and Joan Lampada for typing the manuscript.

References

Broca, P. (1865). Sur le Siège de la Faculté du Language Articulé. Bulletin de la Societé d'Anthropologie, 6, 377-393.

Brown, J.W. and Hécaen, H. (1976). Lateralization and Language Representation. Neurology (Minneap), 26, 1410-1411.

Chui, H.C. and Damasio, A.R. (1980). Human Cerebral Asymmetries Evaluated by Computerized Tomography. J. Neurol. Neurosurg. Psychiatry, 43, 873-878.

Delis, D., Knight, R. and Simpson, C.T. (1983). Reversed Hemispheric Organization in a Left-hander. Neuropsychologia, 1, 13-24.

Eling, P. Broca on the Relation between Handedness and Cerebral Speech Dominance. Brain and Language, 22, 158-159.

Falzi, G., Perrone, P. and Vignolo, L. (1982). Right-left Asymmetry in Anterior Speech Region. Archives of Neurology, 39, 239-240.

Galaburda, A., LeMay, M., Kemper, T.L. and Geschwind, N. (1978). Right-left Asymmetries in the Brain: Structural Differences between the Hemispheres May Underlie Cerebral Dominance. Science, 199, 852-856.

Geschwind, N. and Levitsky, W. (1968). Human Brain: Left-right Asymmetries in the Temporal Speech Region. Science, 161, 186-187.

Gloning, I., Gloning, K., Haub, G. and Quatember, R. (1969). Comparison of Verbal Behavior in Right-handed and Non-right-handed Patients with Anatomically Verified Lesion of One Hemisphere. Cortex, 5, 43-52.

Gloning, I. and Quatember, R. (1966). Statistical Evidence of Neuropsychological Syndromes in Left-handed and Ambidextrous Patients. Cortex, 2, 484-488.

Goodglass, H. and Kaplan, E. (1972). The Assessment of Aphasia and Related Disorders. Lea and Febiger, Philadelphia.

Goodglass, H. and Quadfasel, F.A. (1954). Language Laterality in Left-handed Aphasics. Brain, 77, 523-548.

Graves, R. (1983). Mouth Asymmetry, Dichotic Listening and Tachistoscopic Visual Field Advantage as Measure of Language Dominance. Neuropsychologia, 21, 641-649.

Gur, R.E. , Gur, R.C., Sussman, N.M., O'Connor, M.J. and Vey, M.M. (1984). Hemispheric Control of the Writing Hand: The Effect of Callosotomy in a Left-hander. Neurology (Minneap), 34, 904-908.

Hécaen, H. and Sauguet, J. (1971). Cerebral Dominance in Left-handed Subjects. Cortex, 7, 19-48.

Henderson, V.W., Naeser, M.A., Wiener, J.M., Pieniadz, J.M. and Chui, H.C. (1984). CT Criteria of Hemisphere Asymmetry Fail to Predict Language Laterality. Neurology (Minneap), 34, 1086-1089.

Koff, E., Naeser, M.A., Pieniadz, J.M., Foundas, A.R. and Levine, H. (1986). CT Scan Hemispheric Asymmetries in Right- and Left-handed Males and Females. Archives of Neurology, 43, 487-491.

LeMay, M. (1977). Asymmetries of the Skull and Handedness: Phrenology Revisited. Journal Neurol. Sci., 32, 243-253.

Levy, J. and Reid, M. (1976). Variation in Writing Posture and Cerebral Organization. Science, 194, 337-339.

Naeser, M.A., Alexander, M.P., Helm-Estabrooks, N., Levine, H.L., Laughlin, S.A. and Geschwind, N. (1982). Aphasia with Predominantly Subcortical Lesion Sites - Description of Three Capsular/Putaminal Aphasia Syndromes. Archives of Neurology, 39, 2-14.

Naeser, M.A. and Borod, J.C. (1986). Aphasia in Left-handers: Lesion Site, Lesion Side and Hemispheric Asymmetries on CT. Neurology, 36, 471-488.

Pieniadz, J.M. and Naeser, M.A. (1984). Computed Tomographic Scan Asymmetries and Morphologic Brain Asymmetries: Correlation in the Same Cases Post Mortem. Archives of Neurology, 41, 403-409.

Pieniadz, J.M., Naeser, M.A., Koff, E. and Levine, H. CT Scan Cerebral Hemispheric Asymmetry Measurements in Stroke Cases with Global Aphasia: Atypical Asymmetries Associated with Improved Recovery. Cortex, 19, 371-391.

Spreen, O. and Benton, A.L. (1969). Neurosensory Center Comprehensive Examination for Aphasia. Department of Psychology, University of Victoria, Victoria, Canada.

Strauss, E. Lapointe, J., Wada, J., Gaddes, W. and Kosaka, B. Study of the Relationship between Anatomical Asymmetry in Planum Temporale Region on Angiograms, Dichotic Listening and Wada Test Results. In Preparation.

21
Lateralization of Language in Children with Developmental Dyslexia: A Critical Review of Visual Half-field Studies

Morris Moscovitch

INTRODUCTION

Sensory half-field studies is one of the techniques available to neuropsychology for investigating the functional organization of the normal brain. Compared to other techniques, such as EEG, cerebral blood flow and PET scans, it is by far the cheapest and requires the least expertise to use. Many scientists would consider this a mixed blessing. The very accessibility of the technique makes it possible for anyone with minimal training and equipment to conduct experiments and produce publishable results. As a result, the laterality literature is replete with effects that are small or inconsistent (sometimes both), and with theories that are typically vague or simple-minded. Or so critics from outside and within the field would have us believe.

Of all the laterality techniques, the most readily accessible are tachistoscopic ones that are used to infer the nature of hemispheric organization from visual field asymmetries in perception. One would expect, therefore, that if these negative impressions were by and large accurate, it is in this area that they would be most readily confirmed. Quite the opposite is the case. A critical examination of the literature, and the application of some judicious criteria in weeding out inappropriate studies, reveals, even to a skeptical reader, a body of research that is as consistent as any in neuropsychology and cognitive psychology. Even the size of the effects are not usually smaller than many reported in the cognitive literature. As such, laterality studies can be used to provide a firm basis for testing, confirming, extending, and even challenging neurologically-derived data and hypotheses about the normal brain. It can even contribute to theory and model-building in cognitive psychology rather than merely borrow from it (Coltheart, 1981; Moscovitch, 1979; Zaidel, 1978).

324

To illustrate my points, I will focus on one area that I
initially thought typified the sorry state of laterality research
- face recognition and developmental dyslexia. To the
disinterested observer, and even to the committed investigator,
it has often been difficult to find order amidst the variety of
inconsistent results and competing claims about the nature of
hemispheric specialization in individuals with specific reading
disorders (Naylor, 1980; Young & Ellis, 1981). Before turning to
this literature, I will review briefly a model of hemispheric
specialization that I found provided a useful framework for
organizing and evaluating the available literature. More details
about the model can be found in Moscovitch, 1986a.

INTERHEMISPHERIC TRANSFER AND RELATIVE EFFICIENCY MODELS OF LATERALITY

The model assumes that any tachistoscopic task used to
measure visual half-field differences consists of a set of
component operations that are hierarchically organized. Parallel
top-down processing is also possible, as is hemispheric
activation, but for ease of exposition I will restrict the
discussion to a simple, single hierarchically organized process.
(Those craving greater complexity are referred to Moscovitch,
1986a and Trevarthen, 1984.) Leaving aside scanning habits and
strength of projection pathways, differences in perception
between the visual fields can be attributed to one of two causes:
interhemispheric transfer or callosal relay and relative
efficiency of hemispheric processes. Interhemispheric transfer
applies if one or more component operations are functionally
localized (Moscovitch, 1973, 1976, 1981) to one hemisphere. All
input must then be shunted to that hemisphere for processing
regardless of which hemisphere receives the input initially. The
magnitude of field differences is determined by factors that
govern the interhemispheric transfer of information (see Figure
1). A relative efficiency model applies when the stimuli are
processed by the receiving hemisphere. Visual field differences
occur because the efficiency of operation of the components on
the two sides may differ. The visual field advantage is then
determined by the difference in efficiency of the sum (or
product) of the component operations in each half brain (Figure
2). It is also possible that different components may be
recruited to perform a task in each half brain, a condition that
is sometimes reflected in a different pattern of results in each
visual field as well as in a difference in processing efficiency
(e.g. Moscovitch, 1981, pp55-57; Hellige, this volume, Sergent,
1981; Young & Ellis, 1985).

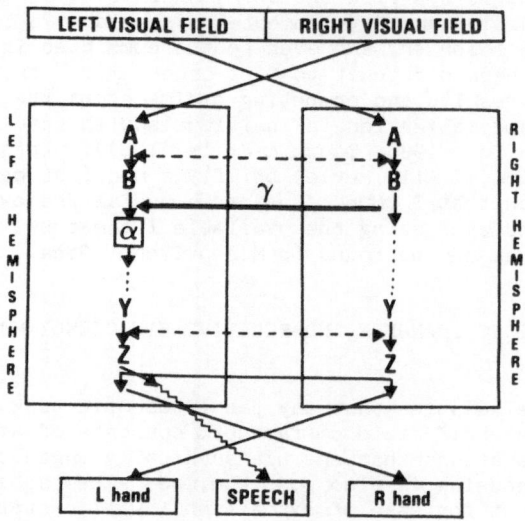

Fig. 1. Interhemispheric transfer. Component process α is
functionally localized (see text) to the left hemisphere. For
simplicity, all other components are represented as identical on
the two sides and only one interhemispheric pathway, α, is
represented as critical apart from the ones involved in response
output. In this instance, the RVF will be favoured by a value
determined by α regardless of which hemisphere initiates the
response (from Moscovitch, 1986a).

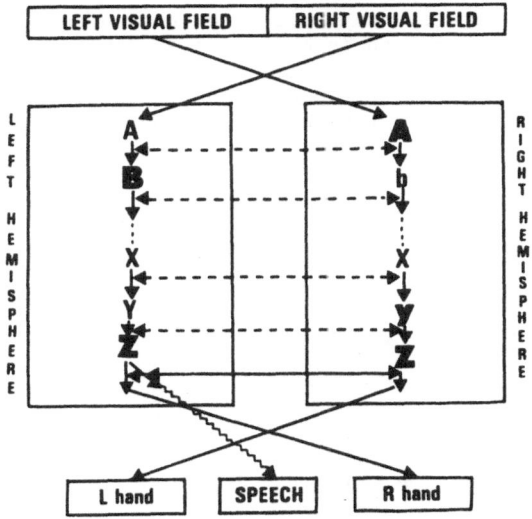

Fig. 2. Relative efficiency. The thickness of the letters
represents the relative efficiency of corresponding component
processes. In this instance, component process A and y are more
efficient in the right hemisphere. B and Z are more efficient in
the left hemisphere, and X is equivalent in the two hemispheres.
Visual field advantage is determined by the efficiency with which
each hemisphere processes the input it receives directly from the
contralateral visual field. The dashed arrows indicate that
interhemispheric transmission is not a determinant of perceptual
asymmetries except, perhaps, at the motor output stage (from
Moscovitch, 1986a).

In most tasks, both interhemispheric transfer and relative efficiency are probably responsible for producing the visual field differences observed (e.g. Figure 3). Without knowing the contribution that relative efficiency or interhemispheric transfer makes to visual field differences in a given task, it is difficult to infer the nature of hemispheric organization from the results.

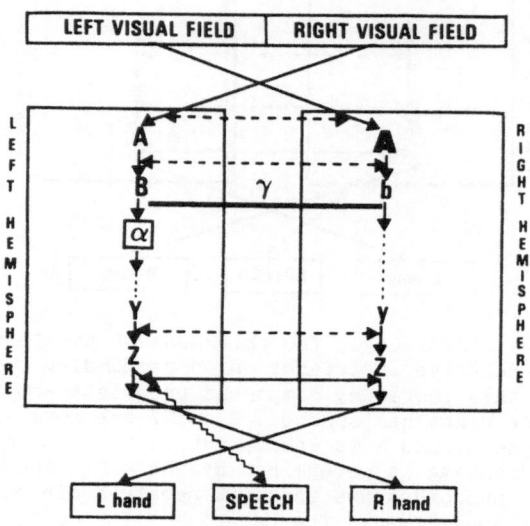

Fig. 3. Mixed case: left-hemisphere localization and right-hemisphere efficiency. Component process α is functionally localized to the left hemisphere. Component process A in the right hemisphere is relatively more efficient than its counterpart in the left. In this case we cannot predict which visual field will be favored. It depends on the advantage that A in the right hemisphere enjoys over A in the left. If the advantage is greater than α then the LVF will be favored whereas if it is less than α then the RVF will be favored. Note that whatever balance of processing efficiency the hemispheres strike subsequent to α it will not affect the magnitude of perceptual asymmetries (from Moscovitch, 1986a).

An examination of figures 1 and 2 indicates that different factors can alter the magnitude and direction of perceptual asymmetries depending on whether the interhemispheric transfer or the efficiency model is applicable. When visual field differences occur only because the hemispheres differ in efficiency, then alterations in the operation of any of the components that constitute a task, or the addition or deletion of components as needed, can affect the visual field advantage that is observed. For example, suppose that both hemispheres can match a pair of identical simultaneously-presented letters or faces (Boles, 1981; Boles & Eveland, 1983). An advantage in favour of the left or right hemisphere can be created by manipulating stimulus variables, such as clarity or spatial frequency composition (Sergent & Hellige, 1986), response factors, such as response hand (see Hellige & Sergent, 1986), or cognitive strategy, such as requiring that the match be based on phonetic features or on structural form (see Moscovitch, 1979). Thus, the advantage can be shifted towards the left visual field by reducing clarity, switching from a right to a left hand response or by adopting a pattern recognition, instead of a phonetic, strategy. In short, if the efficiency model applies, visual field differences can be affected by changes introduced at any locus or component in the information processing sequence from stimulus input to response output.

The conditions over which changes in perceptual asymmetries can be induced are much more restricted when interhemispheric transfer accounts for some of the observed laterality effects. In such cases, visual field differences can be altered by affecting the operation of only the first component that is functionally localized or those components preceding it in the information processing sequence. Changes introduced at subsequent components have no effect on visual field differences.

Examination of Figures 1 and 2 indicates why this is so. When components A and B are equally efficient on the two sides, the advantage in favour of one visual field, the right in the example chosen, is determined only by interhemispheric transfer, α. In the mixed case, both interhemispheric transfer and relative efficiency determine laterality effects. By making the operation of components A or B in the left hemisphere more or less efficient than their counterparts in the right, the right visual field advantage can be increased or decreased, respectively, by a value determined by the difference between left and right-sided processes, $(A + B)_L - (A + b)_R$. The right visual field advantage may even be reversed if the difference is greater than α and favours the right hemisphere.

Having converged on α, the information from either visual field can follow identical or different pathways to the final response output. Assuming that the final output is always determined by the pathway that transmits information most quickly and accurately, then, for all intents and purposes, there really is only one functional pathway from α onwards. Any changes in the component processes beyond α will, therefore, have an identical effect on the input from both visual fields, even if another component that is unique to the right hemisphere is introduced (see Figure 4). Thus, whatever difference exists between the visual fields at α, is maintained at all subsequent points along what is functionally a common pathway.1

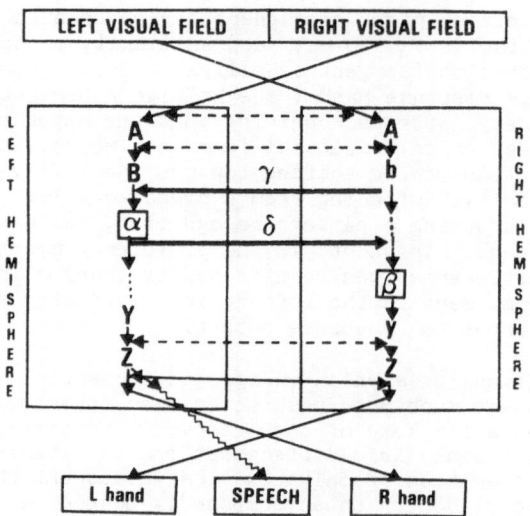

Fig. 4. Mixed case: left- and right-hemisphere localization. Component process α is functionally localized to the left hemisphere whereas β is functionally localized to the right. For simplicity, all other components are equally efficient on the two sides. In this case, the RVF will be favored by a value determined by α (see text). This difference between LVF and RVF stimuli is maintained even as information from α is transmitted across pathway δ to component β in the right hemisphere (from Moscovitch, 1986a).

Though few experiments have been conducted for the express purpose of testing the predictions derived from the model in which interhemispheric transfer applies, the evidence available that bears on this issue is very encouraging. In general, changes in perceptual asymmetry can be brought about more easily by manipulating stimulus parameters or attributes than by altering response modes (see Moscovitch, 1986a, p.100). Presumably the former affects component processes early in the information processing sequence that precede the functionally localized component, α. Changes in response mode, from vocal to manual responses, or from right to left hand responses, necessarily affect the very late stages of processing that occur subsequent to the functionally localized component, α. Thus, decreases in luminance, clarity, spatial frequency, and duration sometimes cause a shift towards the left visual field even for a strongly lateralized task such as word recognition, in which at least one component is believed to be functionally localized to one hemisphere. Changes in response mode from verbal to manual, on the other hand, have little or no effect on perceptual asymmetries on such tasks. In fact, whether or not changes in response mode affect perceptual asymmetries can serve as a diagnostic test for distinguishing those cases in which interhemispheric transfer contributes to the laterality effect from those in which only relative efficiency determines the outcome. In the latter case, changes in response mode are effective in producing changes in laterality whereas in the former case they are not.

Not all changes in laterality tasks are so easily interpreted. Particularly difficult are those manipulations that affect middle level cognitive processes that are sensitive to higher order stimulus attributes such as familiarity and meaningfulness, and that are associated with the initiation and execution of strategies. Unless one can specify the components involved in the task, the order in which they occur, and whether or not any of them is functionally localized, it may be difficult to predict even the direction of perceptual asymmetry, since even small changes in the task may prove to be critical. As might be expected, inconsistencies abound in this aspect of the laterality literature and the interpretations offered to explain them seem suspiciously post hoc. To take but one example, as often as not the typical RVF advantage for words is found to vary with the word's familiarity, concreteness, emotionality, and imageability (see Chiarello, in press, and Moscovitch, 1986a, p.101 and references therein). Those who report positive results cite this as evidence either for a limited right hemisphere lexical contribution to reading, or for the involvement of extra-lexical right hemisphere processes in perception, or both, whereas those who report negative results assert that lexical processes are functionally localized to the left hemisphere in normal people. Similar disputes can be found in the literature on face recognition. Except for rare exceptions, little attempt is made to specify the component processes that are affected by higher-order manipulations and thereby identify the source of the inconsistencies.

This enterprise would be made easier if, for each laterality task, a complete detailed description or model were available of the component processes involved and their sequence of operation. Though current models fall far short of this ideal for any laterality task I can think of, some models of word and face perception are sufficiently well developed to be of service in answering some simple, but fundamental, questions in laterality

research. I now turn to the literature on developmental dyslexia
to illustrate this approach.

DEVELOPMENTAL DYSLEXIA: VISUAL HALF FIELD STUDIES

Overview

     One of the questions that has preoccupied neuropsychologists
interested in developmental dyslexia, often to the exclusion of
other questions that are at least as important, is whether
developmental dyslexia is associated with an abnormal pattern of
hemispheric dominace. Unlike normal children who are typically
right-handed and left-hemispheric dominant for language, many
dylexic children are believed to be weakly lateralized or even to
show reversed dominance.

     Remarkably, the issue is still unresolved despite extensive
research on the topic since the "hemispheric dominance"
hypothesis was proposed by Orton (1937) over fifty years ago. I
will not attempt to review the literature, having neither the
space nor the qualification to do the job adequately. Instead, I
will set myself the much more manageable task of examining only
those studies that compared visual field differences in
perception in normal and dyslexic children. Although these
studies comprise only a fraction of the published
neuropsychological literature on dyslexia, they share a set of
concerns that are common to the entire field. It is hoped that
by examining a circumscribed area carefully, the knowledge gained
will be relevant to other areas in the field.

     In a typical tachistoscopic laterality study of
developmental dyslexia, the difference in perception of stimuli
presented in the two visual fields in dyslexic children is
compared with that of appropriately matched control subjects. In
most of the studies it is assumed, either implicitly or
explicitly, that the visual field difference on a laterality task
provides a direct measure, or at least an index, of hemispheric
dominance or lateralization for such global functions as language
or visual-spatial abilities. The larger the difference in favour
of one visual field, the more that function is presumed to be
lateralized to the hemisphere contralateral to the favoured
field.

     Were the findings consistent, there would be little reason
to question this basic approach. Unfortunately, positive
findings of a laterality difference between a dyslexic and a
normal child, are reported as often as negative findings
(somewhat less often if a simple count is taken of every
published study), leaving the interested reader at a loss in
trying to make sense of the available evidence.

Some recent critical reviews have been very helpful in guiding the reader through the dense and contradictory literature by calling attention to a variety of factors, in addition to hemispheric specialization, that potentially can influence the outcome of laterality studies (Bryden, 1982; Corballis, 1983; Naylor, 1980; Pirozzolo, 1979; Satz, 1976; Witelson, 1976; Young & Ellis, 1981). Among these factors are subject selection on the basis of IQ, reading proficiency and type of reading disorder, stimulus presentation procedures, cognitive processes involved in reading and cognitive strategies used to decode tachistoscopic input. The reviews are useful in making the reader aware both of the complexity involved in conducting tachistoscopic laterality studies on dyslexic children and of the need for more experiments in which the above factors are carefully controlled. One senses, however, that the reviewers, in the end, are unwilling to commit themselves on the issue of language lateralization in dyslexic children.

The available evidence does not warrant such a cautious position. In their effort to promote ideal studies the reviewers may have discounted too quickly some perfectly adequate studies that may have some minor flaws.

While acknowledging a debt to the excellent reviews, I would like to propose a less stringent set of criteria by which to evaluate published studies. As is clear from the reviews, and from the previous discussion on models of laterality, a variety of factors other than hemispheric specialization for language can influence the direction and magnitude of visual field difference. Even if subjects and controls are carefully selected and care is taken that the testing procedures are methodologically sound, successful completion of the tachistoscopic task itself often relies on the operation of a variety of components in addition to the ones specifically concerned with language. Some of the extra-linguistic components may themselves be lateralized and contribute to the overall laterality effect. It is entirely possible that visual field differences in laterality tasks between dyslexic and normal chidren, should they exist, are due entirely to differences in lateralization of these extra-linguistic components rather than in the language component. An equally likely possibility is that dyslexic children adopt a different strategy in decoding verbal stimuli than do normal children (Young & Ellis, 1981). As a result, different components are called into play when a dyslexic child attempts to identify a verbal tachistoscopic stimulus than when a normal child tries to do so. In the absence of precise knowledge of the components involved in tachistoscopic identification and how they are influenced by strategic factors, it is difficult, if not impossible, to interpret the significance of laterality differences, or even similarities, between normal and dyslexic children. Nonetheless, by adopting a less stringent set of criteria that are in keeping with the spirit, if not entirely with the substance, of these excellent reviews, it is possible to find surprising uniformity among the studies that meet these criteria and to suggest, on that basis, a tentative solution to the problem of lateralization in dyslexia. In the following section, the rationale for choosing these criteria will be set forth.

Criteria for Visual Half-field Studies of Reading-Disabled Children

Despite great variability in visual laterality, it is assumed that hemispheric dominance for language is stable - at least over relatively long time periods. The first priority, then, was to find a tachistoscopic test that would reflect that underlying invariance as closely as possible. Put in terms of the models we described earlier, the ideal test should yield a visual field difference that was very sensitive to hemispheric asymmetry in the operation of "language" components and relatively insensitive to the operation of the other components. Though it is prejudging the issue, the most likely task would be one in which the linguistic component is functionally localized since, according to the models, that at least limits the influence of non-linguistic factors to those components that precede the linguistic one. Of the tests available, identification of words and word-like letter strings produces the most consistent right visual field advantage. It can be argued, as I have elsewhere (Moscovitch, 1973, 1976, 1979, 1986a,b), that word identification generally behaves as if at least one of the components is functionally localized in the left hemisphere. Admittedly, the evidence in favour of this view, though compelling, is not conclusive (see Chiarello, in press and Patterson & Besner, 1984, and related commentaries by Rabinowicz and Moscovitch, 1984 and Zaidel and Schweiger, 1984 for a balanced overview). Nonetheless, it is a simplifying assumption that I will adopt because some strong predictions follow from it. <u>The first criterion, therefore, is that the tachistoscopic laterality test involve the identification of words or word-like letter strings</u>.

Adopting this criterion immediately raises an obvious problem: it is precisely in this type of test that dyslexic children are most impaired. Comparison between visual field differences in normal and dyslexic children will be confounded, at the very least, by the different level of skill that both groups bring to the test and, at most, by possibly different strategies that the two groups use in decoding words. In principle, these are legitimate concerns; in practice, their consequence is diminished. To convince the reader of this point it is necessary to examine some tachistoscopic studies of word recognition in normal people.

Even if it is assumed that the linguistic components are functionally localized to one hemisphere, variation in visual field advantage can occur in response to factors that affect earlier components. As we noted previously, these components are likely to be involved in the apprehension of structural or pictorial features of the stimulus. Hemispheric differences in extracting visual sensory features are generally absent, but when they do occur they are small and typically favour the right hemisphere (Davidoff, 1982; Kimura & Durnford, 1974; Moscovitch, 1986a; Sergent and Hellige, 1986). Larger and more consistent hemispheric asymmetries are found when sensory features must be integrated into more abstract perceptual representations. With regard to words, those representations may be primarily orthographic or pictorial thereby favouring the left or right hemisphere, respectively. Thus, the typical right visual field-left hemisphere advantage in word identification may be diminished or even reversed if the subject is required to rely to a greater than usual extent on pictorial component processes in identifying words or letter strings.

The reliance on such pictorial processes is most evident if we consider children who have not yet learned to read. To them, the written word has roughly the same status as, say, semitic script or sanskrit does to anyone who is unfamiliar with either. Even the beginning reader may still rely heavily on these pictorial component processes in trying to identify tachistoscopically-presented words. Consequently, the expected right visual field advantage may not emerge until word identification is relatively automatic in the sense that graphemic access to linguistic, or at least lexical, component processes or representations in the left hemisphere can occur without the type of laborious pictorial decoding in which the right hemisphere specializes. The above analysis is based, in part, on an excellent series of experiments conducted on Isareli children. Published over a period of five years, the experiments constitute a developmental study of visual field differences to Hebrew and English words in children from grade 1 to 11. Carmon, Nachshon, and Starinsky (1976) and Silverberg, Gordon, Pollack, and Bentin (1980) observed that a right visual field advantage in reading Hebrew words is first found at about grade 3. In the earlier grades, there is either no visual field difference or the left visual field is favoured. Carmon et al concluded that before they become familiar with written words, children rely on a pictorial, right-hemisphere strategy as much as, if not more than, a linguistic, left hemisphere strategy to identify words. The shift towards a right-visual field advantage by grade 3 reflects the shift in emphasis from a pictorial to a linguistic strategy.

To test this hypothesis, Silverberg, Bentin, Gaziel, Obler, and Albert (1979) reasoned that the right visual field advantage should be abolished in fluent Hebrew readers if they are

introduced to a new script. In grade 5, Israeli children first
learn to read English. The same children who show a consistent
right visual field advantage for Hebrew words, have a slight left
visual field advantage for English words. By grade 11, the
visual field advantage for English had shifted to the right,
presumably as a result of their increased familiarity with that
written language. As a final demonstration, Bentin (1981) showed
that blurring can reverse the visual field advantage from the
right to the left even for Hebrew words in fluent, grade 7
readers, and that a RVF advantage for English words can be
accelerated by making the subjects familiar with them.

Apart from showing that emphasis on different decoding
strategies can influence the direction of perceptual asymmetries
even for such highly lateralized stimuli as words, the studies on
Israeli children suggest a solution to the problem of equating
dyslexic and normal children for reading ability and provide us
with our second criterion. Though theoretically it would be
ideal to equate dyslexic and normal children on all relevant
dimensions, including reading level, it is impossible to do so.
By equating normal and dyslexic children on reading ability and
IQ, there must necessarily be a discrepancy in their ages. Who
is to say that age is a less important determiner of hemispheric
specialization and, hence, perceptual asymmetry than reading
ability (Brown & Jaffe, 1975)? What the studies on Israeli
children suggest is that once children attain a certain minimal
level of reading proficiency, they display a RVF advantage that
is affected very little by further increments in skill
(presumably so long as common words are used - but see below).
Normal children attain that minimal level in their native
language by about grade 3. This suggests that a sufficient
criterion for comparing visual field differences between normal
and dyslexic children is that they each attain a minimal level of
reading proficiency. The more stringent criterion of equating
the reading level of the two groups aside from generating other
problems, may actually provide few added benefits. Our second
criterion, then, is that normal and dyslexic children must read
at least at a grade 3 level. If the left hemisphere is dominant
for language in dyslexic, as in normal children, they, too,
should have a normal RVF advantage for identifying words once
they are reading at a grade 3 level. Differences in visual field
asymmetry between the two groups might then suggest a difference
in hemispheric organization.

The third criterion is that all subjects must be right-
handed. It is well known that the proportion of left-handers
among dyslexic people is greater than in the general population.
Whether these are natural or pathological left-handers is not
known, so I thought it best to exclude them initially. The issue
of handedness in dyslexia will be taken up later.

The remaining criteria have little to do with hemispheric specialization as such. The fourth criterion is that the full scale IQ should be within or above normal limits for both groups. As with reading level, it may be preferable, but not necessary, that the IQ of the two groups be equated. The fifth criterion concerns the definition of dyslexia. Here, too, I wish to opt for the laxest possible criterion, namely, that a dyslexic child is one who reads below his grade level despite having the emotional and intellectual capacity to read normally. In all but one of the studies reviewed, the dyslexic children read, on the average, 2-3 years below their grade level. In one study, however (Olson, 1973), some of the very youngest children were classified as dyslexic if they read only 6 months below their grade level.

Are dyslexics weakly lateralized? A review of the literature in light of the proposed criteria

In reviewing the literature, seven studies meet all five criteria. Those studies are listed in Table 1 along with the reading grade level, the stimuli used, the visual field advantage reported for dyslexic children, and how that advantage compared with that of normal children. As can be seen from Table 1, there is a striking uniformity in the results. All seven studies reported a significant RVF advantage in dyslexic children. In six, that advantage was as great or greater than in normal children, whereas in the studies by Kershner (1977, 1979), it was smaller. In the Wellman and Allen (1983) study, the RVF advantage was equivalent in normal and dyslexic children when the comparison was between large groups of both types of children. When, however, the top 20 in the normal group were compared with the bottom 20 in the dyslexic group, the magnitude of the RVF advantage was greater in the normal group. Because there is no information on whether the subjects from the two extremes of the distribution differed on other dimensions (a highly likely possibility), it is difficult to determine whether they even met our criteria. Kershner (1979) reported similar results when he compared extreme groups.

Based on the majority of the studies, the simple conclusion would be that language functions, at least those involved in word identification, are as strongly lateralized in dyslexic, as in normal, children. The exceptions, however, caution against endorsing that conclusion hastily. Before a firm conclusion can be reached, it is necessary to be satisfied that the criteria are valid by examining as well the studies that did not meet them, and, then, to provide an account of the anomolous findings that does not invalidate the conclusion that lateralization of function is normal in dyslexic children.

Recall that the same consideration determined the choice of the first three criteria: to ensure that the tachistoscopic test engages the language components of the left hemisphere and produces a RVF advantage that primarily reflects the linguistic superiority of the left hemisphere over the right. Any departure from these criteria increases the likelihood that nonlinguistic operations, such as pictorial ones in which the right hemisphere specializes, will be recruited to process the stimuli. Visual field difference as a measure of language dominance would thereby be contaminated. The likelihood of that occurring is probably greater in dyslexic than in normal readers since the dyslexics' inferior verbal abilities in general (Denkla, Rudel, & Broman, 1978; Gordon, 1980) make them more susceptible to using nonverbal strategies. As a result, those studies that fail to fulfill these criteria are less likely to yield a consistent RVF advantage on ostensibly verbal tests even in normal readers, but especially in dyslexics. Examination of the evidence corroborates this hypothesis.

Table 1. Visual half-field studies on developmental dyslexia that meet all five criteria. Tabulated with each study are reading grade level of the dyslexic children, the stimuli used, the visual field advantage in dyslexic children and whether it was equal, larger, or smaller than in normals.

Study	Reading Grade	Stimulus Type	Visual Field Advantage	Comparison with normals
1. Bouma & Legein (1977)	2-5	Unilateral 3-5 letters	RVF	equal
2. Kershner (1977, 1979)	3-4	Bilateral 4 letters	RVF	smaller
3. McKeever & Huling (1970)	3-8	Bilateral 4 letters	RVF	equal
4. McKeever & VanDeventer (1975)	4-6	Bilateral, Unilateral	RVF	equal
5. Olson (1973)	2-5	Bilateral	RVF	equal
6. Wellman & Allen (1983)	3-5	Unilateral	RVF	equal
7. Yeni-Komshian et al (1975)	3-4	Unilateral	RVF	larger

When trigrams are used instead of words, the results are inconsistent. Aylwood (1984) failed to find a significant visual field difference in both normal and disabled readers whereas Keefe and Swinney (1979) found a RVF advantage in both groups. Using a string of numerals as stimuli, Yeni-Komshian, Isenberg, and Goldberg (1975) found no visual field difference in normal children but a significant RVF advantage in dyslexics. Corballis, Macadie, and Beale (1985) found the reverse when subjects were asked to identify rotated letters. Bouma and Legein (1977) had three conditions in their experiment: single letters that were presented alone, letters that were flanked by Xs, and bilaterally presented words. In the solitary letter condition, there was no visual field advantage for either normal or dyslexic children whereas in the embedded letter condition and in the bilateral word condition there was a significant RVF advantage that was equivalent in the two groups. Gross, Rothenberg, Shottenfeld, and Drake (1978) reported a RVF advantage in threshold for identifying single letters that was greater for disabled readers.

Even if words are used, it may be necessary that the response be one of identification rather than recognition. Pirozzolo and Rayner (1979) presented words bilaterally and then had subjects point to the target items among an array of four distractors. They found a significant right visual field advantage in normal, but not in dyslexic, children. Naylor, Lambert, Sassone and Hardyck (1980) who asked subjects to match two words with each other, reported no consistent RVF advantage even in normal children. It may be that recognition and visual matching encourages the dyslexic child, who may have difficulty encoding information verbally, to rely more on a pictorial than a verbal strategy in this situation. This interpretation is consistent with evidence from studies on normal adults showing that response modes, such as matching and recognition, can reduce, eliminate, or reverse the RVF advantage for words that is typically seen when identification is required (Brand, Van Bekkum, Stumpel, & Kroeze, 1983; Gibson, Dimond, & Gazzaniga, 1972; Klein, Moscovitch, & Vigna, 1976).

For similar reasons, using subjects who read below a grade 3 level will decrease the likelihood of obtaining a RVF advantage even in a word identification test. Marcel and his colleagues' (Marcel, Katz, & Smith, 1974; Marcel & Rajan, 1975) failure to find a RVF advantage for words in dyslexic, but not in normal, children may have occurred because the dyslexic children in those studies were reading below the level of normal children who were themselves only in grade 3. The dyslexic children in Aylwood's study (see above) were also reading below a grade 3 level. Olson (1973) used subjects of different ages in her study. Those dyslexic children who were 8-9 years old were reading below a grade three level. They showed a RVF advantage for word identification when the words were presented bilaterally but not when they were presented unilaterally. The group of 10-11 year old dyslexic children, who were reading at, or above, a grade 3 level, however, favoured the RVF in both conditions. These results replicate exactly those obtained with normal children by Carmon et al (1976).

Data on left-handed dyslexic children are almost not available. Occasionally a small number of left-handers are included in some studies, but their data are not reported separately. The exception is the study by Gross et al (1978) who found that all three non-right-handed dyslexics showed a LVF advantage for single letter thresholds, whereas all the right-handed dyslexics favoured the RVF.

The dearth of information on left-handed dyslexics is puzzling since they are supposed to be overrepresented in the dyslexic population. In fact, the high proportion of left-handers among dyslexic children has, and continues to be, cited as evidence of anomolous lateralization in dyslexia (Geschwind & Galaburda, 1986). This conclusion is unwarranted. That there is increased left-handedness among dyslexics has no bearing on the question of lateralization in right-handers who form the majority even among dyslexic children. What the data on left-handed dyslexics does suggest is that hemisphere damage is more common among left-handed dyslexic children than among normal children. If this is true, one would expect to find a greater incidence of LVF-right hemisphere advantage in word identification among a group of left-handed dyslexic children than among left-handed normal children. To my knowledge, no one has tested this hypothesis.

A review of the evidence seems to justify the choice of criteria and to add weight to the conclusion that language functions necessary for word identification are as strongly lateralized in right-handed dyslexic children as in normal children. What remains is to explain the somewhat anomolous results obtained by Kershner (1977, 1979).

Some recent studies by Young and Ellis (1985) and by Bub and Lewine (in press) on normal adults are relevant. They found that as word length increases, performance in word identification is progressively impaired in the left visual field but remains stable both in the fovea and in the right visual field. As a result, the RVF advantage increases as the words get longer. One interpretation of these results is that words directed to the left hemisphere are processed as a unit in the skilled reader. Words directed initially to the right hemisphere, however, are held in a visual buffer where they are processed serially before gaining access to the left hemisphere for further encoding. Dyslexic children, even when reading at a grade 3 level, may not be as facile as normal children in encoding words that are projected to the left hemisphere. Indeed, Bouma and Legein

(1977) have found that older dyslexic children show word length
effects to foveal presentation whereas normal children do not.
As a result, performance to foveal presentation becomes
progressively impaired with increasing word length in dyslexic
children. If similar effects occur to right visual field
presentation in dyslexic children it implies that the RVF
advantage in dyslexic children, unlike that in normal children,
will remain relatively constant with increasing word lengths
because the scores in both fields will drop as word length
increases. Consequently the dyslexic child who has not yet
reached the level of skill necessary to encode words as a unit
will have a smaller RVF advantage than normal readers who have
acquired that skill. The difference will be due almost entirely
to the RVF score which drops with increasing length only in the
dyslexic child. The left visual field score should be similar in
both groups.

This explanation may account for some of Kershner's results.
He compared disabled readers with gifted and normal children of
close to equivalent reading ability. All groups had a RVF
advantage. When compared to normal readers, the dyslexic child
had a smaller, but not significantly different, RVF advantage.
The main difference between the groups was in the lower RVF score
obtained by the dyslexics. In comparison to that group, the
disabled readers may not have been less lateralized, but may
simply have been encoding words inefficiently in both visual
fields (see Bub and Lewine, in press, for a similar observation).
This explanation does not hold up well when the dyslexic children
are compared with gifted children of above average reading skill.
As expected, the gifted children have a higher RVF score but,
surprisingly, their LVF score is significantly lower than that of
both of the other groups. This finding remains the only
exception among those that meet all five criteria. It is
fitting, perhaps, that it involves only exceptional children.

Conclusion

The preceding critical review of the literature indicates
that when appropriate criteria are applied, tachistoscopic visual
half-field studies produce a consistent and equivalent RVF
advantage in right-handed dyslexic and normal people. Such
results strongly suggest that there is no difference in
lateralization of language function between normal and disabled
readers who are right-handed. In this group of disabled readers
dyslexia is not related to anomolous hemispheric specialization
but to impaired language functions that are lateralized, as in
normal people, to the left hemisphere. In almost all the studies
reviewed, both those that meet our criteria as well as those that
did not, the absolute level of performance on verbal tasks,
irrespective of visual field differences, was lower in disabled
than in normal readers. Disabled readers typically have a lower

verbal IQ and score worse on a variety of verbal tests, including simple naming (Denkla et al, 1978). This conclusion is consistent with Gordon's (1980) observation that dyslexic children and their family members perform poorly on left hemisphere tasks but do well on right-hemisphere tasks, despite showing normal left-hemisphere lateralization of language as assessed by a verbal dichotic listening task. In short, it is the quality of the language functions in the left hemisphere, not their relative lateralization that determines whether reading will be impaired or will develop normally.

The proviso that the child must read at a grade 3 level before visual half-field tests of word-identification can be valid indicators of hemispheric dominance, may be interpreted by some that I favour the developmental lag hypothesis of dyslexia (Satz, 1976). I do not wish to commit myself on that issue. I do, however, believe that linguistic components are functionally localized in the left hemisphere long before a child begins to read, perhaps as early as birth (Moscovitch, 1977). On that basis it is predicted that dyslexic children younger than grade 3 will have a normal pattern of left-hemisphere lateralization of language functions on laterality tests that make no demands on reading, or other abilities that have developmental milestones. Tests such as the dichotic listening test would seem to be ideal candidates. Unfortunately, the literature on dichotic listening tests in disabled readers indicates that this field of research is not without its problems (Bryden, 1982), some of which are similar to those encountered in the tachistoscopic literature and some of which are unique. A proper test of this prediction must await a solution to those problems. As for those children and adults who cannot even attain a grade-3 level of reading proficiency despite extensive schooling, I make no predictions.

The conclusion that lateralization of language functions is normal applies only to right-handed dyslexics. The likely presence of a disproportionate number of pathological left-handers among left-handed disabled readers suggests that, as a group, left-handed dyslexics may have a greater tendency to favour the LVF-right hemisphere than would left-handed normal readers. Because it results from different causes, the reading impairment in pathological left-handers may be different from that observed in other dyslexics. It is among this group of pathologically left-handed dyslexics that the theory of anomolous hemispheric specialization in dyslexia may finally be vindicated.

Acknowledgements

This paper was prepared, in part, while the author was a Fellow at the Institute for Advanced Studies of the Hebrew University in Jerusalem. I thank the staff of the Institute for the help given and the Neuropsychology Group, headed by

I. Nachshon, for stimulating discussions that helped develop the
ideas in this paper. The work on the paper was also supported by
an NSERC of Canada grant to the author. I thank Maureen Patchett
for typing the manuscript.

References

Aylwood, E.H. (1984). Lateral asymmetry in subgroups of dyslexic
    children. Brain Lang., 22, 221-231.
Bentin, S. (1981). On the representation of a second language in
    the cerebral hemispheres of right-handed people.
    Neuropsychol. 19, 599-603.
Boles, D.B. (1981). Variability in letter-matching asymmetry.
    Perc. Psychophys., 29, 285-288.
Boles, D.B. and Eveland, D.C. (1983). Visual and phonetic codes
    and the process of generation in letter matching. J. Exper.
    Psychol.: Hum. Perc. Perf., 9, 657-674.
Bouma, H. & Legein, Ch.P. (1977). Foveal and parafoveal
    recognition of letters and words by dyslexics and average
    readers. Neuropsychol., 15, 69-30.
Brand, N., Van Bekkum, I., Stumpel, M., & Kroeze, J.H.A. (1983).
    Word matching and lexical decisions: A visual half-field
    study. Brain Lang, 18, 199-211.
Brown, J.W. & Jaffe, J. (1975). Hypothesis on cerebral
    dominance. Neuropsychol., 13, 107-110.
Bryden, M.P. (1982). Laterality: Functional Asymmetry in the
    Intact Brain. Academic Press, New York.
Bub, D.N. & Lewine, J. (in press). Different modes of word
    recognition in the left and right visual fields. Brain Cog.
Carmon, A., Nachshon, I., & Starinsky, R. (1976). Developmental
    aspects of visual hemifield differences in perception of
    verbal material. Brain Lang., 3, 463-469.
Chiarello, C. (in press). Lateralization of lexical processes in
    the normal brain: A review of visual half-field research.
    In Studies in Neuropsychology. (eds. H.A. Whitaker and A.
    Caramazza). Erlbaum, New York
Coltheart, M. (1981). Disorders of reading and their
    implications for models of normal reading. Visible Lang.,
    15, 245-286.
Corballis, M.C. (1983). Human Laterality. Academic Press, New
    York.
Corballis, M.C., Macadie, L., & Beale, I.L. (1985). Mental
    rotation and visual laterality in normal and reading
    disabled children. Cortex, 12, 225-236.
Davidoff, J.B. (1982). Studies with non-verbal stimuli. In
    Divided Visual Field Studies of Cerebral Organization, (ed.
    J.G. Beaumont). Academic Press, London.
Denkla, M.B., Rudel, R.G., & Broman, M.B. (1981). Tests that
    discriminate between dyslexic and other learning disabled
    boys. Brain Lang., 13, 118-129.

Geschwind, N. & Galaburda, A.M. (1985). Cerebral lateralization: Biological mechanisms and pathology: I. A hypothesis and a program for research. Arch. Neurol., 42, 428-459.

Gibson, A.R., Dimond, S.J., & Gazzaniga, M.S. (1972). A left field superiority for word matching. Neuropsychol., 10, 463-466.

Gordon, H.W. (1980). Cognitive asymmetry in dyslexic families. Neuropsychol., 18, 645-656.

Gross, K., Rothenberg, S., Shottenfeld, S., and Drake, C. (1978). Duration thresholds for letter identification in left and right visual fields for normal and reading-disabled children. Neuropsychol., 16, 709-715.

Hellige, J.B. and Sergent, J. (1986). Role of task factors in visual field asymmetries. Brain Lang., 5, 200-222.

Keefe, B. and Swinney, D. (1979). On the relationship of hemispheric specialization and developmental dyslexia. Cortex, 15, 471-481.

Kershner, J.R. (1977). Cerebral dominance in disabled readers, good readers, and gifted children: Search for a valid model. Child Devel., 48, 61-67.

Kershner, J.R. (1979). Rotation of mental images and asymmetries in word recognition in disabled readers. Can. J. Psychol., 33, 39-50.

Kimura, D. and Durnford, M. (1974). Normal studies in the function of the right hemisphere in vision. In Hemisphere Function in the Human Brain. (eds. S.J. Dimond and J.G. Beaumont). Elek Science, London.

Klein, D., Moscovitch, M., and Vigna, C. (1976). Attentional mechanisms and perceptual asymmetries in tachistoscopic recognition of words and faces. Neuropsychol., 14, 55-66.

Marcel, T., Katz, L., and Smith, M. (1974). Laterality and reading proficiency. Neuropsychol, 12, 131-139.

Marcel, T. and Rajan, P. (1975). Lateral specialization for recognition of words and faces in good and poor readers. Neuropsychol., 13, 489-497.

McKeever, W.F. and Huling, M.D. (1970). Lateral dominance in tachistoscopic word recognition of children at two levels of ability. Q. J. Exper. Psychol., 22, 600-604.

McKeever, W.F. and Van Deventer, A.D. (1975). Dyslexic adolescents: Evidence of impaired visual and auditory language processing with normal lateralization and visual responsivity. Cortex, 11, 361-378.

Moscovitch, M. (1973). Language and the cerebral hemispheres: Reaction-time studies and their implication for models of cerebral dominance. In Communication and Affect: Language and Thought. (eds. P. Pliner, T.A. Alloway, and L. Krames) Academic Press, New York.

Moscovitch, M. (1976). On the representation of language in the right hemisphere of right-handed people. Brain Lang., 3, 47-71.

Moscovitch, M. (1977). The development of language functions and its relation to cognitive and linguistic development: A review and some theoretical speculations. In Language Development and Neurological Theory. (eds. S.J. Segalowitz and F.Gruber). Academic Press, New York.

Moscovitch, M. (1979). Information processing and the cerebral hemispheres. In Handbook of Behavioral Neurobiology: Vol 2: Neuropsychology. (ed. M.S. Gazzaniga). Plenum Press, New York.

Moscovitch, M. (1981). Right-hemisphere language. Top. Lang. Dis., 1, 41-61.

Moscovitch, M. (1986a). Afferent and efferent models of visual perceptual asymmetries: theoretical and empirical implications. Neuropsychol., 24, 91-114.

Moscovitch, M. (1986b). Hemispheric specialization, interhemispheric codes and transmission times: Inferences from studies of visual masking of lateralized stimuli in normal people. In Two hemispheres: one brain. (eds. F. Lepore, M. Ptitot, and H.H.Jasper). Alan R. Liss, New York.

Moscovitch, M., Scullion, D., and Christie, D. (1976). Early versus late stages of processing and their relation to functional hemispheric asymmetries in face recognition. J. Exp. Psychol: Hum. Perc. Perf., 2, 401-416.

Naylor, H. (1980). Reading disability and lateral asymmetry: An information-processing analysis. Psychol. Bull., 87, 531-545.

Naylor, H., Lambert, N.M., Sassone, D.M., and Hardyck, C. (1980). Lateral asymmetry in perceptual judgments of reading disabled, hyperactive, and control children. Int. J. Neurosci., 10, 135-143.

Olson, M.E. (1973). Laterality differences in tachistoscopic word recognition in normal and delayed readers in elementary school. Neuropsychol., 11, 343-350.

Orton, S.T. (1937). Reading, writing, and speech problems in children. Norton, New York.

Patterson, K. and Besner, D. (1984). Is the right hemisphere literate? Cog. Neuropsychol., 1, 315-341.

Pirozzolo, F.J. (1979). The Neuropsychology of Developmental Reading Disorders. Praeger, New York.

Pirozzolo, F.J. and Rayner, K. (1977). Hemispheric specialization in reading and word recognition. Brain Lang., 4, 248-261.

Rabinowicz, B. and Moscovitch, M. (1984). Right hemisphere literacy: A critique of some recent approaches. Cog. Neuropsychol., 1, 343-350.

Satz, P. (1976). Cerebral dominance and reading disability: An old problem revisited. In The Neuropsychology of Learning Disorders: Theoretical Approaches. (eds. R.M. Knights and D.J. Bakker). University Park Press, Baltimore.

Sergent, J. (1982). About face: Left-hemisphere involvement in processing physiognomies. J. Exper. Psychol.: Hum. Percep. Perf., 8, 1-14.

Sergent, J. and Hellige, J.B. (1986). Role of input factors in visual-field asymmetries. Brain Lang., 5, 174-199.

Silverberg, R., Bentin, S., Gaziel, T., Obler, L.K. and Albert, M.L. (1979). Shift of visual field preference for English words in native Hebrew speakers. Brain Lang., 8, 184-190.

Silverberg, R.., Gordon, H.W., Pollack, S., and Bentin, S. (1980). Shift of visual field preference to Hebrew words in native speakers learning to read. Brain Lang., 11, 99-105.

Trevarthen, C. (1984). Hemispheric specialization. In Handbook of Physiology: Volume III Section 1, The Sensory Systems. Part 1. (ed. I. Darian-Smith). The American Physiological Society, Bethesda.

Wellman, M.M. and Allen, M. (1983). Variations in hand position, cerebral lateralization, and reading ability among right-handed children. Brain Lang., 18, 277-292.

Witelson, S.F. (1976). Abnormal right hemisphere specialization in developmental dyslexia. In The Neuropsychology of Learning Disorders: Theoretical Approaches. (eds. R.M. Knights and D.J. Bakker). University Park Press, New York.

Yeni-Komshian, G.H., Isenberg, D., and Goldberg, H. (1975). Cerebral dominance and reading disability: Left visual field deficit in poor readers. Neuropsychol., 13, 83-94.

Young, A.W. and Ellis, A.W. (1981). Asymmetry of cerebral hemispheric function in normal and poor readers. Psychol. Bull., 89, 183-190.

Young, A.W. and Ellis, A.W. (1985). Different methods of lexical access for words presented in the left and right visual hemifields. Brain Lang., 24, 326-358.

Zaidel, E. (1978). Lexical organization in the right hemisphere. In Cerebral Correlates of Conscious Experience. (eds. B.A. Busser and A. Rougeul-Buser). Elsevier, Amsterdam.

Zaidel, E. and Schweiger, A. (1984) A wrong hypothesis about the right hemisphere: Commentary on K. Patterson and D. Besner, "Is the right hemisphere literate?" Cog. Neuropsychol., 1, 351-364.

Session V
Integration of Hemispheric Functions

Chairman: C. Trevarthen

22
Role of Callosal Connections in the Representation of the Visual Field in the Primary Visual Cortex of the Cat

G. Berlucchi, A. Antonini, G.G. Mascetti and G. Tassinari

INTRODUCTION

Current ideas about the functional significance of the corpus callosum and other commissures are chiefly derived from studies of the behavioral deficits following commissural sections in both man and experimental animals. The main conclusion supported by these studies is that the forebrain commissures are essential for unifying and coordinating cognitive processes that take place separately and independently in the right and left cerebral hemispheres (Sperry, 1982). Attempts at analyzing this conclusion at the anatomical and physiological levels have provided a considerable amount of data, but the interpretation of the findings has often met with the difficult problem of establishing a common principle of commissural organization within an overall theory of the brain. We believe that the 'principle of supplemental complementarity' proposed many years ago by Sperry (1962) is best suited for understanding the available evidence on the anatomy and the physiology of the interhemispheric connections of the cerebral cortex (Berlucchi, Tassinari and Antonini, 1986). In brief, the principle states that the commissural connections are organized in such a way as to allow the activity of each hemisphere to be supplemented with different and complementary information about concurrent activities in the other hemisphere. For example, the representation of each hand in the contralateral somesthetic cortex via the specific sensory pathways can be supplemented and complemented by a representation of the other hand transmitted to the same hemisphere via the corpus callosum. This arrangement is both supplementary, because the representations of the two hands add to one another in each hemi-

350

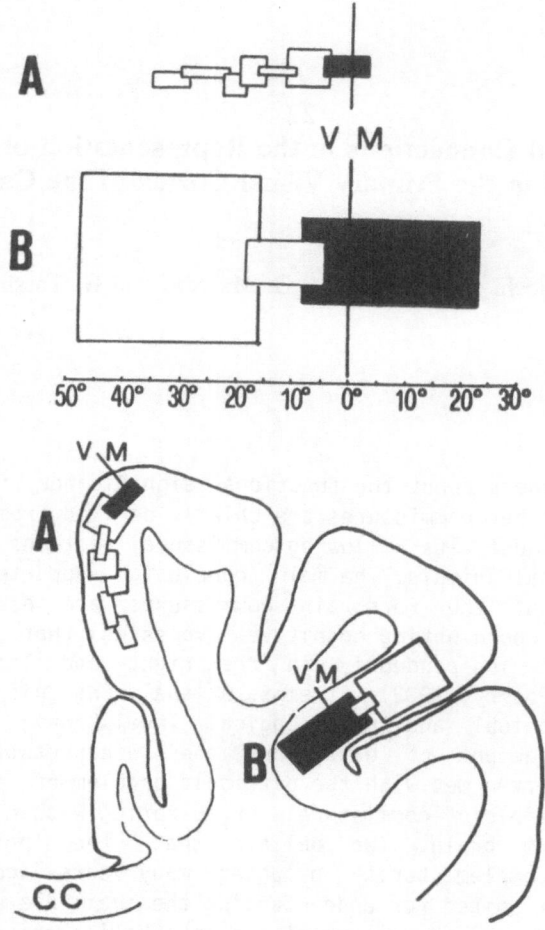

Figure 1. Contribution of the corpus callosum (c.c.) to the visua
field representation in areas 17/18 (A) and in the latera
suprasylvian areas (B). In both cases the corpus callosum may b
involved in extending some receptive fields (black rectangles
across the vertical meridian (V.M.) into the ipsilateral visua
field. Owing to the larger size of receptive fields in the latera
suprasylvian areas, the callosal representation of the ipsilatera
hemifield is much more extended in these areas than in area
17/18.

sphere, and complementary, because this addition can provide a unitary substrate for the control of bimanual activities.

THE PRINCIPLE OF SUPPLEMENTAL COMPLEMENTARITY IN THE VISUAL SYSTEM

For many years we have studied the physiology of the callosal connections of visual cortical areas in cats and have found that their organization in general follows the principle of supplemental complementarity (Berlucchi, Tassinari and Antonini, 1986). More specifically, we have found that neurons in visual cortical areas may receive visual information via both intrahemispheric and interhemispheric afferents. As a rule, the intrahemispheric afferents convey information from the half visual field contralateral to the cortex, whereas the interhemispheric afferents convey information from the ipsilateral visual field. The convergence of the two sets of afferents is patterned so that it occurs only for inputs that are matched at the vertical meridian of the visual field.

Figure 1 shows a schematic representation of the visual field in cortical areas which will be helpful for understanding the general premises for most experiments reported here. The visual receptive fields of cortical neurons usually lie in the contralateral visual field, and as one moves in one particular direction in a given cortical area the receptive field position also moves gradually from the contralateral peripheral visual field toward the central vertical meridian. Some receptive fields stop at the meridian, but a few others may extend wholly or partly into the ipsilateral visual field. This ipsilateral extension depends on the size of the receptive fields: In areas 17 and 18, which may be considered to constitute the primary visual cortex of the cat, receptive fields are usually small and the invasion of the ipsilateral visual field is limited to a narrow strip adjoining the vertical meridian. In other visual cortical areas such as the suprasylvian visual areas, where receptive fields are much larger, some receptive fields may reach the far periphery of the ipsilateral visual field.

The important point, however, is that in both cases the callosal connections of these cortical areas may be involved in providing cortical neurons with information from the ipsilateral visual field which is spatially continuous with that conveyed to the same neurons from the contralateral visual field via intrahemispheric pathways. Further, the continuous receptive fields built up by the appropriate convergence of intrahemispheric and interhemispheric inputs are homogeneous with respect to orientational and directional selectivity, suggesting that the two sets of inputs reaching the same cortical neurons are not only related to contiguous visual field regions on opposite sides of the vertical midline, but also matched for sensitivity to most characteristics of the visual stimulus. If the interhemispheric connections transmitted visual information from regions away from the vertical meridian, or if they converged with intrahemispheric inputs endowed with different response properties, the cortical neurons receiving such mismatched inputs would have receptive fields divided into disjoint portions, one in one hemifield and the other in the opposite hemifield, and their differential sensitivity to the orientation and direction of the visual stimulus would not be the same throughout their receptive field. Further, any mismatching between the ipsilateral and contralateral components would be incompatible with a continuous representation of the visual field over the cortex.

EXPERIMENTAL APPROACHES TO THE PHYSIOLOGY OF VISUAL CALLOSAL CONNECTIONS

Physiological studies of the organization of the callosal connections in the visual system have generally followed two approaches. One approach employs the split-chiasm preparation in which a midsagittal section of the optic chiasm restricts the output of each retina to the ipsilateral hemisphere. Responses of cortical neurons in one hemisphere to photic stimulation of the eye on the same side are mediated by the extant intrahemispheric corticopetal fibers of the visual pathway. However the pathway mediating responses of the same neurons to stimulation of the opposite eye, when present, must involve an interhemispheric component.

This component is the corpus callosum, since the cortical neuronal responses to stimulation of the opposite eye, but not those to stimulation of the ipsilateral eye, are suppressed by a callosal section. Following the initial experiment of Berlucchi and Rizzolatti (1968), this approach has been used repeatedly with good success for analyzing the tranmission of visual information to various cortical areas by the corpus callosum in the cat (Lepore and Guillemot, 1982; Antonini, Berlucchi and Lepore, 1983; Antonini, Di Stefano, Minciacchi and Tassinari, 1985; Cynader, Garner, Dobbins, Lepore and Guillemot, 1986; Lepore, Ptito and Guillemot, 1986; Maffei, Berardi and Bisti, 1986). The results have generally confirmed the view that in spite of the complete elimination of the crossed optic fibers by the midsagittal division of the optic chiasm all visual cortical areas, and even a subcortical visual center such as the superior colliculus (Antonini, Berlucchi and Sprague, 1978; Antonini, Berlucchi, Marzi and Sprague, 1979), can still receive visual information from the contralateral eye via specific callosal connections and combine it with direct visual information from the ipsilateral eye. Further, these binocular interactions seen in physiological experiments on split-chiasm cats have suggested that in principle the corpus callosum can indeed have a role in building up cross-midline receptive fields like those shown in Fig.1.

These possibilities for binocular interactions can explain why various aspects of binocular vision, such as interocular transfer and equivalence (Myers, 1956; Berlucchi and Marzi, 1982) and binocular stereopsis (Lepore, Ptito and Lassonde, 1986), are wholly or partly preserved in split-chiasm cats notwithstanding the surgical separation of the inputs from the two retinae. Since the evidence on split-chiasm cats has been reviewed recently (Berlucchi, Antonini and Tassinari, 1986; Cynader, Gardner, Dobbins, Lepore and Guillemot, 1986; Lepore, Ptito and Guillemot, 1986; Berlucchi and Antonini, in press), this approach to the functional analysis of callosal transmission of visual information will not be considered further.

The second approach to this problem uses animals with intact optic pathways and examines the characteristics of visual field maps in cortical areas. In animals with frontal eyes and partially crossed optic nerves, the connections of the primary visual pathways are such that the right and left visual hemifields are mapped in a crossed fashion onto the cerebral hemispheres. Thus, each visual cortical area in each hemisphere usually contains a more or less complete map of the opposite half-field, but as we have mentioned before such maps can also include an ipsilateral field component. The occurrence of such ipsilateral representations suggests the participation of the corpus callosum to these cortical maps in accordance with Sperry's principle of supplemental complementarity, and this possibility can be investigated by examining the amount of ipsilateral visual field represented in cortical areas before and after a callosal section.

Recording from the crown of the middle suprasylvian gyrus in the cat Dow and Dubner (1971) reported neurons that responded to visual targets moving across large portions of both contralateral and ipsilateral visual fields. Cutting the corpus callosum or destroying the cortex in the opposite hemisphere abolished the response in the ipsilateral but not in the contralateral field. Similarly, Rocha-Miranda, Bender, Gross and Mishkin (1975) and

Gross, Bender and Mishkin (1977) found that the large bilateral receptive fields typically found in the inferotemporal cortex of the macaque monkey became restricted to the contralateral visual field following a section of the splenium of the corpus callosum and the anterior commissure. Marzi, Antonini, Di Stefano and Legg (1982) recorded from single neurons in the postero-medial lateral suprasylvian visual area in the cat and confirmed previous evidence on the extensive binocular representation of the ipsilateral visual field in this area. By sectioning the corpus callosum they were also able to show that this ipsilateral representation was drastically reduced or suppressed, whereas there was no change in the representation of the contralateral visual field. The reduced ipsilateral representation which survived after sectioning the corpus callosum was limited to the eye contralateral to the cortex, and could be accounted for by the crossed contingent of fibers from the temporal retina which is known to exist in the cat (Stone and Fukuda, 1974; Kirk, Levick, Cleland and Wassle, 1976).

These studies clearly show that cortical neurons with large receptive fields crossing the midline of the visual field do indeed receive the ipsilateral components of their receptive fields via the corpus callosum and, in the monkey, via the anterior commissure as well. Yet it remains doubtful whether this is also true of cortical neurons with small receptive fields overlapping the vertical meridian, such as those in areas 17 and 18. It is true that by sectioning one optic tract (Choudhury, Whitteridge and Wilson, 1965) or the optic chiasm (Berlucchi and Rizzolatti, 1968; see also above) one can show that visual information transmitted by the corpus callosum is effective in discharging neurons in areas 17 and 18, but there is no direct evidence in neurally intact animals that the small ipsilateral receptive fields or receptive field components in these areas owe their existence to the corpus callosum.

In fact the only experimenter who compared the ipsilateral visual field representation in areas 17 and 18 of the cat before and after a callosal section reported no difference (Leicester, 1968), and it has been suggested that the callosal connections of these areas are involved in visual functions quite unrelated to the extension of receptive fields into the ipsilateral visual field. One of these functions could be the establishment and the maintenance of binocular interactions for neurons with receptive fields in the vicinity of the vertical meridian (see e.g. Blakemore, Diao, Pu, Wang and Xiao, 1983; Payne, 1986).

Recently we have readdressed this problem by recording from

single neurons in the transition zone between areas 17 and 18 of cats lightly anesthetized with sodium thiopental and paralyzed with pancuronii bromidium. We studied 201 binocular neurons and 27 monocular neurons in 12 neurally intact cats, and 132 binocular neurons and 10 monocular neurons in 8 cats which had undergone a surgical section of the posterior two-thirds of the corpus callosum. We wanted to learn if the receptive fields of some of the neurons in the intact cats lay partially or wholly in the ipsilateral field, and, in the affirmative case, if this required, to some extent at least, the integrity of the corpus callosum.

The line of separation between the ipsilateral and contralateral visual fields is the vertical or zero meridian, which runs through the fixation point in the area centralis of the retina (Nikara, Bishop and Pettigrew, 1968; Cooper and Pettigrew, 1979). Since the area centralis is difficult or impossible to locate ophthalmoscopically in the cat retina, the position of the vertical meridian in the visual field cannot be assessed directly by a projection technique. Instead, it can be determined indirectly by two independent methods, as we have done in the present experiments.

The first method is based on the assumptions that the centers of the two monocular receptive fields of the average binocular cortical neuron are exactly superimposed when the vertical meridians of the two eyes are also exactly superimposed, and that the two eyes are perfectly symmetrical. In practice, the receptive fields of several binocular neurons are mapped on a tangent screen at a fixed distance from the cat's eyes along with the projections of the two optic disks. The perpendicular distance between the zero meridian and the optic disk center in each eye is equated with the half algebraic sum of the distance between the two optic disk centers and the average (median) separation between the centers of pairs of receptive fields. The latter measure is taken as negative when the position of the two receptive fields relative to the eyes is crossed (i.e. the receptive field in the right eye is to the left of the receptive field in the left eye), and as positive in the reverse case (Nikara, Bishop and Pettigrew, 1968).

The other method requires that one maps receptive fields of cortical neurons in both hemispheres through one eye and finds out the width of overlap between receptive fields from different hemispheres (Hubel and Wiesel, 1967; Leicester, 1968). Obviously recordings must be made from cortical regions of the two hemispheres where receptive fields have the same altitude, since an overlap between two partial hemispheric representations of the visual field at different altitudes would be physically impossible. The vertical meridian is then traced through the centerpoint of the area of overlap and the procedure is repeated for the other eye.

It is crucial with both methods to check for changes in eye position during the recording session. This was obtained in our experiments by plotting repeatedly the optic disks and by making appropriate corrections in the receptive field positions when an occasional eye drift was detected. Such drifts were in all cases rare and quite small. Further we consistently recorded from cortical areas representing visual field altitudes within 10 degrees of the horizontal meridian in each hemisphere in the overall group , including intact and callosotomized cats, the median altitudinal separation between receptive field centers of the two hemispheres being at most 3.5 degrees in all cases.

In both intact and callosotomized cats the distance between the optic disk center and the vertical meridian estimated with the first method was highly correlated with that estimated with the second method. The coefficient of correlation was .93 for the intact cats (p<.005) and .83 for the callosotomized cats (p=.005) The strong concordance betwen the two sets of estimates, which were arrived at with independent methods, satisfied ourselves that our assessments of the position of the vertical meridian were reliable.

In the callosotomized cats the time between the callosal section and the first electrophysiological recording ranged from one hour to several weeks, and the results were not affected by the length of this interval. The portion of the corpus callosum which was sectioned is known to contain the interhemispheric connections of all visual cortical areas.

The results can be summarized as follows.
1) We found with the first method that in all cats with an intact corpus callosum a majority of receptive fields were located partly or totally within the ipsilateral hemifield. This was not unexpected, since we limited our recordings to the 17-18 border which is connected with the vertical meridian region of the visual field. In all, 67.9% of 412 receptive fields occupied portions of the ipsilateral visual field, 9% being entirely ipsilateral. The mean across cats of the mean receptive field extension into the ipsilateral visual field, calculated in each cat as the mean distance between the vertical meridian and the farther receptive field border in the ipsilateral field, was 1.5 degrees. The mean across cats of the maximal extension into the ipsilateral visual field, calculated in each cat as the distance between the vertical meridian and the farther border of the receptive field which

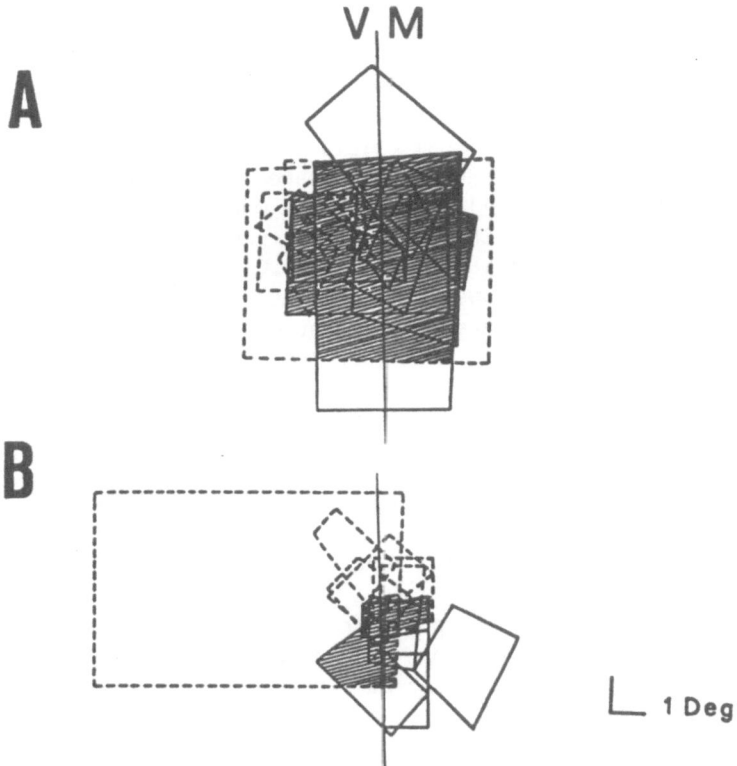

Figure 2. Receptive fields of the left eye in the left hemisphere (continuous lines) and in the right hemisphere (dashed lines) of the same cat before (A) and after (B) a posterior callosal section. All receptive fields are referred to the center of the optic disk of the left eye and the width of the zone of the visual field in which they are superimposed corresponds to the horizontal visual angle represented in both hemispheres (overlap). The vertical line passing through the midpoint of the overlap coincides with the vertical meridian of the visual field. Note that callosotomy affects the width of the overlap symmetrically on both sides of the vertical meridian.

extended most into the ipsilateral visual field, was 3.7 degrees.
2) In fair agreement with the results obtained with the first
method, the average maximal width of the overlap between the
visual field representations of the two hemispheres was 5.8
degrees, giving a mean ipsilateral representation of 2.9 degrees
for each hemisphere.
3) After the callosal section, the percentage of fields extending
into the ipsilateral visual field was 43.5% of 255 receptive
fields, a significant reduction compared with the 67.9% value of
the intact cats (p<.001). Further, the mean ipsilateral extension
was .9 degrees, and the .6 degrees reduction relative to the
intact cats was significant at the p=.03 level; and the maximal
ipsilateral extension was 2.1 degrees, with a significant 1.8
degrees reduction compared to the intact cats (p=.01).
4) The average maximal width of the overlap between the two
hemispheric visual field representations in the callosotomized
cats was 3.6 degrees, suggesting that each hemisphere had an
ipsilateral representation of 1.8 degrees, in good accord with the
measure obtained with the other method. The 2.2 degrees reduction
in the width of overlap relative to the intact cats approached
significance (p=.07).

Some of these findings are illustrated in graphic form in
Figs.2 and 3.

The results on the presence of the ipsilateral receptive
fields in the transition zone between areas 17 and 18 of the cat
are in keeping with those of many previous similar studies (see
e.g. Leicester, 1968; Blakemore, 1969; Tusa, Palmer and
Rosenquist, 1978; Tusa, Rosenquist and Palmer, 1979; Harvey, 1980;
Albus and Beckmann, 1980; Whitteridge and Clarke, 1982). All these
studies indicate that ipsilateral receptive field centers do not
extend beyond 2 or 3 degrees from the vertical meridian, with the
exception of the study by Whitteridge and Clarke which reports
ipsilateral receptive fields centered as much as 12 degrees from
the vertical meridian. In our present experiment we have been
unable to find such far ipsilateral receptive fields, and in view
of our extensive exploration of the cortical region studied by
Whitteridge and Clarke (1982) it seems unlikely that we have
completely missed their band of strongly ipsilateral
representation. This discrepancy therefore remains unresolved.

On the other hand, our present results on the extent of
ipsilateral visual field represented at the boundary between areas
17 and 18 are consistent with those on the overlap between the
visual field representations in the two hemispheres which we
independently demonstrated in the same experiment, as well as with

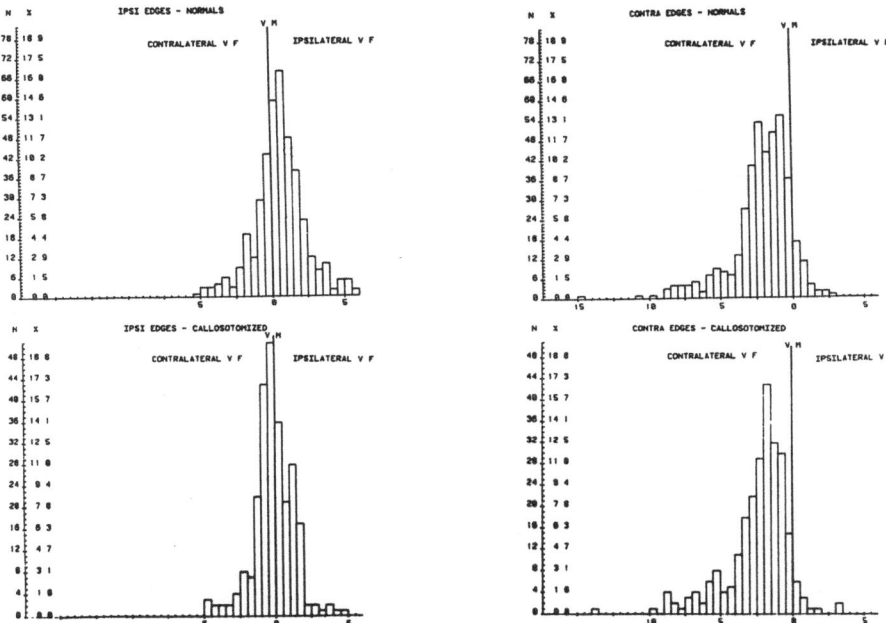

Figure 3. The histograms on the left show the overall distribution of the ipsilateral receptive field borders (left receptive field borders in the left hemisphere and right receptive field borders in the right hemisphere). The histograms on the right show the distribution of the contralateral borders of the same receptive fields (right borders of receptive fields in the left hemisphere and left borders of receptive fields in the right hemisphere). In the top row are represented the data for intact cats, in the bottom row those for callosotomized cats. The position of the vertical meridian (0°) was determined in each cat using the median horizontal separation of receptive field centers of binocular neurons. Callosotomy shifts the peak of the distribution of the ipsilateral fields borders toward the contralateral visual field, leaving the distribution of the contralateral visual field borders unaffected.

previous evidence on such overlap obtained by Hubel and Wiesel
(1967) and Leicester (1968).

The main point of the results is that the ipsilateral field
representation was significantly reduced, both in extent and in
number of neurons contributing to it, by a posterior callosal
section. Only one previous experiment examined this possibility
(Leicester, 1968), and the conclusion was that a callosal section
had no effect on the bilateral representation of the vertical
meridian region in areas 17 and 18. However, this conclusion was
based on a few receptive fields recorded in a single cat after a
callosal section, a piece of evidence which on the basis of our
findings must be considered insufficient for reaching such a
conclusion.

We agree with Leicester (1968) that the corpus callosum is not
necessary for the existence of ipsilateral receptive fields in
areas 17 and 18, as well as for the overlap between the visual
field representations in the same areas of the two hemispheres.
The pattern of crossing of the optic nerve fibers in the chiasm is
apt to explain these properties of visual field mapping in areas
17 and 18, because there is a retinal strip around the vertical
meridian which projects to both hemispheres. More specifically,
since in our results the ipsilateral representation was the same
in the ipsilateral and contralateral eyes, both before and after
the callosal section, we can suggest that it was mainly due to the
pattern of decussation of the X-fibers of the retina which is
symmetrical about the vertical meridian (Stone and Fukuda, 1974;
Kirk, Levick, Cleland and Wassle, 1976).

However it also appears that callosal connections contribute
to both ipsilateral representation and hemispheric overlap by
increasing the number of cortical neurons involved in these
functions, and by slightly but significantly expanding the amount
of visual field represented in areas 17 and 18 of the two
hemispheres. The possible functional significance of this callosal
participation to the organization of areas 17 and 18 must be
discussed in the context of findings of other studies.

THE ROLE OF THE CORPUS CALLOSUM IN BINOCULAR INTERACTION IN AREAS
17 AND 18

In relating binocular stereopsis along the midsagittal plane
to receptive field disparities in the visual cortex, Blakemore
(1969) reasoned that cortical neurons which code for large
binocular disparities in the central visual field must have a
receptive field in one eye and one visual hemifield coupled with a

receptive field in the other eye and the other hemifield. In principle this kind of binocular interaction could result from the convergence on a single cortical neuron of a intrahemispheric input from one eye and the contralateral visual field, and an interhemispheric (callosal) input from the other eye and the ipsilateral visual field. Neurons receiving these combined inputs should obviously become monocular following a callosal section.

The effects of a callosal disconnection on binocular interaction in areas 17 and 18 are controversial. The percentage of binocular neurons at the border between areas 17 and 18 was observed to fall significantly after an ablation of the contralateral corresponding cortical regions in some experiments (Dreher and Cottee, 1975; Blakemore, Diao, Pu, Wang and Xiao, 1983) but not in others (Cynader, Gardner, Dobbins, Lepore and Guillemot, 1986). Studies reviewed by Payne (1986) indicate that a callosal section causes a permanent reduction (from about 80% to about 40%) in the percentage of binocular neurons in portions of areas 17-18 that represent visual field regions within 4 degrees of the vertical meridian. On the contrary, Minciacchi and Antonini (1984) and Elberger and Smith (1985) reported that section of the corpus callosum in adult cats had no effect on binocular interactions in areas 17 and 18, although in the latter study a reduction in the percentage of binocular neurons could be observed if the callosal section was performed before 19 days of age.

The data of the experiment we have described in the previous section lend themselves to a re-examination of this issue. Considering neurons with receptive field centers within 5 degrees of the vertical meridian, we found that neurons that could be activated through either eye were 189 out of 216 (87.5%) in intact cats, and 113 out of 122 (92.6%) in callosotomized cats. Clearly these results run counter the hypothesis that the corpus callosum has a role in binocular interaction in areas 17 and 18.

This finding does not mean by itself that callosal connections of areas 17 and 18 are not involved in binocular stereopsis. The callosal input might participate in this function by generating horizontal receptive field disparities in the midline region for neurons that however receive binocular information via intrahemispheric afferents. We looked into this possibility by analyzing receptive field correspondence in binocular neurons of both intact and callosotomized cats. This was done by superimposing the estimated vertical meridians of the two eyes and measuring the horizontal disparity between the centers of each pair of receptive fields of binocular neurons. Crossed and uncrossed disparities were symmetrical in both intact and callosotomized cats. The mean crossed disparity was 0.88 degrees (S.D.=0.65) in the intact cats and 0.97 degrees (S.D.=0.81) in the callosotomized cats; the mean uncrossed disparity was 0.88 degrees (S.D.=0.9) in the intact cats and 0.88 degrees (S.D.=0.78) in the callosotomized cats.

Thus we can argue that the pattern of horizontal disparities of receptive fields in the central visual field is hardly modified by a callosal section. Yet, the significance of this finding for the interpretation of the importance of the corpus callosum in binocular stereopsis is unclear, because the distribution of receptive field disparities does not necessarily reflect stereoscopic functions (see e.g.von der Heydt, Adorjani, Hanny and Baumgartner, 1978). However, it should be mentioned that in behavioral studies on binocular depth perception (Timney, Elberger and Vandewater, 1985) and binocular stereoperception (Lepore, Ptito and Lassonde, 1986) in cats severing the corpus callosum had no effect on the animal's performance.

CONCLUSION

If the callosal connections of the primary visual cortex are not involved in binocular stereopsis, then their function could be that of linking the two halves of the visual field in perception. An object in the center of the visual field is perceived as a unified whole in spite of the fact that its right and left halves are projected to different hemispheres. Sperry (1952) argued that to this unitary perception must correspond a brain pattern of activity which is also unitary, and we maintain that the interplay of intrahemispheric and interhemispheric visual inputs which occurs in the visual cortex is important in allowing unitary patterns of cerebral organization to include neurons in both hemispheres (Berlucchi and Antonini, in press). However, this proposition is difficult to test directly, and it is disappointing to realize how ignorant we are about the relations between the striking disconnection syndrome shown in visual perception by callosotomized patients (Sperry, 1982) and the physiology of visual callosal connections. It is obvious that future experiments in this area need guidance from new insights into the problem.

REFERENCES

Albus, K. and Beckmann, R. (1980). Second and third visual areas of the cat: interindividual variability in retinotopic arrangement and cortical location. J.Physiol. (London) 299, 247-276.

Antonini, A., Berlucchi, G. and Lepore, F. (1983). Physiological organization of callosal connections of a visual lateral suprasylvian area in the cat. J.Neurophysiol. 49, 902-921.

Antonini A., Berlucchi, G., Marzi, C.A. and Sprague, J.M. (1979). Importance of corpus callosum for visual receptive fields of single neurons in cat superior colliculus. J.Neurophysiol. 42, 137-152.

Antonini A., Berlucchi, G. and Sprague, J.M. (1978). Indirect, across-the-midline retinotectal projections and representation of ipsilateral visual field in superior colliculus of the cat. J.Neurophysiol. 41, 285-304.

Antonini, A., Di Stefano, M., Minciacchi, D. and Tassinari, G. (1985). Interhemispheric influences on area 19 in the cat. Exp.Brain Res. 59, 179-186.

Berlucchi, G. and Antonini, A. (in press). The role of the corpus callosum in the representation of the visual field in cortical areas. In Brain Circuits and Functions of the Mind. (ed. C.Trevarthen) Cambridge University Press, Cambridge.

Berlucchi, G. and Marzi, C.A. (1982). Interocular and interhemispheric transfer of visual discrimination in the cat. In Analysis of Visual Behavior (eds. D.J.Ingle, M.A.Goodale and R.J.W. Mansfield) MIT Press, Cambridge MA.

Berlucchi, G. and Rizzolatti, G. (1968). Binocularly driven neurons in visual cortex of split-chiasm cats. Science 159, 308-310.

Berlucchi, G., Tassinari, G. and Antonini, A. (1986). The organization of the callosal connections according to Sperry's principle of supplemental complementarity. In Two Hemispheres - One Brain (eds. F.Lepore, M.Ptito and H.H.Jasper) Alan R.Liss, New York.

Blakemore, C. (1969). Binocular depth discrimination and the nasotemporal division. J.Physiol. (London) 205, 471-497.

Blakemore, C., Diao, Y., Pu, M., Wang, Y. and Xiao, Y. (1983). Possible functions of the interhemispheric connexions between visual cortical areas in the cat. J.Physiol. (London) 337, 331-349.

Choudhury, B.P., Whitteridge, D. and Wilson, M.E. (1965). The function of the callosal connexions of the visual cortex. Quart.J.Exp.Physiol. 50, 214-219.

Cooper, M.L. and Pettigrew, J.D. (1979). A neurophysiological determination of the vertical horopter in the cat and owl. J.Comp.Neurol. 184, 1-26.

Cynader, M., Gardner, J., Dobbins, A., Lepore, F. and Guillemot, J.P. (1986). Interhemispheric communication and binocular vision: Functional and developmental aspects. In Two Hemispheres - One Brain (eds. F.Lepore, M.Ptito and H.H.Jasper) Alan R.Liss, New York.

Dow, B.M. and Dubner, R. (1971). Single-unit responses to moving stimuli in middle suprasylvian gyrus in the cat. J.Neurophysiol. 34, 47-55.

Dreher, B. and Cottee, L.J. (1975). Visual receptive-field properties of cells in area 18 of cat's cerebral cortex before and after acute lesions in area 17. J.Neurophysiol. 38, 735-750.

Elberger, A. and Smith, E.L. (1985). The critical period of corpus callosum section to affect cortical binocularity. Exp.Brain Res. 57, 213-223.

Gross, C.G., Bender, D.B. and Mishkin, M. (1977). Contribution of the corpus callosum and the anterior commissure to visual activation of inferior temporal neurons. Brain Res. 131, 227-239.

Harvey, A.R. (1980). A physiological analysis of subcortical and commissural projections of areas 17 and 18 in the cat. J.Physiol. (London) 302, 507-534.

Hubel, D.H. and Wiesel, T.N. (1967). Cortical and callosal connections concerned with the vertical meridian of the visual field. J.Neurophysiol. 30, 1561-1573.

Kirk, D.L., Levick W.R., Cleland, B.G. and Wassle, H. (1976) Crossed and uncrossed representation of the visual field by brisk-sustained and brisk-transient cat retinal ganglion cells. Vision Res. 16, 225-231.

Leicester, J. (1968). Projection of the visual vertical meridian to cerebral cortex of the cat. J.Neurophysiol. 31, 371-382.

Lepore, F. and Guillemot, J.P. (1982). Visual receptive field properties of cells innervated through the corpus callosum. Exp.Brain Res. 46, 413-424.

Lepore, F., Ptito, M. and Guillemot J.P. (1986). The role of the corpus callosum in midline fusion. In Two Hemispheres - One Brain (eds. F.Lepore, M.Ptito and H.H.Jasper) Alan R.Liss, New York.

Lepore, F., Ptito, M. and Lassonde, M. (1986). Stereoperception in cats following section of the corpus callosum and/or the optic chiasma. Exp.Brain Res. 61, 258-264.

Maffei, L., Berardi, N. and Bisti, S. (1986). Interocular transfer of adaptation after effect in neurons of area 17 and 18 of split chiasm cats. J.Neurophysiol. 55, 966-976.

Marzi, C.A., Antonini, A., Di Sefano, M. and Legg C.R. (1982). The contribution of the corpus callosum to receptive fields in the lateral suprasylvian visual area of the cat. Behav.Brain Res. 4, 155-176.

Minciacchi, D. and Antonini, A. (1984). Binocularity in the visual cortex of the adult cat does not depend on the integrity of the corpus callosum. Behav.Brain Res. 13, 183-192.

Myers, R.E. (1956). Function of corpus callosum in interocular transfer. Brain 79, 358-363.

Nikara, T., Bishop, P.O. and Pettigrew, J.D. (1968). Analysis of retinal correspondence by studying receptive fields of binocular single units in cat striate cortex. Exp.Brain Res. 6, 353-372.

Payne, B.R. (1986). Role of callosal cells in the functional organization of cat striate cortex. In Two Hemispheres - One Brain (eds. F.Lepore, M.Ptito and H.H.Jasper) Alan R.Liss, New York.

Rocha-Miranda C.E., Bender, D.B., Gross, C.G. and Mishkin, M. (1975). Visual activation of neurons in inferotemporal cortex depends on striate cortex and forebrain commissures. J.Neurophysiol. 38, 475-491.

Sperry, R.W. (1952). Neurology and the mind-brain problem. Am.Sci. 40, 291-312.

Sperry, R.W. (1962). Some general aspect of interhemispheric integration. In Interhemispheric Relations and Cerebral Dominance (ed. V.B.Mountcastle) The Johns Hopkins Press, Baltimore.

Sperry, R.W. (1982). Some effects of disconnecting the cerebral hemispheres. Science 217, 1223-1226.

Stone, J. and Fukuda, Y. (1974). The naso-temporal division of the cat's retina re-examined in terms of Y-, X- and W-cells. J.Comp.Neurol. 155, 377 394.

Timney, B., Elberger, A. and Vandewater, M.L. (1985). Binocular depth perception in the cat following early corpus callosum section. Exp.Brain Res. 60, 19-26.

Tusa, R.J., Palmer, L. and Rosenquist, A.C. (1978). The retinotopic organization of area 17 (striate cortex) in the cat. J.Comp.Neurol. 177, 213-236.

Tusa, R.J., Rosenquist, A,C. and Palmer, L. (1979). Retinotopic organization of areas 18 and 19 in the cat. J.Comp.Neurol. 185, 657-678.

Von der Heydt, R., Adorjani, C., Hanny, P. and Baumgartner, G. (1978). Disparity sensitivity and receptive field incongruity of units in cat striate cortex. Exp.Brain Res. 31, 523-545.

Whitteridge, D. and Clarke, P.G.H. (1982). Ipsilateral visual field represented in the cat's visual cortex. Neurosci. 7, 1855-1860.

23
Reversed Brain Anatomical Asymmetries in Schizophrenia: A Search for Contributing Variables

M.S. Myslobodsky and D.R. Weinberger

The anatomical asymmetry of the brain known today under the label of 'typical' asymmetry, i.e. the predominance of the left parietooccipital area, was recognized in the beginning of this century and was related to a 'specialization of the two hemispheres' (Smith, 1907). The possibility that cognitive disturbances in the mentally ill might be associated with some sort of 'atypical' brain anatomy was entertained in the early twenties. Some hemisphere asymmetries in psychotic patients, including patients with dementia praecox, had been observed (Fortig, 1922; Inglessis, 1925), but they seemed to be within the range of the non-psychotic population.

More recently, an interest in comparing the two hemispheres was kindled by a theory attributing aberrant behaviors to abnormalities of the brain laterality balance (Flor-Henry, 1969). After LeMay (1976) pioneered the quantification of brain asymmetries in vivo, using CT scanners, an opportunity has arisen for a modern reassessement of brain asymmetries in schizophrenia (Table I). In 1979, Luchins, Weinberger & Wyatt reported that a subgroup of schizophrenic patients had predominantly reversed hemispheric asymmetry, i.e the predominance of the right occiput over its counterpart on the left side. This 'atypical' hemisphere asymmetry was encountered in the normal population, but seemed to be related to left-handedness (LeMay, 1976; LeMay & Kido, 1978). Soon, reversed asymmetries were replicated in a different sample of schizophrenic patients by Luchins, Weinberger, & Wyatt (1982) and by other groups (Table 2). And yet, within a single year, several studies were published reporting conflicting or inconsistent data (Table 2).

Could it be that in spite of this disagreement, all sides to the dispute are right? If so, how? Among the most likely explanations for these discordant results is the heterogeneity of schizophrenic patients along clinical lines, premorbid adjustment,

Table 1. Parietooccipital width asymmetries ("normative" data)

Reference	N	Age (Yrs)	Sex	Population	Distribution (%) L	E	R
LeMay	60[a]	?[c]	M	patients	67	28	5
(1977)	70	?	F		65	22	13
Chui	22-28[a]	44.2	M	patients	39	32	29
(1980)	22		F		32	59	9
Nyback (1982)	46	27	M,F	n. volunt.[c]	54	20	26
Luchins (1982)	100	30	?	med & neurol patients	46	49	5
Jernigan (1982)	32	22-60	M	n. volunt.	75	3	22
Tsai (1982)	18	3-18	M,F	n. popul.	22[d] (28)	72 (50)	6 (22)
Andreasen (1982)	40	matched to x̄ 29.9	M,F	neurol. patients	25	68	8
Luchins (1983)	62	?	?	headache controls	55	31	15
Koff	64-69	16-80	M	'neurol. normal'	64	23	13
(1986)	61-63		F		53	23	25

[a] Only the data for right-handers were isolated in this and the rest of the studies.
[b] "?" denotes the missing data.
[c] No information on handedness.
[d] Measurements according to LeMay & Kido (1978) vs. Chui & Damasio (1980) (in brackets).

the progress of the disorder, response to treatment, evidence of brain anatomical pathology, etc. Another, and a more prudent possibility, attributes these discrepancies to poorly controlled handedness, eye dominance, sex, age, race, neuroleptic medication, techniques of measurement, and other variables.

Gender was seldom examined as a variable since male/female differences in vulnerability to schizophrenia are typically attributed to social factors (Lewin, 1981; Leventhal, Schuck & Rothstein, 1984). However, indications that brain hemisphere balance in human and even in subhuman animals is sexually dimorphic (McGlone, 1980; Robinson, Becker, Camp et al. 1985) and that younger females may be somewhat protected by estrogens acting as neuroleptic-like substances (Duffy, Vincent, Fleury et al., 1979) invites biological explanations of gender differences in schizophrenia (Seeman, 1982).

Inglessis (1925) was probably the first to mention that brain hemispheres in normal and psychotic patients are more asymmetric in males. The examination of CT scans failed to confirm the gender-related differences in the pattern of asymmetries (LeMay &

Table 2. Occipital Asymmetries in Schizophrenia

Reference	N	Age	Sex	Distribution (%)		
				L	E	R
Luchins (1979)	57	19-49	?[c]	53	25	22
Luchins (1982)	66	30	?	52[d]	30	18
non-atrophic subgroup	45	29	?	38	40	22
Naeser (1981)[a]	16	?	?	50	19	31
Jernigan (1982)	31	33+9	M	84	3	13
Andreasen (1982a,b,)	43	29.9	M,F	37	47	28
Nyback (1982)	46	32[b]	M,F	44	17	39
Luchins (1983)	45	?	?	60	24	16
Tsai (1983)	53	?	M	33	36	31
Present sample	37	18-42	M,F	49.5	43	7.5
				(24)[e]	(56.5)	(19.5)

a: Patients with bilateral prefrontal leukotomy. One patient is left-handed.
b: Older than controls at p <.05.
c: "?" denotes the missing data.
d: Combined results for right- and left-handers.
e: Measurements according to Chui & Damasio (1980) with values corrected for falx deviations in brackets.

Kido, 1978; Chui & Damasio, 1980; Pieniadz & Naeser, 1984; Koff, Naeser, Pieniadz et al., 1986). Consequently, the male/female ratio of the samples is not always specified (Table 1, 2) even though the examination of Table 1 suggests that the typical direction of asymmetry is somewhat more frequently encountered among the males.

Race. Taking note of the fact that blacks, notably black males, may be over-represented among process schizophrenics (Allon, 1971), race may become an additional factor to be considered in brain asymmetry studies. The data that brain asymmetries may be less common among blacks were mentioned by Smith (1907). In line with the mood of his time he has asserted that "In the white Races there seems to have been a greater specialization of the two cerebral hemispheres than in the case of the Negro and in the former the resulting dissimilarity of shape in the cerebral hemispheres produces a cranial asymmetry (Smith, 1907, p. 578)". Strauss, Kosaka & Wada (1983) mentioned one unpublished study dealing with the "differential pattern of asymmetry according to race" but its precise role remained obscure.

Neuroleptic medications and tardive dyskinesia. The association between tardive dyskinesia and radiological brain abnormalities remains uncertain. CT scan abnormalities were described by some workers (Bartel & Themelis, 1983) whereas others reported negative results (Jeste, Wagner, Weinberger et al. 1980).

Handedness. Several writers found an increase in left-handedness in schizophrenia (Wahl, 1976; Gur, 1977; see, however, Taylor, Dalton & Fleming, 1980; Merlin, 1984). According to LeMay & Kido (1978) left-handers have a reversed pattern of petalias and of cerebral width. These two group of findings suggested that either reversed dominance may be associated with an abnormal pattern of brain laterality or it is a result of excessive sinistrality in schizophrenia, or both. Chui & Damasio (1980) and more recently Koff et al. (1986) have refuted this possibility being unable to detect a different pattern of brain asymmetry in normal sinistrals. Moreover, Naeser and associates (Pieniadz & Naeser, 1984; Naeser & Borod, 1986) indicate that radiologically established asymmetries are hardly reliable predictors of handedness and the site or the side of speech representation. There was, however, a significant increase in reversed asymmetries in right-handed females with a family history of sinistrality (Koff et al., 1986), suggesting that the issue is worthy of a re-examination. In fact, most researchers have felt that it is not safe to disregard handedness and, by and large, have tended to deal with right-handed subjects (see, however Luchins et al., 1982).

Age should be specified because mixed cerebral dominance had been found more often in a younger population (Oddy & Lobstein, 1972).

Techniques of measurements are unlikely to be solely responsible for the encountered variance. Luchins & Meltzer (1983), adopting a modified technique of Chui and Damasio (1980) failed to replicate the original finding of Luchins et al. (1979, 1982). However, Tsai, Jakoby, Stewart, and Beisler (1983) using the error-prone method of LeMay (1976) were able to confirm reversed asymmetries in schizophrenia. Paradoxically, the same technique in the hands of Jernigan, Zatz, Moses, and Cardinello (1982) revealed a higher incidence of reversed asymmetries in normal volunteers rather than in schizophrenic patients.

Nevertheless, problematic methodology of CT quantification undeniably plagues numerous research efforts.

Crossed sighting dominance. An excessive number of patients with crossed dominance has been reported in schizophrenia (Piran et al., 1982; see, however, Merrin, 1984). There is a possibility that individuals with reversed asymmetries are recruited from a population with crossed ocular dominance (Myslobodsky, 1985), but it was impossible to isolate this factor inasmuch as the items for sighting dominance have often been used in some inventories to assess handedness.

What it is that makes sighting dominance worthy of consideration when dealing with brain laterality is not readily apparent. One possibility suggested by Piran et al. (1982) is based on the assumption that left hemisphere functions are impaired in schizophrenia (Flor-Henry 1969), an event which presumably takes place in early ontogeny. Consequently, the right hemisphere would have a larger share of control over otherwise 'left hemisphere functions'. Assuming further that sighting dominance is controlled contralaterally one might expect that the crossed pathway from the left nasal retina (in the right-handers) would become in such cases a preferential sighting channel. This implies that left hemisphere deficiency might have caused a relative functional advantage (and a relative increase in size?) of the visual centers in the right hemisphere. The latter is expected to be more noticeable among the right-handers with left sighting eyes.

Unfortunately, no point of this suggestion rests on solid evidence. Even if one assumes that the left hemisphere is 'deficient' in schizophrenia, the locus of this deficit and its relation to visual centers has yet to be specified (Myslobodsky, 1983, for review). Moreover, it is not clear if there is some coupling between ocular and cerebral dominance (Porac & Coren, 1981, for review) and if such coupling does exist, whether the dominant eye is controlled contralaterally. One may, in fact, entertain a possibility of the homolaterally controlled dominant eye. Some tests of sighting dominance require to align the near and remote targets by placing the former off the Vieth-Muller horopter. Under such conditions, disparate bi-temporal retinal points would be stimulated thereby causing physiological diplopia. The latter is a fleeting experience because normally only one image is selected. According to Ogle (1962), "The eye whose half-image tends to be suppressed in physiological diplopia is the non-dominant eye (p. 410)". Put a different way, the dominant eye is the eye whose temporal retina is preferred or whose crossed disparity detectors are maximally sensitive to stimuli. Apparently, partial homolateral control of the dominant eye is not unlikely.

Finally, crossed dominance may not be associated with psychopathology, let alone schizophrenia. A recent study of Merrin (1984) failed to replicate increased frequency of crossed dominance in schizophrenia. And yet, this hypothesis offers a strong testable prediction which makes schizophrenia a useful arena where the interrelations between brain asymmetries and sighting dominance can be examined.

METHODS and RESULTS

This study was conducted with the volunteer inpatients at the NIMH research wards of St.Elizabeths Hospital, Washington, DC who met DSM-III diagnostic criteria for schizophrenia. Only strongly right-handed subjects were selected. Miles (1929) ABC test was used for establishing the sighting dominance. On the basis of this test, subjects with the right dominant eye were assigned to the homolateral dominance subsample (HD), whereas patients with the

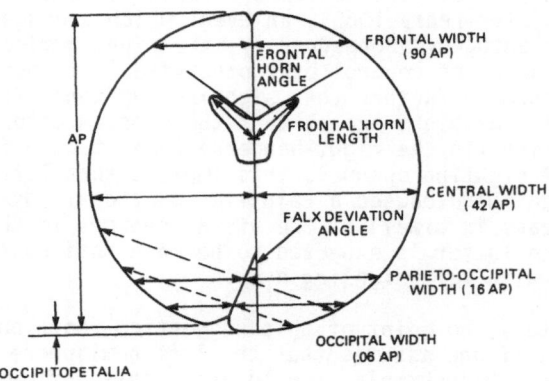

Fig. 1. Illustration of measurements of hemispheres as well as
frontal horns length and angles. AP-maximal antero-posterior axis
drawn through anterior midline landmarks. Perpendiculars at .065,
.16, .42, and .90 along the midline represent occipital,
parietooccipital, central, and frontal width-levels, respectively.
Interrupted lines drawn orthogonally to the falx denote corrected
measurements. The central width was determined at the level of the
pineal gland when it was visualized (LeMay and Kido 1978) or at .42
of the maximal hemisphere distance. The frontal horn length (FHL)
was determined by measuring the distance from the midline through
the middle of the horns to their tips. Other auxiliary measures and
their origin are shown in Table I.

left sighting eye, to the crossed dominance subsample (CD). The two
groups were matched for age, race, and cephalic index but there was
a noticeable preponderance of the males and whites in both
subsamples (Table 4). The CT measurements were conducted on the
uppermost slices of the low ventricular series and the lower slices
of the high ventricular series as shown in Fig. 1 and Table 3.

Table 4 summarizes major brain asymmetries for HD and CD
patients. It shows virtually no differences between the two, with
the exception of a significant group differences on corrected
occipital width and a greater left occipitopetalia in HD patients.
It should be noted, however, that a systematic falx deviation may
contribute to the detection of occipitopetalia due to an accepted
practice of drawing the midline though the rostral landmarks. When
the midline is drawn through the caudal landmarks, occipitopetalia
is obscured whereas the asymmetry of frontopetalia is more
apparent. In more than 50 % of cases there was a clear deviation of
the falx. This may explain the less consistent reports of
frontopetalia and the conviction that radiologically, frontal lobe
asymmetries are less demonstrable than occipital asymmetries
(Weinberger, Luchins, Morihisa et al. 1982).

Table 3. Variables examined in the study

Measurements	References
1 Occipitofrontal length (occipito- & frontopetalia)	LeMay & Kido (1978)
2 Cepahlic index	Montagu (1960)
3 Falx deviation	Fig. 1, 2
4 Occipital (O) width	Pieniadz & Naeser (1984)
5 Corrected O width	Fig. 1
6 Parietooccipital (PO) width	Chui & Damasio (1980)
7 PO width corrected for Falx deviation	Fig. 1
8 Temporal (central) width	LeMay & Kado (1978)
9 Frontal area width	Chui & Damasio (1980)
10 Cerebroventricular indices	
Frontal level[a]	Hahn & Rim (1976)
Caudate level [b]	Hahn & Rim (1976)
Frontal horn length (FHL)	Glydenstedt (1977)
Frontal horn angle (FHA)	Heinz et al. (1980)

Note: measures 3 through 8 were represented as Asymmetry index (AI)
i.e. the difference between the values of the two sides over
their sum $[(R-L/R+L) \times 100]$.
[a] Right and left anterior horn tip distances over the maximal
hemisphere width at this level;
[b] Right and left width of the cellae mediae at the level of
caudate just anterior to the third ventricle over the maximal
hemisphere width at this level.

Jernigan et al. (1982) have justifiably pointed out that the
term 'midline' as applied to the occiput is ambiguous since the
anatomical midline is often displaced. Chui and Damasio (1980)
detected straight sinus deviation in 28 % of their cases; more
frequently to the right (20%). In accord with their observations,
in the present study, the falx was mostly deviated to the right,
but this trend was noted about twice as often: in 38 % (against 23
% to the left) in HD subsample and also in 38 % (against 16 %) in
CD subsample. Whether a higher incidence of falx deviation can be
attributed to schizophrenia remains to be examined. But it
certainly can contribute to an over-estimation of the contralateral
hemisphere width when it is measured from the anatomical midline
and perpendicular to its 'geometrical' (i.e. rostrally justified)
midline. This is a possible explanation of why Jernigan et al.
(1982) have reported increased frequency of normal asymmetry among
schizophrenics than normal volunteers. In fact, Table 2 shows that
the left/right ratio of occipito-parietal width obtained in the
present study is practically identical with Jernigan et al.
Moreover, when occipital width was measured from the anatomical
midline ('corrected occipital width') there was a considerable
reduction of cases with asymmetric width values. Also, after this
correction the difference between the two subsamples appeared to be
marginally significant (Table 4).

Table 4
Characteristics of the sample and hemispheric asymmetry data

	GROUPS		
	Homolateral (n= 21)	Crossed (n= 16)	Total (n=37)
Age	28.8+ 4.7	28.6+5.9	28.7
Sex (male/female)	15/6	12/4	27/10
Race (white/blacks)	19/2	15/1	34/3
Cephalic index (CI)*	78	78	78
Falx deviation (%)	62	50	56
Occipitopetalia (%)**			
Left	43	31	38
Equal	57	56	57
Right	0	13	5
Occipital width***			
occipital (O) level	-3.8+ 6.04	-1.0+ 5.71	-2.4+6.05
corrected O level****	-2.8∓ 6.85	1.8∓ 6.00	-0.5∓6.42
parieto-occipital (PO) level	-1.6∓ 3.40	0.2∓ 6.93	-0.8∓5.23
corrected PO level	-1.2∓ 3.48	1.1∓ 4.28	-0.2∓3.96
Central width (CW)	0.9∓ 2.20	1.8∓ 2.20	1.3∓2.33
Frontal width (FW)	3.8∓ 4.68	1.4∓ 4.06	2.8∓4.52
Cerebroventricular indices			
frontal level	-5.5+ 9.21	-7.9+19.20	-6.6+14.22
caudate level	3.8∓14.88	-1.7∓11.28	1.5∓13.84
frontal horn length	-2.3∓ 8.91	-0.1+12.97	-1.3∓10.75
frontal horn angle	0.9∓ 5.35	2.7∓ 6.07	1.7∓ 5.65

* Cephalic index (CI) of the skull (width/length X 100);
dolichocephalic - 75.9 or less; mesocephalic - 76-80.9;
brachycephalic - 81-85.4; hyperbrachycephalic - more than 85.5
** Chi^2 = 9.8, df=2, p<.01
*** This and other values are expressed as asymmetry indices
(R-L/R+L X 100).
**** $F_{1,35}=4.51$; p<.04 (ANOVA, subsample X level)

When the asymmetry scores of the hemispheres were further
compared, two things emerged. One is a general reduction of the
asymmetry index as the measurements were placed more rostrally. The
other is a noticeable (albeit non-significant) trend for the
occipital width to be skewed to the left whereas the frontal width
is shifted to the right, albeit chiefly in HD patients. In
contrast, the CD patients, especially CD females, had reversed
asymmetry index or (HD females) consistently smaller asymmetry
values suggesting more symmetric hemispheres.

The asymmetry indices tabulated in Table 4 are presented in
Fig. 2 except that the averages for male and female subjects were
plotted separately. This shows that the dynamics of the
occipitofrontal asymmetries are primarily related to sex rather
than to dominance subcategory. Analysis of variance (subsample X
sex X level) yielded a highly significant interaction between sex
and level ($F_{4,128}=7.84$; p <.001). The difference between the
subsamples was not significant. When analysis of variance was

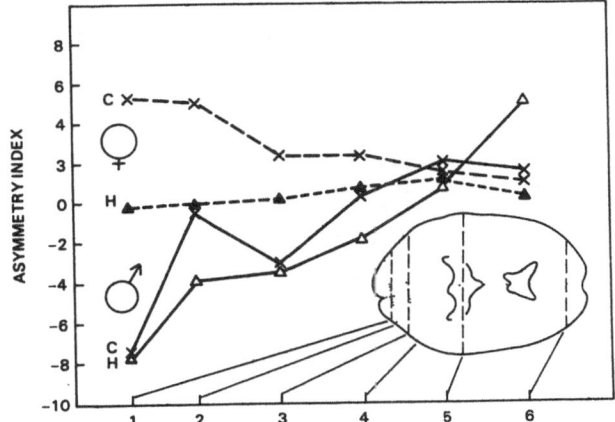

Fig. 2. Hemispheric width asymmetry indices at different levels. Methods of measuring hemispheric width are schematized in the insert. The gender scores in HD and CD subsample ('H' and 'C', respectively) are plotted separately.

conducted for each level separately, a highly significant effect of sex was obtained only for occipital width ($F_{1,33}=16.05$; p < .0003). Predictably, occipital width corrected for falx deviation reduced the asymmetry index; the effect contributed to a marginally significant group effect ($F_{1,33}=4.35$; p<.05). However, the effect of sex was lost after this correction ($F_{1,33}=2.88$, ns).

COMMENT

Two variables examined in the present study, sighting dominance and gender seem to contribute to the pattern of hemisphere asymmetry. It is not surprising that sighting dominance was not detected so far. It was not factored in strongly in the present study. Admittedly, the hypothesis justifying our interest in crossed dominance and brain asymmetry may be based on completely erroneous assumptions (see Introduction). Crossed dominance may be no more than a factor of sinistrality. In the absence of a biologically based theory of sighting dominance this question cannot be adequately resolved.

However, whatever the solution of the foregoing problem might be, the contribution of gender-related factors affecting the morphological brain balance clearly overrides the role of sighting dominance. The reason why the weight of this variable has not been detected or emphasized so far is not certain. It seemed to be noticable in several studies published in the past. De Lasocte-Utamsing and Holloway (1982) reported a wider posterior portion of the corpus callosum (splenium) in normal females, which one might expect in a case of symmetric and hyperconnected hemispheres. This finding was not replicated by a more recent study of Witelson (1985). She noted, however, that mixed-handers had a larger anterior and posterior part of the corpus callosum than consistent right-handers (Witelson 1985). Using magnetic resonance imaging technique, Nasrallah, Andreasen, Coffman et al. (1986) reported that the corpus callosum is wider in this region only in schizophrenic female patients. The reversed ('atypical') asymmetry of the brain in females, notably females with a family history of sinistrality can also be derived from the recent study by Koff, Naeser, Pieniadz et al. (1986) conducted in normal population. Hence, the present findings may be conceived as basically suggesting that reversed asymmetry requires at least two factors to emerge, gender(female) and some components of sinistrality.

While at first glance, our data are in line with the views of more symmetric brain functions in the human (McGlone 1980) and infrahuman female species (Robinson et al. 1985), it should be recalled that even though the notion of the sexually dimorphic brain is widely accepted, the correlation of structural imbalance with functional and cognitive asymmetries is by no means conclusive. For example, in rodents, females show more consistent directionality preferences in spontaneous locomotion and choice tasks than males while asymmetries of the neocortex in the rat are chiefly seen in males (see, Robinson et al., 1985 for review). The danger of drawing facile parallels is apparent as many cognitive functions are not necessarily more symmetrically represented in the female brain (McGlone 1980; Kimura & Harshman, 1984).

Almost no data are available relating anatomical brain asymmetries in schizophrenia to gender. However, such a possibility might be entertained because right-handed females as a group show a somewhat higher incidence of crossed dominance (Poren and Corac 1981) whereas a recent study by Shang-Ming, Flor-Henry, Dayi, et al. (1985) found excessive crossed dominance in schizophrenia only in females. It is possible that hemisphere asymmetries examined concurrently with changes of the callosal anatomy using magnetic resonance imaging would assist in separating subgroups with different laterality patterns.

SUMMARY
Some evidence bearing on variables contributing to radiological brain asymmetry data is reviewed and new findings are presented. While methodology of CT scan quantification remains wanting, it was possible to notice that gender and sighting dominance may determine the type of brain asymmetry. From a more general perspective this

result adds to the opinion that the 'right-left' dichotomy is too
general to be correct and should give way to a more fragmented view
of brain duality. Such a concept of brain asymmetry would be in
accord with the representation of sighting dominance and gender as
components contributing to individual differences in brain
organization (Kimura & Harshman, 1984).

REFERENCES

Allon, R. (1971). Sex, race, socioeconomic status, social mobility,
and process-reactive rating of schizophrenics. J Nerv Ment Dis 153,
343-350.

Andreasen, N.C., Dennert, J.W., Olsen, S.A., Damasio, A.R.,
(1982a). Hemispheric asymmetry and schizophrenia. Am. J. Psychiat.
139, 427-430.

Andreasen, N.C., Smith, M.R., Jacoby, C.G. et al. (1982b).
Ventricular enlargement in schizophrenia: definition and
prevalence. Am. J.Psychiat., 139, 292-296.

Bartels, M., Themelis, J. (1983). Computerized tomography in
tardive dyskinesia: Evidence of structural abnormalities in the
basal ganglia system. Arch. Psychiat. Neurol. Sci., 233, 371-379.

Chui, H.C., Damasio, A.R. (1980). Human cerebral asymmetries
evaluated by computed tomography. J. Neurol. Neurosurg. Psychiat.
43, 973-878.

Damasio, H., Maurer, R.G., Damasio, A.R., Chui, H. (1980).
Computerized tomographic scan findings in patients with autistic
behavior. Arch. Neurol. 37, 504-510.

Dufy, B., Vincent, J.-D., Fleury, H., Du Pasquier, P., Grourdji,
D., Tixier-Vidal, A. (1979). Dopamine inhibition of action
potentials in a prolactin secreting cell line is modulated by
oestrogen. Nature 282, 855-857.

Flor-Henry, P. (1969). Psychosis and temporal lobe epilepsy. A
controlled investigation. Epilepsia 10, 363-395.

Fortig, H. (1922). (cited by Hadziselimovic & Andelic, 1963 &
Hadziselimovic & Cus, 1966).

Glydensted, C. (1977). Measurements of the normal ventricular
system and hemispheric sulci of 100 adults with computed
tomography. Neuroradiology, 14, 183-192

Gur, R.E. (1977). Motoric laterality imbalance in schizophrenia.
Arch. Gen. Psychiat., 34, 33-37.

Hadziselimovic, H. and Andelic, M. (1963). On the appearance of some interior brain structures in relation to the exterior configuration of the brain. Acta Anat. 52, 260-268.

Hadziselimovic, H. and Cus, M. (1966). The appearance of internal structures of the brain in relation to configuration of the human skull. Acta Anat. 63, 289-299.

Hahn, F.J.Y., Rim, K. (1976). Frontal ventricular dimensions on normal computed tomography. Am. J. Radiol. 126, 593-596.

Heinz, E.R., Ward, A., Drayer, B.P., Dubois, P.J. (1980). Distinction between obstructive and atrophic dilation of ventricles in children. J. Comp. Assist. Tomogr. 4, 320-325.

Hier, D.B., LeMay, M., Rosenberg, P.B. and Perlo, V.P. (1978). Developmental dyslexia. Evidence for a subgroup with a reversal of cerebral asymmetry. Arch. Neurol., 35, 90-92.

Inglessis, M. (1925). Uber Kapazitatsunterschiede der linken und rechten Halfte am Schadel bei Menschen (insbesondere Geisteskranken) und uber Hirnasymmetrien. Z. ges. Neurol. Psychiat. 97, 354-373.

Jeste. D.V., Wagner R.L., Weinberger D.R., Reith K.G., Wyatt R.J. (1980). Evaluation of CT scans in tardive dyskinesia. Am. J Psychiat., 137, 247-248

Kimura, D., Harshman R.A. (1984). Sex differences in brain organization. Progr. Brain. Res. 61, 423-441.

Koff, E. Naeser M.A., Pieniadz, J.M., Foundas, A.L., Levine, H.L. (1986). Computed tomographic scan hemispheric asymmetries in right- and left-handed male and female subjects. Arch. Neurol., 43, 487-491

LeMay, M. (1976). Morphological cerebral asymmetries of modern man, fossil man, and non-human primates. Ann. N.Y. Acad. Sci. 32, 243-253.

LeMay, M., Kido, D.K. (1978). Asymmetries of the cerebral hemispheres on computed tomograms. J. Comp. Assist. Tomogr. 2, 471-476.

Leventhal, D.B., Schuck, J.R., Rothstein, H. (1984). Gender differences in schizophrenia. J.Nerv. Ment. Dis. 172, 464-467

Lewine, R. (1981). Sex differences in schizophrenia: Timing of subtypes. Psychol. Bull. 90, 432-444

Luchins, D.J., Meltzer, H.Y. (1983). A blind, controlled study of occipital cerebral asymmetry in schizophrenia. Bsychiat. Res. 10, 87-95.

Luchins, D.J., Weinberger, D.R., Wyatt, R.J. (1979).
Schizophrenia. Evidence of a subgroup with reversed cerebral
asymmetry. Arch. Gen. Psychiat. 36, 1309-1311.

Luchins, D.J., Weinberger, D.R., Wyatt, R.J. (1982). Schizophrenia
and cerebral asymmetry detected by computed tomography. Am. J.
Psychiat., 139, 753-757.

McGlone, J. (1980). Sex differences in human brain asymmetry: a
critical survey.Behav. Brain Sci 3, 215-263.

Merrin, E.L. (1984). Motor and sighting dominance in schizophrenia
and affective disorder. Br. J. Psychiat., 146, 539-544.

Michaels, D.D. (1985). Visual optics and refraction. A clinical
approach. The C.V.Mosby Co.: ST.Louis.

Miles, W.R. (1929). Ocular dominance demonstrated by unconscious
sighting. J Exp Psychol., 12, 113-126.

Montagu, M.F.A. (1960). An introduction to physical anthropology.
C.C. Thomas, Springfield.

Myslobodsky, M. (1986). An attempt at localizing the soft signs in
schizophrenia. C.Shagas et al. (eds), Biological Psychiatry,
Elsevier, Amsterdam, pp 1130-1132

Myslobodsky, M. (Ed.) (1983). Hemisyndromes: Psychobiology,
Neurology, Psychiatry, Academic Press,New York. pp. 347-388.
Naeser, M.A., Levine, H.L., Benson, D.F., Stuss, D., Weir, W.S.
(1981). Frontal leukotomy size and hemispheric asymmetries on CT
scans of schizophrenics with variable recovery. Arch. Neurol. 38,
30-37.

Nasrallah, H.A., Andreasen, N.C., Coffman, J.A., et al (1986). A
controlled magnetic resonance imaging study of corpus callosum
thickness in schizophrenia. Biol. Psychiatry, 21, 274-282.

Nyback, H., Wiesel. F.A., Berggren, B.M., Hindmarsh, T. (1982).
Computed tomography of the brain in patients with acute psychosis
and in healthy volunteers. Acta Psychiat. Scand., 65, 403-414.

Oddy, H.C. and Lobstein, T.J. (1972). Hand and eye dominance in
schizophrenia. British Journal Psychiatry, 34, 33-37.

Ogle, K.N. (1962). Ocular dominance and binocular rivalry. In:
Davson, H. (ed), The eye, v. 4. New York: Academic Press, pp 408-417

Pieniadz, J.M., Naeser M.A. (1984). Computed tomographic scan
cerebral asymmetries and morphologic brain asymmetries. Arch.
Neurol. 41, 403-409.

380

Piran, N., Bigler, E.D., Cohen, C. (1982). Motoric laterality and eye dominance suggest unique pattern of cerebral organization in schizophrenia. Arch. Gen. Psychiat., 39, 1006-1010.

Porac, C., Coren, S. (1981). Lateral preferences and human behavior. Springer-Verlag: New York.

Robinson, T.E., Becker, J.B.,Camp, D.M., Mansour, A. (1985). Variation in the pattern of behavioral and brain asymmetries due to sex differences. In Cerebral Lateralization in Non-human Species. (ed. S. Glick) Academic Press: New York, pp.185-231.

Seeman, M.V. (1982). Gender differences in schizophrenia. Can. J. Psychiat. 27, 107-11.

Shang-Ming, Y., Flor-Henry, P., Dayi, C., Tiangi, L., Shuguang, Q., and Zenxiang, M. (1985). Imbalance of hemisphere functions in the major psychoses: A study of handedness in the Peoples Republic of China. Biological Psychiatry, 30, 906-917.

Smith, G.E. (1907). On the asymmetry of the caudal pole of the cerebral hemispheres and its influence on the occipital pole. Anat. Anz. 30, 574-578.

Strauss, E., Kosaka, B., & Wada J. (1983). The neurobiological basis of lateralized cerebral function. A review. Human Neurobiol., 2, 115-127

Taylor, P.A., Dalton, R., Fleminger, J.J. (1980). Handedness in schizophrenia. Br. J. Psychiat., 136, 375-383.

Tsai, L., Jackoby, C.G., Stewart, M.A., Beisler, J.M. (1982). Unfavorable left-right asymmetries of the brain and autism: A question of methodology. Brit. J. Psychiat., 140, 312-319.

Tsai, L., Nasrallah, H.A., Jacoby, C.G. (1983). Hemispheric asymmetries on computed tomographic scans in schizophrenia and mania. Arch. Gen. Psychiat. 40, 1286-1289.

Wada, J., Clarke, R., Hamm, A. (1975). Cerebral hemispheric asymmetry in humans. Cortical speech zones in 100 infant brains. Arch. Neurol. 32, 239-246.

Wahl, O.F. (1976) Handedness in schizophrenia. Percept. Mot. Skills, 42, 944-946.

Weinberger, D.R., DeLisi, L.E., Perlman, G.P., et al (1982). Computed tomography in schizophreniform disorder and other acute psychiatric disorders. Arch. Gen Psychiat 39, 778-791.

Weinberger, D.R., Luchins, D.J., Morihisa, J. et al. (1982b). Asymmetric volumes of the right and left frontal and occipital regioms of the human brain. Ann. Neurol. 11, 97-100

24
Subcortical Influences on Cortical Processing in 'Split' Brains

Colwyn Trevarthen

INTRODUCTION

The most difficult, but crucial, aspect of higher brain function concerns 'centrencephalic' regulations, from inside the brain. They can change how cortical cells associate sensory information, how different sectors of cortical tissue, including the two hemispheres, interact and complement each other in psychological function, and how efferent influences from cortex and basal ganglia impinge on final common path motoneurones of stem and cord. We must understand these facilitatory and organizing processes if we are to explain the arousal of consciousness, selective attention, intentional movement, motiviation and a variety of active memory functions. Here the experimental virtues of the split-brain preparation meet advances in knowledge of anatomy, physiology and pharmacology of the interneuronal networks and nuclei of the core of the brain, and their upward projections through the thalamus, basal ganglia and limbic system. Findings in this area, and comparisons to lower vertebrates in which cortical processes are less dominant, will help us dispel obscurity in our conception of the emotional, purposeful and aware states of our minds. They have immense importance for comprehending disorders of human motivation and emotion and of cognitive processes as well. They help balance what may be a rationalist bias to treat concrete or factual, environmentally-driven behaviour, memory, awareness and reasoning as belonging to a separate computational level of processing that might be carried out autonomously in the neocortex.

I will review studies with split-brain subjects, including some rather old ones, to illustrate the diversity of contributions from brain stem motive machinery to the information processing, rationalizing and motor programming that we normally attribute to the neocortex.

The Classic Split-Brain

The first experiments with split brain cats, made by Myers and Sperry at the University of Chicago in the 1950s, were to find out if perception and learning could be divided by cutting the corpus callosum and anterior commissure down the sagittal midline (Myers and Sperry, 1953; Myers, 1956). They were done against the background of Lashley's theory that perception and learning were products of 'mass action' in the brain; i.e., that the associations involved can be made in a multiplicity of ways through an unlimited range of neural topographies (Lashley, 1929, 1950). The results of Myers and Sperry were viewed by Lashley as a serious blow to his theory, because they proved that learning could be trapped in one part of the brain by surgery to one set of axons.

For some years after 1954, work in the group with Sperry, now Hixon Professor of Psychobiology at Caltech, was still aimed to pin down the cerebral circuitry of perception and learning, but their attention soon turned to exploiting the split-brain to study more general aspects of cerebral organization. Attempts were made to understand the control of response movements, which seemed surprisingly normal after an operation that removed a good proportion of all the long integrative connections of the brain. Schrier and Sperry (1959) tested if each of the hemispheres of a split-brain cat could direct coordinated movements of the whole cat, including movement of either forepaw being used to obtain reward by moving covers off food wells. The operation seemed to leave motor coordination fully unified and all combinations of hemisphere and paw worked equally well. Tactile discrimination learning using one forepaw at a time was shown to be split in two by callosotomy (Stamm and Sperry, 1957). Perhaps, the refined movements for feeling the stimulus pedal with the pads on the digits were also divided. This question has still not been fully investigated in the cat.

In 1961, Sperry described the divided hemispheric control over brain stem integrative mechanisms in the split brain as 'bilateral hegemony over lower centers'. His diagram (Sperry, 1961a, p.1754) shows only descending arrows, from cortex to stem and cord.

By progressive ablation, cutting away tissue around the primary projection zones in one hemisphere, Sperry's group exploited the fact that the split brain could provide one hemisphere for experiment while the other was left intact for the cat to regulate essentially normal coordinations. They found that, while a small island of cortex around the central sulcus could suffice for 'excellent' tactile discrimination learning and memory (Sperry, 1959), a large posterior territory extending from occipital into parietal and temporal cortex was required to sustain any visual pattern memory (Sperry, Myers and Schrier, 1960). Sperry's conclusion was that the engram for a cat's object discrimination was built around the system that controlled the motor pattern of response, at least the exploratory and final executive part (Sperry, 1967). For touch, the forepaw was both effector, applying careful pressure on a response pedal, and organ or reception. For vision, the whole cat was guided by either or both eyes to the 'correct' stimulus pattern. This and other studies, in which split brain cats and monkeys with motor cortex removed on one side and visual cortex removed on the other were shown able to aim the homolateral limb to stationary or moving objects seen (Myers, Sperry and McCurdy, 1962), drew attention to the possible contribution to awareness and memory of the preparation for response, in line with Sperry's previously enunciated motor theory of perception (Sperry, 1952) and his concept of facilitatory sets (Sperry, 1958).

In the split-brain experiments, to explore 'the problem of the cerebral correlates of conscious experience', the animal's behaviour was tightly constrained in a test apparatus. After pretraining the animal in the act of response, many trials were given repeating a two-choice problem - pushing one kind of stimulus with nose or forepaw, and not the other different stimulus, gained the subject a fragment of food. Most importantly, the hemispheres were tested one at a time; by covering one eye with a mask, or by holding back one forepaw, with a halter or in a specially shaped box with paw holes. Even when eyes or paws were switched every few trials, there were many contextual experiences available to the subject, inseparable from the experimental controls, that could cue use of one hemisphere at a time. Motivation was assumed to be available in equal measure for the two half brains, and, until the restraining devices were changed, responses were taken to be the same in every trial. The learning was scored as a proportion of correct choices in so many trials given. In short, each hemisphere was being tested as a passive retentive vessel, one that was lined up to collect stimuli and to base present choices on past successes. Dynamic variation in processing in the hemispheres was not studied, except for the changes due to learning.

Early Evidence of Brain-Stem Input: Upstream Effects of a Set to Respond with one Hand

The failure of interhemispheric transfer of learning after commissurotomy proved securely that the corpus callosum was a bridge, perhaps the only bridge, by which the experience for pattern discrimination learning through eye or paw could pass from one neocortex to the other. Then evidence came that some experience of visual stimuli could 'leak' through subcortical systems. Meikle and Sechzer (1960) showed interocular transfer of a brightness discrimination in chiasm-callosum sectioned cats. Voneida (1963) demonstrated that classical conditioned responses to simple stimuli could by-pass the divide in the forebrain.

To test the autonomy and completeness of learning in the two disconnected hemispheres of split-brain monkeys, and for possible involvement of undivided subcortical systems, I made double simultaneous learning experiments (Trevarthen, 1962; 1963; 1965). By means of polarized light, I projected contradictory discrimination tasks overlapping onto a single pair of response screens, transmitting one task to each of the two halves of the brain in every trial through polarizing filters in the eye holes of a mask into which the monkey inserted his face (Figure 1,A). With this technique I discovered that internal cerebral processes, affecting the hemispheres differently, could have a decisive influence over learning and remembering when the input of stimulus information and reward was as perfectly equal as I could make it on the two sides of the brain.

The major intrinsic factor deciding the effects of experience in memory in these double learning tests was some neural activity setting correlated with the arm and hand that the monkey used for responding. If the activities of the arms were changed by holding the preferred limb back, the interhemispheric balance of learning and retention switched within a few trials to the other hemisphere (Figure 1,B). Seeing and remembering were allocated to left or right cortex according to the laterality of an impulse to move. This put learning of a discrimination in a different light. If, of two hemispheres receiving the same input of experience and the same motivational enducement (peanuts taken from below the response screens and put in the mouth), one hemisphere should learn and the other not, then learning must require an intrinsic pattern of facilitation. Some intracerebral condition related to moving one arm must meet perceptual experience and fix it in a form that can be made available to guide future choices. Not just the idea or taste of the food reward, but some kind of 'intention' to move in a certain way had a decisive role to play in retention.

My tests confirmed that the perception of different kinds of stimuli were not always confined in a cortical process of perceptuo-motor coordination and memory formation. Some kinds of stimuli even led to formation of a memory that, while it had come in through one eye and up the optic tract on one side of the brain, was afterwards accessible to both eyes in spite of midline division of the optic chiasma and all the interhemispheric commissures. This violated the accepted principle that input to the cortex, viewed as the sole location for pattern perception and learning, was by the geniculo-striate system that permitted no crossing of visual input above the two optic tracts. It indicated that an alternative visual system, presumably one involving the superior colliculi (optic tectum of the midbrain), could hold some forms of visual learning, or that corticifugal pathways could transfer perception or the engram between the cortices through the brain stem (Trevarthen, 1968a).

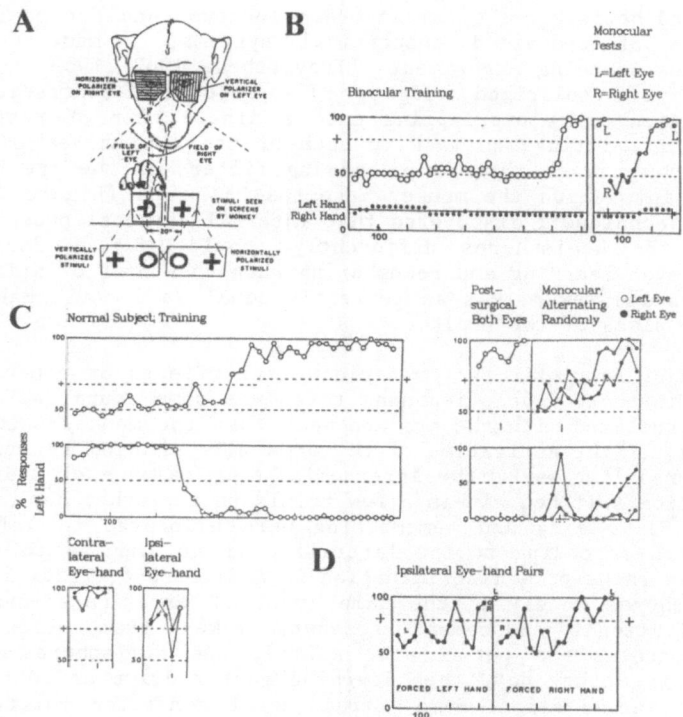

Figs 1.
A: Split-brain monkey performing a double visual discrimination test with conflicting stimuli. (Trevarthen, 1968).

B: Learning and spontaneous choice of hands; split-brain monkey with conflicting stimuli; binocular training and monocular retention tests.

C: Normal monkey learns a visual discrimination without conflict. Postsurgical test with both eyes shows good retention; monocular test with random alternation between the eyes causes fall of score, 90% of errors being due to neglect of stimuli contralateral to the eye in use. Note preference for contralateral eye-hand pairs after the split-brain operation.

D: Fluctuating performance of ipsilateral eye-hand pairs; same subject and same test as in B, i.e. after both hemispheres had been shown fully retentive with free use of contralateral eye-hand combinations, after split-brain surgery.
(B, C and D from Trevarthen, 1965)

All learning curves show percentage correct responses. Dashed line, p = 0.01.

Complementary Visuomotor Systems: Two Visions Theory

Considering the evolution of visuo-motor systems in invertebrates and influenced by the findings of Sperry, Glickstein and Sperry (1960), Downer (1959) and Voneida (1963) with split-brain cats and monkeys, and by Kuyper's theory of the parallel central motor systems for proximal and distal musculature (Kuypers, 1964), I was led to formulate a Two Visual Systems Theory. Sperry's motor theory of perception (Sperry, 1952), and similar or complementary approaches to perceptual processing and active information pickup of von Holst and Mittlestaedt, (1950), Teuber (1960) and J.J. Gibson (1957), led me to seek more than one level of vision because movements of different kinds need different afferent information, (different visual invariants) to control them (Trevarthen, 1968a).

The cardinal distinction was between whole-body orienting and locomotion in the ambient spatial arena of surfaces, spaces and large masses round the body, and focal manipulation by hand or in the mouth of small identified objects. The ambient body-centred field was the terrain (layout) or earth, vegetation, buildings, air and water, structured by gravity and sunlight and moved by the flow of wind or water. Focal vision concentrated on the identity of objects taken out of the geographical context for the sake of their intrinisic or constitutional properties or uses.

I analyzed the evolution of ambient and focal visual systems in terms of ancient brain stem arrangements for control of orienting and consummatory action in the behaviour field. I found evidence in the literature for one map of the body-centered visual 'behavioural field' on the optic tectum common to all vertebrates. Physiological and anatomical studies continue to confirm this uniformity of organization in the midbrain tactile and visual maps and show that different parts of the posterior cortex are in close reciprocal communication with the midbrain system at different depths (Gaither and Stein, 1979; Ogasawara, McHaffie and Stein, 1984). A common somatotopic frame is provided for perception of the animals surroundings in vision and several other modalities, all in orderly relationship to the neural mechanisms that aim movements in different directions (Meredith and Stein, 1986). I call this field for orienting telotopic, rather than somatotopic, because it maps directions round a centre on the polarized axis of body action, not body surface form.

The Organization of Motives to Pick-up Visual Experience

The learning studies with split-brain monkey lead away from a linear information-processing or behaviouristic account of perception and learning that this research had begun with, towards one that saw both perceiving and learning as organized to assist the solving of motor problems and to satisfy action in pursuit of predictive aims that were formulated deep in the CNS. When perceiving and responding to stimuli in a discrimination test, the subject is collecting information to resolve rival patterns within complex internal motivational and associative processes involving many systems of the brain.

It became my aim to ask questions about the effects of splitting the brain on the growth and use of perceptions and memories under the influence of intrinsic processes that would include undivided functions in integrative and regulatory centrencephalic systems in the brain-stem. These functions are presumed to be concerned with the fundamental 'motives' of intelligence, including: arousal to action, or withdrawal into sleep; attention and selective orientation to particular sensory stimuli; coordinating of integrated and reciprocal motor patterns; automatic guidance of motor synergies for posture, locomotion and prehension; autonomic regulation of body physiology; evaluation of goal objects in relation to vital needs and basic emotional response; and, finally, with organization of emotional expression for communication between subjects by means of signal movements. They could have a powerful voice in the changing shape of consciousness and memory at the level of the neocortex, and profound effects on the cerebral planning of action. They would normally work in cooperation with the forebrain commissures (Trevarthen, 1972).

For visual awareness, oculomotor orienting to left or right integrated at the midbrain level is likely to be associated with activation of one or other half of the brain stem and the corresponding hemisphere. My binocular training method (Trevarthen, 1965) allowed me to begin experimentation on this question, later developed into a general neuropsychological theory of attention by Kinsbourne (1972, 1974). I showed that a chiasm-callosum sectioned monkey could fail in intermittent monocular tests of retention because of double, mirror oculomotor neglect, failing to make orientation to one of two response screens placed side-by-side and missing trials in which the rewarded screen was on the neglected side (Figure 1C). The direction of neglect was presumed to be to the ipsilateral side of the hemisphere activated. In other words, activation of a hemisphere would bias visual orienting to the contralateral half of the visual field. As in earlier tests, the activated hemisphere was mainly the one contralateral to the hand used for response.

That hemispheric mechanisms for orienting attention to left or right are tightly coupled to equivalent brain stem systems had been elegantly demonstrated by Sprague (1955). He showed that unilateral visual neglect caused by a lesion on an occipital lobe of the forebrain of a cat could be 'cured' by removing the opposite Superior colliculus. Welsh and Stuteville (1958) had found the same reciprocal relationship between frontal and occipital cortices in control of visual orienting - neglect caused by a frontal lesion on one side could be relieved by excising the opposite occipital cortex. Recent physiological and anatomical studies of rats, cats and monkeys, and clinical evidence from humans, indicate that the cerebral mecahanisms of directed attention couples (1) polymodal sensory representations in the posterior panetal ortex, linked to the midbrain telotopic maps; (2) motor representations in the frontal cortex, linked to the basal ganglia and brain stem motor system; and (3) a motivational representation in the limbic cortex (cingulum), linked to the hypothalamus and long ascending monoaminergic pathways (Mesulam, 1981). The neocortical commissures are only one component of this extensive, reciprocally connected system that controls pick-up of sensory information from the environment in relation to states of motivation and preparations for movement.

Mechanisms of Self-control and Emotional Expression

The brain-stem incorporates emotional mechanisms, centered on the hypothalamus, the midline reticular system and limbic structures (Heilman and Satz, 1983). These, in addition to social signalling functions, attend to body maintenance and defend it and the cognitive machinery against damage or disorganization.

Incidental observations on frantic emotional displays by a normal monkey presented with conflicting visual stimuli, in comparison with the calm behaviour of the chiasm-callosum sectioned subjects with the same stimuli as one hemisphere took over from the other when the hand in use was changed, led to the idea that emotions associated with cognitive decision-making and 'self-evaluation' had been entirely divided by commissurotomy. The normal animal was in an awkward perceptual impasse, struggling to disambiguate two overlapping, contradictory, two-choice discrimination tasks, each task being polarized and the two tasks being made visible to different eyes through appropriately oriented polarizing filters in the eye holes of the training box. Both eyes had been trained to choose the same stimulus, then the cue value of the stimuli was reversed for one eye. For brief periods the problem was resolved by the animal paying attention, without closing the eye-lids, to input from one eye and ignoring the one that had the stimuli switched; but this internal lateralization of attention (as in interocular rivalry) could not be maintained. Cyclic phases of emotional display each time the score fell to chance were expressed in wrist biting, circling, compulsive attention to something of no significance and violent shaking of the training cage. The angry and impatient animal, a tame and normally calm crab-eating macaque, also grimaced at her reflection in the glass front of the cage and refused proffered food.

The concept of separate hemispheric emotional reactions to experience in the split brain was supported by evidence that unilateral ablation of the pole of the temporal lobe in a split-brain monkey could give rise to a unihemispheric Kluver-Bucy syndrome, the animal being tamed, losing visual object recognition and tending to indiscriminate orality when one eye was open, but unaffected when the other was open. All of these are preliminary reports, so the distribution of emotional processes in split brain monkeys remains uncertain. The evidence suggests that while the whole animal is coordinated in an emotional state and expresses itself as a whole, emotional coordination with the external world and affective evaluation of stimuli is divided between the disconnected hemispheres (See Sperry, 1961a, pp 1756-1757; 1961b, p 615).

Mishkin (1982) has shown that the anterior infero-temporal area TE is essential for visual object recognition, and that the amygdala and hippocampus play a part in establishment of both recognition memory and associative memory. Mishkin's group have used a partial split-brain method, as recommended by Sperry (1916b), to demonstrate the part of the limbic structures, which are cross-connected through the anterior commissure, in formation of visual associations and memories in monkeys (Mishkin and Phillips, 1986). There is further evidence of a contribution of the anterior and dorsal medial thalamus to recognition memory in conjunction with the limbic structures. Given the importance of meso-limbic, meso-frontal systems in emotional states and affective illness, it becomes significant that this work on memory systems in monkeys opens the way to a theory of convergence and reciprocal interaction between environmental input ascending to the hemispheres from the sensory projection systems and an evaluative input from the core motivating mechanisms that permeate the cortex via the limbic system and non-specific (intrinsic) thalamic relays.

How Hand Movements are Controlled in Split-Brain Monkeys

Split-brain monkeys can coordinate any eye-hand combination well for reaching and for catching moving objects (Mark and Sperry, 1968). However, if they are forced to use the left eye (connected via the geniculo-stiate projection to the left hemisphere after division of the optic chiasma) with the left hand (mostly coordinated from the right hemisphere), or the right eye with the right hand, they are apraxic when they try to adjust the digits to grasp a small object of a particular shape (Downer, 1959; Trevarthen, 1962, 1965; Brinkman and Kuypers, 1972). The visuo-spatial control of proximal arm and shoulder muscles is intact; but control of distal forelimb muscles of the hand by visual perception of pattern on a target object, or of local structure in the surroundings of a small object sought for, is divided into two efficient intrahemispheric contralateral eye-limb systems, and two inefficient ipsilateral systems. The latter are either using a weak representation for the ipsilateral limb in each hemisphere, or an unreliable brain-stem mechanism that permits bridging between the functions of the hemispheres.

In discrimination tests, the operated animals learned at a slower rate when they were forced to use ipsilateral eye-hand combinations, as compared to the favoured contralateral pairs, and the ipsilateral combinations were also very unstable (Figure 1D). Some of the problem was due to a fluctuating neglect of one of the two horizontally arranged response screens, but when vertical pairs of response locations were used it became clear that there was also an intermittent blindness or forgetting of the target meaning. Sometimes, even though both halves of the brain knew a solution, the animal just stared at the stimuli as if nothing had appeared. This, in fact, was the first evidence we obtained of blank periods in cortical perception related to a forced use of an unfavoured motor system for response (See also Downer, 1959).

Sperry, Glickstein and Mark used orientation tests to show how complementary channels of afferent guidance for the two forelimbs may be associated in the brain. Mark's study (see Sperry, 1964, p50) asked, could a split brain monkey bring the two hands to meet at one point in extracorporal space when both hands are invisible? One hand had to push a peanut through a hole in a horizontal partition so that the nut could be caught by the other hand underneath. The results suggested that neither cutting the corpus callosum, nor sectioning through the roof of the midbrain could break this coordination, though combining the two operations seemed to do so. Coordination of the arms of a monkey to reach out in one body action space does not require callosal connections.

Hamilton (1967) trained one eye-hand combination of chiasm-callosum sectioned monkeys to compensate reaching for prismatic displacement of the visual field. Then reaching to visual targets was tested with all eye-hand combinations, trained and untrained. Even deep midline surgery extending through forebrain commissures, midbrain roof and cerebellum failed to divide the mechanism of this adaptation. The acquired deviation or recalibration of reaching transfered between the eyes, so this was not a cortical visual learning like pattern recognition or object identification. If the movements were restricted to one limb during adaptation, then transfer between the limbs could be incomplete. However, this kind of failure of motor learning could also be observed in normal, intact monkeys and humans, too; it was not a hemispheric deconnection effect. It appears that adaptation to prismatic displacement is a body sense change, perhaps recalibration of some 'felt position of the limbs', or of a kinematic image of the moving body and 'felt displacement'.

To investigate where the brain learned to control precisely timed and precisely aimed action of the hands, I used a test of eye-hand reaction time and a problem box task designed to encourage young baboons to make rapid alternating moves of the two hands.

First, reaction time tests with a right-handed baboon indicated an asymmetry for triggering of a precise aiming of the index finger to touch a 2mm metal knob at the center of a light that had turned on in the vertical meridian while the subject was oriented and highly attentive (Trevarthen, 1978; Figure 2A). After midline chiasm-callosum surgery, the animal showed a reaction time slowed by 30 milliseconds and occasional visual neglect (functional blindness) when forced to initiate response of the right hand to visual input to the right hemisphere. All other hand-hemisphere combinations were unaffected by surgery in this test. The apparent loss of vision at the moment of manual response, the monkey staring at the illuminated spot without reacting, resembles visual neglect or 'perceptual erasure' of the image of the target object reported from the left speaking hemisphere of a commissurotomy patient immediately after initiation or a movement towards the target by the left hand (see below).This kind of experiment has since been followed up by Beaubaton, Requin and Guiard in the Marseille laboratory where my work with baboons was done under the guidance of Professor Jacques Paillard (Beaubaton and Requin, 1973; Guiard, 1980; Guiard and Requin, 1978; Paillard and Beaubaton, 1976). Their results support the conclusion that split-brain surgery divides visually 'guided', but not 'elicited' eye-hand coordination, and that it reduces allocation of motor preparatory sets between the hands.

The above-mentioned baboon was one of two tested after split-brain surgery for retention of a symmetric problem box task involving four manipulative steps that had to be performed in a particular sequence (Trevarthen, 1978). The task was designed to encourage rapid alternating moves of the two hands, to explore the formation of a learned executive program capable of generating precisely timed, and precisely measured, reciprocal action of the two hands (Figure 2B). Both animlas showed clear but opposite manual dominance in presurgical learning, one hand acting as leader and the other as assistant. Midline forebrain surgery temporarily broke the most efficient two-handed strategies. In one subject tested with all eye-hand combinations, the motor program for the precise alternation of moves on the visually perceived task appeared to be retained for both hands about two times more strongly in the hemisphere opposite the dominant right hand. Immediately after surgery the subdominant hand was akinetic, leaving its steps unperformed even when both eyes, and thus both hemispheres, had sight of the box. This is evidence for cerebral dominance in control of skilled bimanual manipulation.

It has recently been argued that monkeys may show consistent innate manual asymmetry in more delicate manipulations where object support and selective displacement of parts of the object are divided between the hands. This evolutionary step led to specialization of the left hemisphere for control of more rapid sequences of digital movement and hence, in humans, to lateralization of 'frame-content' speech control to that

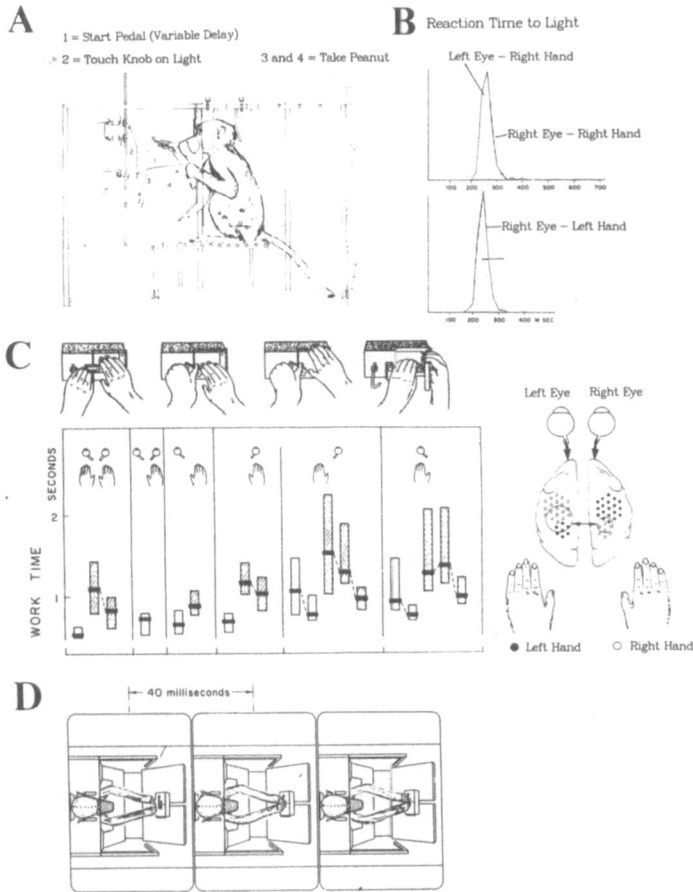

Figs 2.
A: Reaction time test with split-brain baboon and performance with all four eye-hand pairs after training completed. (Trevarthen, 1978).

B: Problem box task; patterns of two-handed coordination learned before surgery. Speed for opening box, for all four eye-hand combinations, before surgery (white bars) and after surgery (cross-hatched bars) (Trevarthen, 1978).

C: Double reaching movement with problem box. Drawn from stroboscopic film, 40 msec between frames.

hemisphere (MacNeilage, 1986).

Paradoxically, commissurotomy also caused both baboons, when they had their posture and attention balanced in confrontation to a goal, to make occasional precisely synchronous, slavishly coupled, mirror reaching movements of the two hands to items in the center of visual attention, (Figure 2C). These bisymmetric moves contrasted with the omissions of one hand's participation in many trials to test the learned bimanual manipulative skill (see Trevarthen, 1965, discussion pages 103-106). The animal had to see the problem-box to know the orientation and spatial configuration of the goal on a given trial, but much of the skill in opening the box would have been independent of vision. The seemingly contradictory kinds of response after surgery, the deficient hand either moving redundantly or failing to play its part, give evidence of two motor control processes in the hemispheres that resemble the hypothesised functions of the medial and lateral motor control systems recently proposed by Goldberg (1985).

Goldberg, citing data on effects of unilateral lesions of the supplementary motor area (SMA) and of subsequent commissurotomy in monkeys, proposes that precisely alternated moves of the two hands require commissural linking of activity of the two SMA's. My data indicate that commissurotomy can, at least initially, lead to the same decoupling of control for alternation of hand moves as does unilateral SMA lesion. Both split animals regained a partially coordinated bimanual response after a period of clumsy conflict between the hands. Activation of the subordinate hand led to increased clumsiness of the dominant hand. At first the bimanual pattern was a slow ambling kind of swinging of moves between the hands, which would entail an increased involvement in more proximal musculature and increased mechano-proprioception.

When we consider the role of sensory input while action is in course, and the intimate relation it must have to central assimilation of this input into the motor program, the boundary between feed-forward and feed-back dissolves. This regulation of movement must involve basal ganglia and subhemispheric systems, including the ventral midbrain and the cerebellum. Each of the latter has its own visual input different from that to the visual cortex, and each projects to parietal and frontal cortices to contribute to setting up of cortical motor schemata. Brain stem sensory-motor systems will be particularly involved in attentional selections preparatory to action and in maintaining an exproprioceptive visual-spatial frame for body coordination, drawing on input from both peripheral and foveal parts of the retinae.

To summarize: research with split-brain cats and monkeys shows that, while most perceptual processing and memory and a part of motor programming related to the distal musculature can be divided into two independent cortical domains by callosotomy, the activities of the neocortex are continuously modulated through reciprocal connections with the brain stem mechanisms of orientation, postural adjustment and emotional state. These subhemispheric integrations receive clarification from later work with human commisurotomy patients.

Tests for Hemispheric Modes of Human Consciousness Separated by Commissurotomy

With human commissurotomy subjects we encounter cerebral dominance for language and a new kind of deconnexion effect in perception due to different kinds of cognitive process in the hemispheres. The hemispheres of a human being sustain different mental strategies or programs which access different records of experience; they have associative fields that are organized differently to deal with a variety of problems in understanding potentially meaningful stimuli; they also operate with complementary principles of motor coordination. Despite post-surgical adaptations, commissurotomy patients keep consistent and enduring differences in mental processes in their two hemispheres, with input to completely separated perceptual and learning functions leading to the accumulation of separate, if largely duplicate, lines of experience.

The tests that have been done to bring to light the separation and the different modes of consciousness have involved restriction of the mobility and orienting of the subjects, and a toning down of motivation as well. Experimenters learned to present tasks in which the reward or confirmation of response is immediately available to a lateralized process that discriminates between significant stimuli, and to direct responses into forms that will not provide intrusive bilateral stimulation that can give rise to reafferent cross-cuing (Bogen, 1986). The picture which emerges from such testing has been vividly described by Sperry and colleagues in a number of now classical papers (Sperry, 1966; 1968; 1970; 1974; Sperry, Gazzaniga and Bogen, 1969).

Attention and volition can be captured in either half-brain by directing stimuli to vision from left or right of the fixation point, to left or right hand, to left or right nostril, or, with dichotic interference stimulation, to left or right ear. Usual response modes are speech (which engages the left hemisphere of all right handed subjects to the virtual total exclusion of any output from right hemisphere consciousness) and pointing or grasping with left or right hand. The subject may be asked to draw a representation, to spell out a word with plastic letters, to write a word, to imitate a posture with the hand, to show how a familiar cipher or tool is used, etc. Cross-modal perception is tested by requiring the subject to choose an object after matching experiences, on some preset criterion, between suitably lateralized input to two different modalities.

Many studies have demonstrated that commissurotomy patients can retain rich and detailed images of different visual stimuli presented one each side near the vertical meridian. Processes of discrimination and recognition go on independently in the two hemispheres. Similar results are obtained with auditory stimuli presented with dichotic interference, a procedure which suppresses ipsilateral projection of sound stimuli to the cortex in brain stem networks that are capable of passing input from each ear to both hemispheres (Milner, Taylor and Sperry, 1968). Tactual perceptions of the hands (Nebes, 1974; Sperry, Grazzaniga and Bogen, 1969) and olfaction through the two nostrils (Gordon and Sperry, 1969) are also divided in two.

After about 1970, important modifications in technique allowed the subject to compare or choose between two stimuli presented simultaneously, one to each hemisphere; or the subject was given tasks that involved moving both his or her hands at the same time. Tests of imitation for various positions of the fingers of a hand, for comprehension and performance of symbolic gestures, for comprehension of how objects are to be used by the hands, have given us a new perspective on levels of motor control and their relation to cognition (Zaidel and Sperry, 1976).

Zaidel has developed a contact-lens method of restricting vision to half of the visual field while the eye moves, the other eye being covered. He fitted two of the commissurotomy patients, permitting them to make normal visual scan of a test display with one eye while input is kept to one side of the fovea and thus confined to one geniculo-striate pathway and one hemispheric visual system. He gained a richer picture than hitherto available of the awareness, intelligence, self-consciousness and world knowledge of each of the two hemispheres, especially the previously less accessible right hemisphere (Sperry, Zaidel and Zaidel, 1979; Zaidel, 1977; 1978).

Recordings of eye movements have shown that visual orienting is under dual convergent control by the two hemispheres, each hemisphere being capable of following cues to look left or right (Trevarthen, 1974a; 1975). But each hemisphere has, as expected, a preattentive bias to seek new visual information in the contralateral perifoveal field; activity of the left hemisphere biases the subject to look to the right, and vice versa. Tests with large kinetic displays extending into the far periphery of the retinal fields have uncovered unitary or undivided cerebral systems for awareness of events round the body that may have profound facilitatory or inhibitory effects over immediate awareness in more finely perceived parts of the visual field that rely on input taken up from nearer the fovea (Trevarthen and Sperry, 1973).

Finally, tests of vigilance, cognitive processing capacity, speed of sorting and memory have brought to light marked impairments that had previously been overlooked (Dimond, 1976, Ellenberg and Sperry, 1979; Kreuter, Kinsbourne and Trevarthen, 1972; Kinsbourne, 1974)).

The new methods for examining the cerebral processes of split-brain subjects have culminated in a new interest in how the central regulatory systems in subcortical parts of the brain may modulate, or even direct, neocortical processing of perception, cognition, learning and coordination of responses. Sperry's review of 1974 is more cautious than the previous ones concerning the extent to which cerebral regulatory processes and sensory uptake can be bisected by neo-cortical commissurotomy, as the following indicates:

'Separate parallel performance on different tasks, though possible under special facilitating conditions, seems not to be the general rule. The fact that attention on many tests seems to become focussed in one separate hemisphere and simultaneously become repressed in the other would appear to reflect lateral differentiation in the thalamo-brain stem attentional mechanisms. The sleep-wake cycle and associated arousal mechanisms also retain strong bilateral unity after commissurotomy, presumably centered in mid brain structures ... With most of our research interest naturally concentrated on the divided aspects of brain function it is easy to underemphasise the many components of behavior that remain unified.' (Sperry, 1974, pp 8 and 10).

Bilateral Coordination of the Hands and Drawing after Commissurotomy

Early tests of imitated hand positions indicated that the commissurotomy patients had two separate systems for coordinating individuated finger movements, and that the left hemisphere was better at imitating hand signs (Gazzaniga, Bogen and Sperry, 1967). Simpler arm and hand movements of one limb could be directed from either hemisphere. Reaching tests showed that the patients retained one undivided system representing the space on both sides of the body for both limbs (Trevarthen, 1974a; 1975).

D.Zaidel and Sperry (1977) gave the same patients a large battery of standardized tests of motor performance 5-10 years after surgery. They confirmed that the two hands were both quick and accurate when working separately, as long as there were no signs of unilateral cerebral pathology impairing motor control on one side. The patients were slower than normal controls and sometimes they exhibited duplicate volition, but this was felt to be a minor and infrequent disorder; most conflicting movements were attributed to 'forced reflex movements and other largely involuntary responses'. For the most part the patients were acting as single coherent agents.

There was severe qualitative and quantitative impairment of bimanual coordination in tasks that demanded alternation of the hands and independent control. This agrees with my findings with baboons (Trevarthen, 1978) and with Kreuter, Kinsbourne and Trevarthen (1972) who imposed double index finger tapping at a maximum rate as a background task to test for duplication of cerebral processing capacity after commissurotomy. In fact, commissurotomy reduced the capacity of the subject to keep tapping with the fingers synchronized while doing mental arithmetic or reciting every third letter of the alphabet. Right hand tapping was interfered with by moderate mental concentration on these verbal tasks. When the subject made a mistake or could not work out how to respond, both hands faltered, then stopped.

Ellenberg and Sperry (1979) had commissurotomy patients sort small objects for half-an-hour with 2 hands making synchronized movements as rapidly as possible. Most achieved good rates with the two hands (16-38 objects/minute) but both rates and errors showed evidence for two kinds of interference between control from the disconnected hemispheres. When one hemisphere had cortical damage affecting hand movements, this led to slowing of both hands when they were moved together, and more errors. When both cortices were in good condition, activity of the right hemisphere (left hand) was interfered with by an idle left hemisphere (right hand doing nothing). When both hands were active, the sorting movements of the left hand accelerated. Subjects with complete commissurotomy frequently lost synchrony of hand movements, but anterior partials (anterior commissure and anterior 5cm of the corpus callosum sectioned) did not. Evidently transmission of tactual information through the posterior callosum was sufficient to maintain coordination in this task.

Preilowski (1975) used a bimanual visual tracing task, a new skill in which the hands had to learn to crank on two handles collaboratively to guide a cursor along a diagonal line. Both complete and anterior partial commissurotomy patients performed badly, indicating that the genu region connecting the frontal lobes is important for bimanual coordination in a difficult joint visual guidance. Well-practised bimanual coordinations, such as tying shoe laces, may be little affected by commissurotomy, possibly because the skill is in united subcortical systems such as the cerebellum, or because a coherent motor program is established in one hemisphere or in duplicate on both sides.

In spite of good verbal comprehension in the right hemisphere, the patients are inept in making symbolic gestures on command with the left hand (left ideomotor apraxia). The right hemisphere can recognize common objects, such as a toothbrush or spoon, grasped by the left hand, and can make proper use of them (ideational praxis), but this hand and hemisphere is poor at writing (left dysgraphia). Conversely, the right hand is inferior to the left in copying drawings (right dyscopia) (Bogen, 1969; Gazzaniga, Bogen and Sperry, 1967; Zaidel and Sperry, 1977).

D. Zaidel and Sperry (1973) made a visuo-tactile version of the Raven's progressive matrices test with images etched as raised metal lines as for a printing block. A pattern seen in free vision was matched to one of three fragments felt out-of-sight in the left or right hand. Scores of commissurotomy patients were low. The left-hand (right-hemisphere) performance was superior in accuracy and 25% quicker, but it was interfered with by a perservering left hemisphere set that was shown up by the subject verbalizing and using a sequential reasoning strategy to describe elements of the pattern. Subjects with anterior partial commissurotomy had higher scores, showing that normal performance is sustained by interhemispheric communication integrating complementary search and recognition strategies in the two hemispheres.

A clear demonstration of an interhemispheric shift of awareness, or recent visual memory for a model, with change of hands for drawing, is given by the performance of the commissurotomy patient NG in Figure 3A. She has been shown to have excellent visual form and colour perception in her right hemisphere. She was asked to reproduce what she saw in a series of chimeric figures presented in a tachistoscope with left or right hand. With change of hands, she switched to draw the side of the chimera homolateral to the hand, and she also changed the style of drawings. Another commissurotomy patient, LB, drew left halves of chimeras with either hand, demonstrating an overriding superiority of the right hemisphere for this task (Figure 3A).

More complex changes of 'facilitatory cognitive set' are illustrated by the performances of a patient, HD, who exhibited both a left-field neglect (Trevarthen, 1974) and a profound inability to recognize people by sight of their faces (i.e. prosopagnosia). This disorder persisted for years after an operation for a right parieto-occipital abscess. Here a selective aversion for faces was capable of causing the usually neglected right half of a chimera to be perceived and drawn (Figure 3B).

Fluctuations in a Partially Divided Ambient Visual Awareness: Perceptual Erasure: Faulty Vigilance

In tests for unity of vision over 180°, and for transfer of awareness of the occurrence, shape and direction of displacement of events in the peripheral visual field with two commissurotomy patients, changing sets to respond, or spontaneous fluctuations in visual attention or vigilance, could lead the subject to neglect one or other half of awareness, or to occlusion of one object (Trevarthen, 1970; Trevarthen and Sperry, 1973). For example, when presented with two lights that were moving about simultaneously one on each side at about 45° from the fovea, the subjects said

400

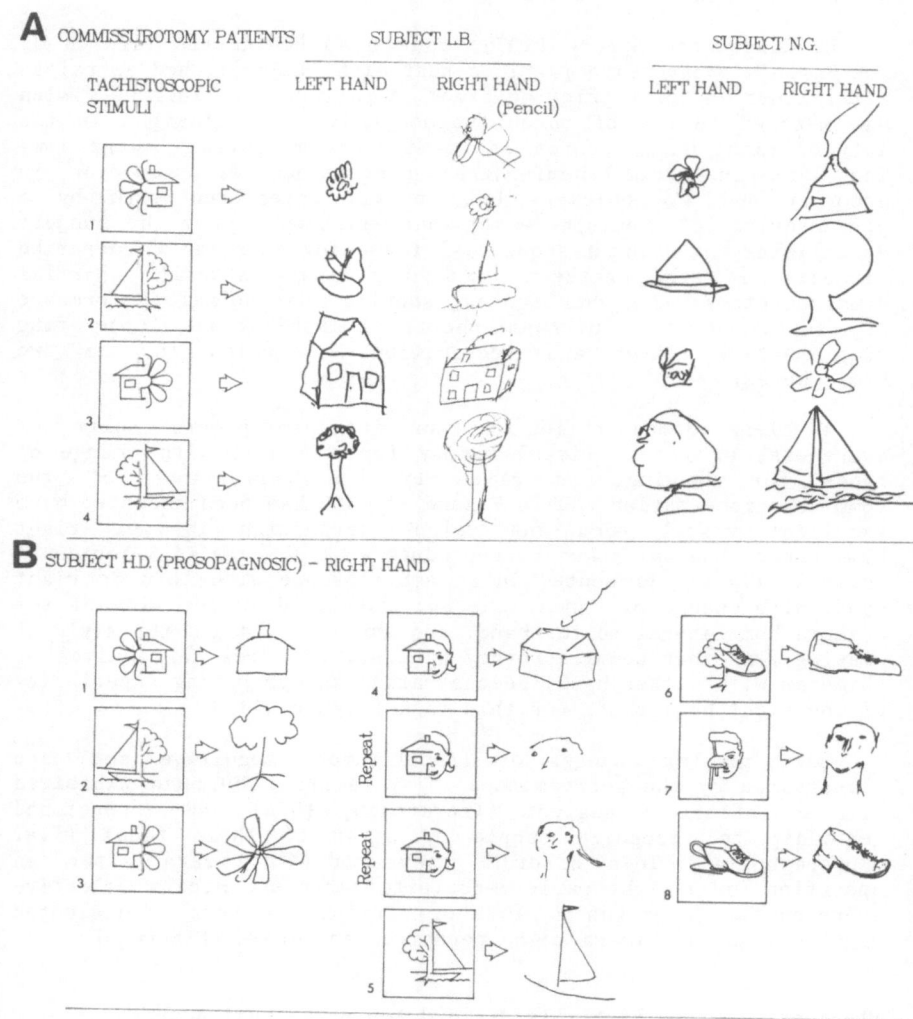

Figs 3.
A: Drawings made by commissurotomy patients from chimeric figures presented tachistoscopically.

B: Prosopagnosic patient with right posterior lesion; drawings from chimeric stimuli presented in order shown.

that they saw either the left or the right light disappear. Sometimes it was the right stimulus that the subject said had vanished. This is evidence for elimination of an element of visual consciousness in the left, speaking hemisphere.

Descriptions the patients made of the fluctuating appearances of peripheral events in ambient visual awareness corresponded closely with the experiences normal people had, though to a less conspicuous degree, with stimuli located at the same off-center locations in the visual field while their gaze was kept steady. Disappearances and reappearances in the periphery resemble the changes experienced in interocular rivalry.

Patients with parieto-occipital or frontal lesions may exhibit similar anomalous variations in peripheral awareness (Teuber, 1960). Visual neglect can result from lesions in parietal, frontal or cingulate cortex, or subcortical structures linked to these areas (Mesulam, 1981). In all these cases, vision is profoundly influenced by factors other than the stimuli received by the eye.

Commissurotomy subjects could accurately describe moving parallactic shadows cast in the left field by a point-source lamp on a back-projection screen. They spoke in single words, quietly, with concentration, and responded quickly. In this way they showed unity in 'ambient' visual awareness, mediated in the brain stem or by extra-geniculo-striate visual input systems to the cortex. When making the efficient bilateral field judgements in the visual periphery, the patients tended to miss the right stimulus more, and they made bursts of involuntary flicks of gaze to the right, indicative of disinhibition of the left hemisphere preattentive visual orienting systems. They were afterwards silent and subdued, their muteness confirming that the left hemisphere had 'gone to sleep' (Trevarthen and Sperry, 1973).

In these subjects, signs of activation of the right hemisphere, favouring more efficient peripheral visual awareness and passive judgements about spatial arrangements round their bodies, are associated with a solemn mood. This applied most strikingly to subject LB who was typically lively, voluble and humorous when he was using the left hemisphere.

The above indications favor the view that the patients possess two integrated cerebral systems, one left, one right, that involve different cortico-subcortical relations affecting attentive, expressive and emotional functions. It is still an open question to what extent the left/right differences in brain processes are originating at the cortical level. The brain-stem components, of attention, mood, etc., may also be asymmetric.

Subject NG gave us dramatic report of vanishing of a percept from the consciousness of the left hemisphere for a large object entirely located in the right visual field (Trevarthen, 1974a,b).

This disappearance happened at the moment when her left hand began to make a mark, as accurately as possible, at the centre of the object with a felt tip pen while her gaze remained fixed at a central point on the table top (Figure 4A). The object was kept in motion, but it remained unseen until one corner passed over the vertical meridian of her visual field. Then her rigidly immobile left hand struck at the target ballistically. The perceptual 'erasure' was coincident with initiation of a reaching movement, prepared to culminate in an accurate, visually monitored approach to a target, to mark it at the center with the tip of a pen held in the fingers. The object remained invisible to the right of the vertical meridian and the stored preparation of the left hand to reach the target was explosively released as soon as the object crossed the vertical meridian into the left visual field.

In vigilance tests, where the patients have to indicate by hand movements the appearance of tachistoscopically presented stimuli either side, or both sides, of the vertical meridian of the visual field, it has been found that commissurotomy causes unstable visual neglect (Gazzaniga and Hillyard, 1973; Kinsbourne, 1974; Trevarthen, 1976; Teng and Sperry, 1974; Dimond, 1976). Target stimuli are generally eliminated in toto on one side by some gating mechanism. The process can occur when the hemispheres are evenly matched for the task, but will be biased by any lateral inequality in cognitive capacity.

Figure 4B shows the effect on reaction time of requiring commissurotomy patients to make two different discriminations at once to control two digital responses. The fingers moved to left or right to release a light-activated switch and stop the reaction timer. Each finger had to move towards a stimulus of a given colour (red or green) or form (circle or cross) and the instructions were different for the two halves of the visual fields and the two hands. This is a most unpleasant task for normal subjects, but their reactions are quicker than the patients', though the latter experience no perceptual conflict.

Ellenberg and Sperry (1979) required commissurotomy patients to do a tedious signal detection task. Their middle fingers rested on a key that vibrated and they had to signal whenever the vibration was interrupted. They did much worse than normal subjects, became sleepy and agitated, and they complained of headaches. Their loss of vigilance was more marked for the right hand (left hemisphere), which supports findings that the left hand (right hemisphere) is superior at tactual discriminations.

Duplicate Awareness in Central Vision: Unilateral Facilitation from Inside the Brain

The tachistoscopic visual chimera technique, developed from double learning experiments with monkeys, has given us a way to study the effects of task instructions and response mode on

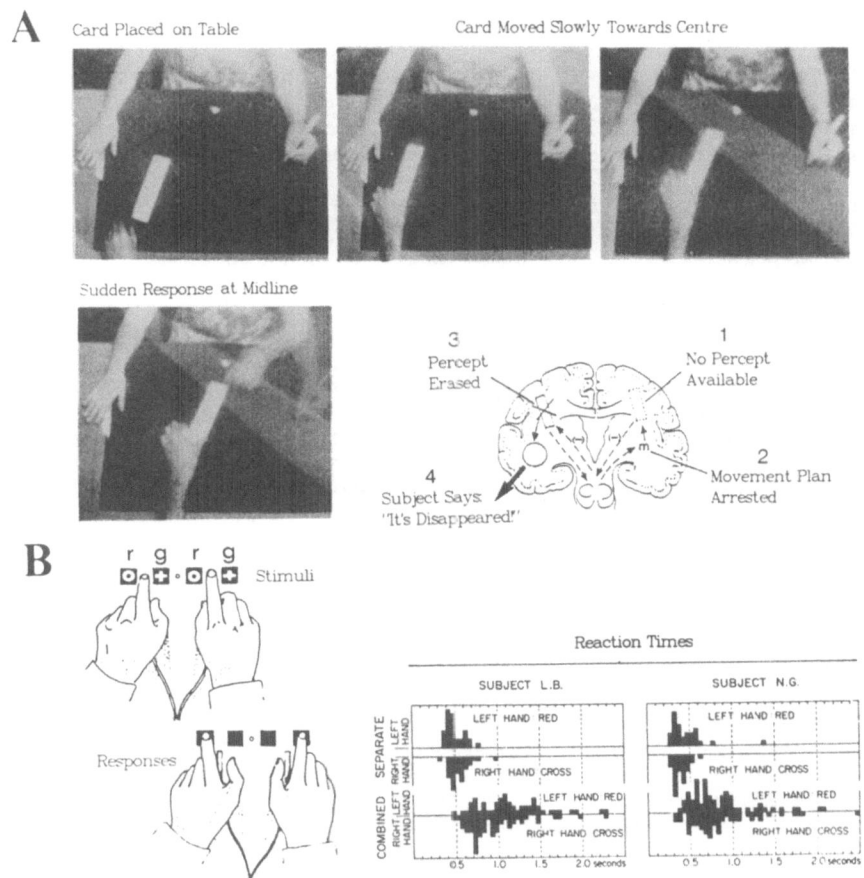

Figs. 4.
A: Commissurotomy patient attempts to place mark with pen at centre of white object on black table, but experiences and reports disappearance of target in right visual field as the left hand starts to move. Evidently an 'interhemispheric specific inhibition' is generated by the absence of perceptual information in the right hemisphere from which the left hand response is to be monitored.

B: Commissurotomy patients' reaction times in a bilateral discrimination test with conflict are slower than when each half of the test is presented separately. Subject sees crosses and circles in red or green light projected onto black squares. Left finger moves to red cross or circle; right finger moves to red or green cross.

allocations of processing in the separated cortical visual systems (Levy, Trevarthen and Sperry, 1972; Levy and Trevarthen, 1976; 1977; 1981). It proves that the two disconnected hemispheres can be aware of different things briefly presented at the same place and at the same time, something which cannot occur in a normally intact brain. Furthermore, it confirms that both hemispheres become aware of a complete meaningful object even though each is receiving sensory information via the striate cortex about only half of the object (Trevarthen, 1974a,b). The subject's choice of which percept to use for selecting a response depends on the kind of response required. Speaking, to say what the experience was, powerfully biased choices in favour of what the left hemisphere saw. Visual matching, requiring free visual scan of a display of potentially matching objects so the subject can pick out and then point to (with left or right hand) the one that is the same as the one seen in the tachistoscope, leads to a strong bias in the opposite direction, favouring the right hemisphere. The subjects performed these tasks easily with almost no sign of confusion and without negative reactions to the choices made, in spite of the contradiction set up by the different halves of the chimera.

When a chimeric stimulus, for example one made of two different faces, is flashed at the fixation point for 1/10 second in a tachistoscope, the commissurotomy patient can be primed to respond to either side with near 100% probability of giving a correct answer. If, after the stimulus has been given, but before the response has started, the instruction or question is reversed, changing the response to a mode favouring the other half of the stimulus, the subject generally shifts over with no difficulty (Levy, Trevarthen and Sperry, 1972). He or she chooses the half stimulus that would have been ignored if the instructions had not changed. This shows that the divided cortex can retain 2 different short term or 'iconic' images (Neisser, 1967), one in each hemisphere, for awareness of the single object.

It is remarkable that only very rarely did a commissurotomy patient seem to perceive the alternative stimulus when instructions were not reversed. Furthermore, on those occasions when awareness was switched from one hemisphere to the other, by change of instructions, subjects showed no consternation and noted no discrepancy of experience. It may be that when a response is made to either side the brain instantly wipes out the trace of any other object in that location. This would require a central decision to erase the percept in the other hemisphere. However, if both kinds of response are caused to occur at once, by double instructions, then the impossibility of the contradictory pair of choices becomes obvious to one or both half brains, and this can cause the patient confusion and anxiety comparable to that described above for the normal monkey experiencing a discrepancy between the objects seen by the two eyes. All the evidence is in favour of the existence of brief perceptual stores in the two hemispheres which are kept alive or extinguished depending upon the distribution of facilitatory processes related to the response set or intention. Some unifying coordinative system is required to ring the changes.

Meta-control: Commissures vs. Brain Stem in the Allocation of Cognitive Resource

In a majority of tests with stimulus chimeras, the factors determining which half-stimulus a commissurotomy patient will choose can be presumed to be intrahemispheric, and to reflect stable differences in cognitive aptitude of the disconnected hemispheres (Levy, Trevarthen and Sperry, 1972; Levy and Trevarthen, 1977). That is what the chimeras were designed to test. However, sometimes the performance of a commissurotomy subject has shown evidence for a temporary disposition in the brain to lateralize cognition at variance with hemispheric abilities. There is evidence for the influence of a factor that overrides the expected bias of hemispheric specialization for cognitive processives. Could such anomalous allocations of cerebral resources, 'tipping the balance' against normal hemispheric dominance, be due to activity of a sub-hemispheric director of cortical processing, like that presumed to switch perception and learning in split-brain monkeys where a change of hands is imposed?

Levy and I performed a study with a set of chimeric stimuli, every half of which could be perceived in two ways. (Levy and Trevarthen, 1976). Only one kind of response movement was used throughout, viz. pointing with a hand to an appropriate matching stimulus, and no speech was permitted. Effects of the side of hand used were controlled. Two kinds of instruction were intended to favour, respectively, the cognitive systems of the left and right hemispheres. The first, called Function Matching, paired drawings by a semantic criterion that coupled the objects represented with reference to the subject's knowledge of their use, occurrence of socially approved meaning. Even though there was no speech, and no word imagery or linguistic processing was required, this kind of match was expected to be allied to language psychologically, and to favour the left hemisphere, and, indeed, this turned out to be the case for all 4 subjects tested. The other condition, called Appearance Matching, requires the subjects to retain a memory of the form of the stimulus for a few seconds, and, disregarding its meaning, to match this image with a drawing that had a similar shape, whatever was represented. As expected, this direct matching of look-alike forms strongly favoured the right hemisphere.

Despite well-marked and stable difference in hemispheric ability, commissurotomy subjects showed some testing runs in which they were making choices with many errors because they were working with the right rule but with the 'wrong' hemisphere. Our conclusion was that the hemisphere is sometimes chosen by a bias of facilitation or activation, which we called 'meta-control', set up in the brain stem.

Levy (1968) proposes that brain-stem sets or dispositions channelling activity into a particular cortical system, even against the relative inferiority of that system, may contribute to many classical neurological syndromes. Sperry, in his Nobel address (Sperry, 1982) also argues that the severity of such highly specific symptoms as aphasia, alexia, colour agnosia, acalculia, apraxia, following local lesions restricted to the left hemisphere may be due to allocation of activity to those damaged parts at the expense of systems with some of the required competence elsewhere in the brain, and particularly in the right hemisphere. Conversely, local lesions of the right hemisphere can cause severe deficit (eg. prospagnosia) that are not shown with right hemispherectomy. Anosognosias, perceptual neglects and completions, alexithymia and such like conditions also may result from narrowing down of the cortical processes of association or motivation, causing diminished awareness of the body, loss of parts of the space of awareness or of potential orientation, or alteration of feelings of pain or of awareness of 'self' and body state. When hemispheres are disconnected, the right hemisphere is released from constraint of this kind and it appears not to be dyslexic, aphasic, or acalculic as classical neuropsychology would predict. Commissurotomy patients also tend to have lowered vigilance and lowered self-awareness. While they have lost some interference between the hemispheres, by the same token they have also lost a measure of vitality in awareness.

Memory, Emotion, Self-Recognition and Social Awareness after Commissurotomy

Commissurotomy patients are forgetful, which causes them considerable daily inconvenience. Formal tests of their memory yield clearly subnormal scores (Zaidel and Sperry, 1974). This was interpreted as due more to the operation than to epilepsy. First, deconnection of the hemispheres prevented collaboration between the specialized memory systems of the two hemmispheres (for example objects seen better by the right hemisphere could not be talked about); and there is evidently an overall depletion of the motivating influences important in both initial uptake and sorting of experiences for storage and in subsequent retrieval. Anterior partial commissurotomy also causes memory loss, especially for highly verbal and semantically complex material.

Using E. Zaidel's scleral contact lens to obtain lateralized visual input, Sperry, Zaidel and Zaidel (1979) gave stimuli to one or other hemisphere to assess reactions to personal, affect-laden and socially or politically significant material. They concluded that both hemispheres were emotional, self aware and socially experienced. They also found evidence for spread between the hemispheres of aspects of the emotional and mental aura generated by recognition of key items:

'In addition to general emotional changes, the central transfer appeared to include also subtle cognitive effects that enabled categorical distinctions like those between government and personal, domestic vs. foreign, historical vs entertainment, etc. Common experience suggests that such emotional and conservative auras play an important orientational role in normal brain function as for example in mnemonic retrieval' (Sperry Zaidel and Zaidel, 1979, p. 165).

This opens the way for explanation of meta-control in terms of a motivating mechanism capable, in the split-brain, of generating subconscious sets that favour activation of particular cognitive mechanisms and memory systems in the hemispheres (Levy and Trevarthen, 1976; 1981). There is evidence 'that commissurotomy causes alexithymia (Hoppe and Bogen, 1977); that is, it dissociates affectively loaded images, weakening both experience of self and memory for a meaningful story.

It would appear that these motive systems, with strong access to emotional states and attention directing mechanisms, have asymmetric affiliations with the hemispheres. There is a considerable amount of evidence to support the conclusion that in general a context for selection and combination of critical items in awareness and thought is set up more by the right hemisphere, which also appears to have stronger emotional evaluation and reference to self-state (Mesulam, 1981; Heilman and Satz, 1983; Ross, 1984)

Conclusion

There clearly is sufficient evidence from split-brain research to encourage vigorous enquiry into the contribution that brain stem systems can have to the associative functions of the neocortex. Motor coordination, directed attention, motivational and emotional evaluation, all appear to rely upon an interhemispheric coordination which is far from eliminated by cutting all the direct neural links between neocortical neurones on the two sides of the cerebrum. Reciprocal connections with the modulatory systems of the brain stem keep segments of the cortex in states of activity that reflect whole brain self-regulatory processes. We need to know much more about the differentiations that so-called reticular systems of the brain's core may introduce into cortical processing.

It seems likely that the earlier developed brain stem neurone networks may play a vital role in regulating the responses of growing cortex to environmental input, and to practice with the consequences of acting during childhood. These systems may contain the germ of lateral asymmetry of cortical function and play a decisive part in embryo and fetal stages when the anatomical basis for bihemispheric mental activies is laid down, before sensory contact with a distal environment is established.

Information processing theories of perception and learning will need to take account of intrinsically regulated programs that develop in phylogenetically older parts of the brain and that modulate the associative functions of the neocortex.

REFERENCES

Beaubaton, D. and Requin, J. (1973). The time course of preparatory processes in split-brain monkeys performing a variable fore period reaction-time task. Physiology and Behavior, 10, 725-730.

Bogen, J.E. (1969). The other side of the brain - I. Dysgraphia and dyscopia following cerebral commissurotomy. Bulletin of the Los Angeles Neurological Society, 34 73-105.

Brinkman J,and Kuypers, H.G.J.M. (1972) Split-brain monkeys: Cerebral control of ipsilateral and contralateral arm, hand and finger movements. Science, 176, 536-539.

Dimond, S.J. (1976). Depletion of attentional capacity after total commissurotomy in man. Brain, 99, 347-356.

Downer, J.L. deC. (1959) Changes in visually guided behaviour following mid-sagittal division of optic chiasma and corpus callosum in monkeys (Macaca mulatta) Brain, 82, 251-259.

Ellenberg, L. and Sperry, R.W. (1979). Capacity for holding sustained attention following commissurotomy. Cortex, 15, 421-438.

Gaither, N.S. and Stein, B.E. (1979). Reptiles and mammals use similar sensory organizations in the midbrain. Science, 205, 595-597.

Gazzaniga, M.S., Bogen, J.E. and Sperry, R.W. (1967).Dyspraxia following division of the cerebral commissures. Archives of Neurology, 16, 606-612.

Gazzaniga, M.S. and Hillyard, S.A. (1973). Attention mechanisms following brain bisection. Attention and Performance, 4, 221-238.

Gibson, J.J. (1957). Optical motions and transformations as stimuli for visual perception. Psychological Review, 64, 288-295.

Glickstein, M. and Sperry, R.W. (1960). Intermanual somesthetic transfer in split-brain rhesus monkeys. Journal of Comparative and Physiological Psychology,53, 322-327.

Goldberg, G. (1985). Supplementary motor area structure and function: Review and hypotheses. The Behavioural and Brain Sciences, 8, 567-616.

Gordon, H.W. and Sperry, R.W. (1969). Lateralization of olfactory perception in the surgically separated hemispheres of man. Neuropsychologia, 7, 111-120.

Guiard, Y. (1980). Cerebral hemispheres and selective attention. Acta Psychologica, 46, 41-61.

Guiard, Y. and Requin, J. (1978). Between-hand vs. within-hand choice-RT: A single channel of reduced capacity in the split-brain monkey. In Attention and Performance, VII. (ed. J. Requin). Erlbaum, Hillsdale, N.J. 391-410.

Hamilton, C.R. (1967. Effects of brain bisection on eye-hand coordination in monkeys wearing prisms. Journal of Comparative and Physiological Psychology, 64, 434-443.

Heilman, K.M. and Satz, P. (eds.) (1983). Neuropsychology of Human Emotion, Guildford Press, London.

Hoppe, I.D. and Bogen, J.E. (1977). Alexithymia in twelve commissurotomized patients. Psychotherapy and Psychosomatics, 28, 148-155.

Kinsbourne, M. (1972). Eye and head turning indicates cerebral lateralization. Science, 176, 539-545.

Kinsbourne, M. (1974). Lateral interactions in the brain. In Hemispheric Disconnection and Cerebral Function (eds. M. Kinsbourne and W.L. Smith). C.C. Thomas, Springfield, Ill. 239-259.

Kreuter, C., Kinsbourne, M. and Trevarthen, C. (1972). Are deconnected cerebral hemispheres independent channels? A preliminary study of the effect of unilateral loading on bilateral finger tapping. Neuropsychologia, 10, 453-461.

Kuypers, H.G.J.M. (1964). The descending pathways to the spinal cord, their anatomy and function. In Progress in Brain Research, Vol. II. Organization of the Spinal Cord. (eds. J.C.Eccles and J.P. Schade). Elsevier, Amsterdam, 178-200.

Lashley, K.S. (1929). Brain Mechanisms and Intelligence. University of Chicago Press, Chicago.

Lashley, K.S. (1950). In search of the engram. Symposia of the Society for Experimental Biology, 4, 454-482.

Levy-Agresti, J. and Sperry, R.W. (1968). Differential Perceptual Capacities of Major and Minor Hemispheres. Proceedings of the National Academy of Sciences, U.S.A., 61, 1151.

Levy, J. (1986). Regulation and Generation of Perception in the Asymmetric Brain. In Brain Circuits and Functions of the Mind (ed. C. Trevarthen) Cambridge University Press, New York (in press)

Levy, J. and Trevarthen, C (1976). Metacontrol of hemispheric function in human split-brain patients. Journal of Experimental Psychology: Human Perception and Performance, 2, 299-312.

Levy, J. and Trevarthen, C. (1977). Perceptual, semantic and phonetic aspects of elementary language processes in split-brain patients. Brain, 100, 105-118.

Levy, J. and Trevarthen, C. (1981). Color-matching, color-naming and color-memory in split-brain patients. Neuropsychologia, 19, 523-541.

Levy, J., Trevarthen, C., and Sperry, R.W. (1972). Perception of bilateral chimeric figures following hemispheric deconnection. Brain, 95, 60-78.

MacNeilage, P. (1986). Bimanual coordination and the beginnings of speech. In Precursors of Early Speech (Wenner-Gren International Symposium Series, Vo. 44). (eds. B. Lindblom and R. Zetterstrom). Macmillan, Basingstoke, Hants, 189-201.

Mark, R.F. and Sperry, R.W. (1968). Bimanual coordination in monkeys. Experimental Neurology, 21, 92-104.

Meikle, T.H. and Sechzer, J.A. (1960). Interocular transfer of brightness discrimition in split-brain cats. Science, 132, 734-735.

Meredith, M.A. and Stein, B.E. (1986). Spatial factors determine the activity of multisensory neurons in cat superior colliculus. Brain Research, 365, 350-354.

Mesulam, M. -M. (1981). A cortical network for directed attention and unilateral neglect. Annals of Neurology, 10, 309-325.

Milner, B., Taylor, L.B., and Sperry, R.W., (1968). Laterally suppressed dichotically presented digits after commissural section in man. Science, 161, 184-186.

Mishkin, M. (1982). A memory system in the monkey. Philosophical Transactions of the Royal Society, London, Series B, 298: 85-95.

Mishkin, M. and Phillips, R.R. (1986). A cortico-limbic memory path revealed through its disconnection. In Brain Circuits and Functions of the Mind. (ed. C. Trevarthen). Cambridge University Press, New York (in press).

Myers, R.E. (1956). Function of corpus callosum in interocular transfer. Brain,79, 358-363.

Myers, R.E. and Sperry, R.W. (1953). Interocular transfer of a visual form discrimination habit in cats after section of the optic chiasum and corpus callosum. Anatomical Record, 115, 351.

Myers, R.E., Sperry, R.W. and McCurdy, N.M. (1962). Neural mechanisms in visual guidance of limb movement. Archives of Neurology, 7, 195-202.

Nebes, R.D. (1974). Dominance of the minor hemisphere in commissurotomized man for the perception of part-whole relationships. In Hemisphere Disconnection and Cerebral Function (eds. M. Kinsbourne and W.L. Smith) Thomas, Springfield, Ill, 155-164.

Ogasawara, K. McHaffie, J.G. and Stein, B.E. (1984). Two visual corticotectal systems in the cat. Journal of Neurophysiology, 52, 1226-1245.

Paillard, J. and Beaubaton, D, (1976). Triggered and guided components of visual reaching: their dissociation in split-brain studies. In Motor System: Neuropsychology and Muscle Mechanism. (ed. M. Shahani). Elsevier, Amsterdam, 333-347.

Preilowski, B. (1975). Bilateral motor interaction: perceptual-motor performance of partial and complete 'split-brain' patients. In Cerebral Localization. (eds. K.J. Zuelch, O. Creutzfeldt and G.C. Galbraith). Springer, Berlin, 115-132.

Ross, E.D. (1984). Right Hemisphere's Role in Language, Affective Behavior and Emotion Trends in Neuroscience, 7, 342-346.

Schrier, A.M. and Sperry, R.W. (1959). Visuo-motor integration in split-brain cats. Science, 129, 1276-1276.

Singer, W. (1984). Learning to see: Mechanisms in experience-dependent development. In The Biology of Learning. (eds. P. Marler and H.S. Terrace). Springer, Berling, 461-477.

Sperry, R.W. (1952). Neurology and the mind-brain problem. American Scientist,40, 291-312.

412

Sperry, R.W. (1958). Physiological plasticity and brain circuit theory. In Biological and Biochemical Bases of Behavior. (eds. H.F.Harlow and C.N. Woolsey). University of Wisconsin Press, Madison, 401-424.

Sperry, R.W.(1959). Preservation of high order function in isolated somatic cortex in callosum-sectioned cat. Journal of Neurophysiology, 22, 78-87.

Sperry, R.W. (1961a). Cerebral organization and behavior. Science, 133, 1749-1757.

Sperry, R.W. (1961b). Some developments in brain lesion studies of learning. Federation Proceedings, 20, 609-616.

Sperry, R.W. (1964). The great cerebral commissure. Scientific American, 210, 42-52.

Sperry, R.W. (1966). Brain bisection and mechanisms of consciousness. IN Brain and Conscious Experience. (ed. J.C.Eccles). Springer, New York, 298-313.

Sperry, R.W. (1967). Split-brain approach to learning problems. In The Neurosciences. (eds. G.C. Quarton, T. Melnechnuk and F.O. Schmitt). The Rockefeller University Press, New York, 714-722.

Sperry, R.W. (1968). Hemispheric deconnection and unity in conscious awareness. American Psychologist, 23, 723-733.

Sperry, R.W. (1970). Perception in the absence of the neocortical commissures. Research Publication of the Association for Research in Nervous and Mental Diseases, 48, 123-138.

Sperry, R.W. (1974). Lateral specialization in the surgically separated hemispheres. In The Neurosciences: Third Study Program. (eds. F.O. Schmitt and F.G. Worden). M.I.T. Press, Cambridge, Mass. 5-19.

Sperry, R.W. (1982). Some Effects of disconnecting the cerebral hemispheres. Science, 217, 1223-1226.

Sperry, R.W. and Gazzaniga, M.S. (1967. Language following surgical disconnection of the commissures. In Brain Mechanisms Underlying Speech and Language (ed. F.L. Darley) Grune and Stratton, New York, 108-121.

Sperry, R.W., Gazzaniga, M.S. and Bogen, J.E. (1969). Interhemispheric relationships: the neocortical commissures; syndromes of hemispheric disconnection. In Handbook of Clinical Neurology, Vol.4. (eds. P.J. vinken and G.W. Bruyn). North-Holland, Amsterdam, 273-290.

Sperry, R.W., Myers, R.E. and Schrier, A.M. (1960). Perceptual capacity of the isolated visual cortex in the cat. The Quarterly Journal of Experimental Psychology, 12, 65-71.

Sperry, R.W., Zaidel, E. and Zaidel, D. (1979). Self-recognition and social awareness in the deconnected minor hemisphere. Neuropsychologia, 17, 152-166.

Sprague, J.M. (1966). Interaction of cortex and superior colliculus in mediation of visually guided behavior in the cat. Science, 153, 1544-1547.

Stamm, J.S. and Sperry, R.W. (1957). Function of corpus callosum in contralateral transfer of somesthetic discrimination in cats. Journal of Comparative and Physiological Psychology, 50, 138-143.

Teng, E.L. and Sperry, R.W.(1974). Interhemispheric rivalry during simultaneous bilateral task presentation in commissurotomy patients. Cortex, 10, 111-120.

Teuber, H.L. (1960). Perception. In Handbook of Physiology, Section 1, Neurophysiology, vol. III. (eds. J.Field, H.W.Magoun and V.E.Hall). American Physiological Society, Washington, D.C., 1595-1668.

Trevarthen, C. (1962). Double visual learning in split-brain monkeys. Science, 136, 258-259.

Trevarthen, C. (1965). Functional interactions between the cerebral hemispheres of the split-brain monkey. In Functions of the Corpus Callosum (Ciba Foundation, Study Group No. 20). (ed. E.G.Ettlinger). Churchill, London.

Trevarthen, C. (1968a). Two mechanisms of vision in primates. Psychologische Forschung, 31, 299-337.

Trevarthen, C. (1968b). Vision in fish: the origins of the visual frame for action in vertebrates. In The Central Nervous System and Fish Behaviour. (ed. D. Ingle). Chicago University Press, Chicago.

Trevarthen, C. (1970). Experimental evidence for a brain-stem contribution to visual perception in man. Brain, Behaviour and Evolution, 3, 338-352.

Trevarthen, C (1972). Brain bisymmetry and the role of the corpus callosum in behavior and conscious experience. In Cerebral Interhemispheric Relations. (eds. J. Cernacek and F. Podovinsky). Slovak Academy of Sciences, Bratislava.

414

Trevarthen, C. (1974a). Analysis of cerebral activities that generate and regulate consciousness in commissurotomy patients. In Hemisphere Function in the Human Brain. (eds. S.J. Dimond and J.G. Beaumont). Paul Elek, London.

Trevarthen, C. (1974b). Functional relations of disconnected hemispheres with the brain stem and with each other: monkey and man. In Hemispheric Disconnection and Cerebral Function. (eds. M.Kinsbourne and W.L. Smith). Thomas, Springfield, Ill., 187-207.

Trevarthen, C. (1975). Psychological activities after forebrain commissurotomy in man. Concepts and methodological hurdles in testing. In Les Synodromes de Disconnection Calleuse chez l'Homme. (eds. F.Michel and B. Schott). Hopital Neurologique, Lyon, 181-210.

Trevarthen, C. (1978). Manipulative strategies of baboons and the origins of cerebral asymmetry. In The Asymmetrical Functions of the Brain. (ed. M. Kinsbourne). Cambridge University Press, New York and London.

Trevarthen, C. (1984). Biodynamic structures, cognitive correlates of motive sets and development of motives in infants. In Cognition and Motor Processes. (eds. W. Prinz and A.F.Saunders). Springer-Verlag, Berlin-Heidelberg-New York, 327-350.

Trevarthen, C. (1984). Hemispheric specialization. In Handbook of Physiology; (Section 1, The Nervous System); Volume 2, Sensory Processes (Section ed. I. Darian-Smith). American Physiological Society, Washington, 1129-1190.

Trevarthen, C. and Sperry, R.W. (1973). Perceptual unity of ambient visual field in human commissurotomy patients. Brain, 96, 547-570.

Voneida, R.J. (1963). Performance of a visual conditioned response in split-brain cats. Experimental Neurology, 8, 493-504.

von Holst, E. and Mitelstaedt, H. (1950). Das Reafferenzprinsip (Wechselwirkungen zwischen Zentral Nervensystem und Peripherie). Naturwissenshaften, 37, 464-476.

Welch, K and Stuteville, P. (1958). Experimental production of unilateral neglect in monkeys. Brain, 81, 341-347.

Zaidel, D and Sperry, R.W.(1973). Performance of Raven's colored progressive matrices test by subjects with cerebral commussurotomy. Cortex, 9, 34-39.

Zaidel, D and Sperry, R.W.(1977). Some long-term motor effects of cerebral commissurotomy in man. Neuropsychologia, 15, 493-504.

Zaidel, E. (1977). Unilateral auditory language comprehension on the Token Test following cerebral commissurotomy and hemispherectomy. Neuropsychologia, 15, 1-18.

Zaidel, E. (1978). Auditory language comprehension in the right hemisphere following cerebral commissurotomy and hemispherectomy. A comparison with child language and aphasia. In Language Acquisition and Language Breakdown: parallels and Divergences. (eds. A. Caramazza and E.B. Zurif). John Hopkins University Press, Baltimore, MD.

25
Specificity and Plasticity in Interhemispheric Integration: Evidence from Callosal Agenesis

Malcolm A. Jeeves and A. David Milner

INTRODUCTION

The failure of growth of the corpus callosum as a commissural bridge between the cerebral hemispheres is a rare condition (Ettlinger, 1977), and it is particularly rare for such individuals to grow to maturity and to be otherwise essentially normal in cognitive and neurological terms. However the systematic behavioural study of these unusual individuals may contribute to both of the issues which constituted the subtitle of this conference: namely, the "unified functioning" and "specialization" of the hemispheres.

We wish to consider each of these aspects in turn in relation to the congenitally acallosal brain. First, we address the question of interhemispheric integration at maturity. A discussion of some recent data on the scope and limits for both perceptual integration and motor integration will thus form the next section of this paper.

We then turn to the role of the corpus callosum in the ontogeny of cerebral function, and in particular, its participation in the development of cerebral specialization, which has been speculated upon by a number of theorists (e.g. Gazzaniga, 1970; Moscovitch, 1973). It is obvious that a crucially important source of evidence for evaluating such ideas is provided by the 'experiments of nature' which congenital absence of the corpus callosum gives us. We will discuss this question in the light of recent observations. We shall also describe some recent data which confirm that the corpus callosum participates in other

*The authors gratefully acknowledge the support of the UK Medical Research Council and the Wellcome Trust for these investigations. They also thank Professor G. Ettlinger and Drs M. D. Rugg and P. H. Silver for their helpful comments on the draft of this paper.

aspects of cerebral development, as argued by Dennis (1976), particularly in the refinement of manual prehension.

We shall restrict ourselves in the present report to presenting data on two totally acallosal subjects, the young woman KC and the young man BF. Further details of these patients may be obtained from our previous reports (Reynolds and Jeeves, 1977; Milner, 1982; Rugg, Milner & Lines, 1985).

INTERHEMISPHERIC INTEGRATION AT MATURITY

We wish to conceptualise integrated action as requiring two interacting factors. First the convergence of afferent information such that environmental events of a given significance, regardless of the particular location or modality of the receptors upon which they impinge, can give rise to a common central representation. Second the central organization of motor outputs such that a coordinated plan can be flexibly executed by means of whichever effectors are required.

Perceptual Integration

Simple visual information. The simplest possible task in which sensory integration may participate is one in which the onset of a punctate light source lateralized to left or right visual hemifield (VHF) is responded to with an invariant finger response by the contralateral hand. It has long been assumed that such "crossed" reaction times (RTs), e.g. of the right hand in response to a left VHF stimulus, will require an extra neuronal relay across the corpus callosum, as compared with uncrossed RTs (Poffenberger, 1912). This supposition is strengthened by the observation that acallosal patients exhibit greatly elevated crossed RTs. Thus the crossed-minus-uncrossed difference (CUD) in their RTs is usually in excess of 20 msec, as compared with mean values of only 2-3 msec in normal subjects (Jeeves, 1969).

Evidence that these lengthened crossed RTs are indeed mediated by visually-coded relay neurones has been afforded by separate investigations of two young adults with total callosal agenesis, one female (KC) and one male (BF). In both, significant effects on the CUD have been found as an inverse function of varying the intensity of the lateralized visual stimulus (Milner, 1982; Milner et al, 1985). Such intensity effects have not been found in normal subjects (Milner and Lines, 1982), and indeed it is generally believed that the normal CUD is based on nonsensory callosal transmission (Berlucchi, 1978; Rizzolatti, 1979). If so, variations in the stimulus would not be expected to influence the interhemispheric transmission time. Interestingly, the completely commissurotomized patients LB and NG have recently been studied in a similar fashion, and their CUDs have been found to be even longer than those of acallosals (Sergent and Myers, 1985). However this effect was not influenced by stimulus intensity, suggesting that the 'acallosal route' was not available to the split-brain subjects. One way of understanding these results would be to suppose that crossed RTs in our acallosal patients exploited the (probable) existence of the anterior commissure (which indeed is enlarged in some acallosals). This route is known to carry visual information in monkeys (Hamilton and Vermeire, 1986), but is believed to have been cut in LB and NG. It is possible that they have been forced into using even-less-efficient ipsilaterally-controlled movements on crossed trials. If this is so one should predict from animal studies (e.g. Brinkman & Kuypers, 1973) that they would have great difficulty in making crossed responses if a precise finger-thumb apposition movement was required rather than the closure of a Morse key which could be achieved by means of a crude hand movement (N.B. such a variation does not affect CUDs in normal subjects: Milner et al, 1986).

We have attempted to follow up our findings on KC and BF by recording visual evoked potentials (VEPs) during performance of RT tasks by these two acallosal subjects. Previous work showed that in normal subjects a lateralized light onset gives rise to a number of consistent components in the averaged VEPs which can be recorded from electrode sites over both hemispheres. Of particular interest is a negative deflection occurring at around 160 msec post-stimulus. This "N160" component occurs later (and with smaller amplitude) at ipsilateral electrode sites than on the contralateral (presumably directly-stimulated) side. These latency differences amount to around 12-15 msec between symmetrical occipital sites and around 3 msec at central sites (Rugg, Lines & Milner, 1984, 1985). To identify the latter difference with the behavioural CUD in normal subjects would be consistent with the fact that in both cases the difference values are invariant as a function of stimulus light intensity (Milner and Lines, 1982; Lines et al, 1984). On the other hand the contralateral/ipsilateral difference in VEP latency at the occipital sites does so vary (Lines et al, 1984).

Our intention then was to try to elucidate further the transmission of elementary visual information in the acallosal brain by identifying and studying the ipsilateral "N160" in our two patients. Unfortunately, this has proved impossible, despite the unambiguous presence of this N160 component, in essentially normal form, at the contralateral electrode sites (Rugg, Milner & Lines, 1985). Nonetheless, the abnormal nature of the ipsilateral recordings (whichever VHF is stimulated) clearly help to validate the VEP technique we have developed as a tool for measuring callosal transmission in normal, and potentially other neurological, subjects. These findings also do not negate our previous inference that elementary visual information does get across in the acallosal brain, e.g. through the anterior commissure (pars interhemispherica). For example, it could be that such transmission is of variable latency; for events to be detected in averaged VEP waveforms they need to be time-locked to the external stimulus. Alternatively, the transmitted information might activate neuronal populations whose anatomical disposition makes their activity undetectable with scalp electrodes.

Spatial versus pattern information. It has been known for some years that more complex perceptual information can be integrated in the acallosal brain. Many experimenters have observed that objects felt in either hand can be vocally named by acallosal subjects, and likewise that words or pictures presented to either VHF can be vocally identified. (There is a recent exception to this in the report by Donoso & Santander, 1982, who describe an acallosal patient who was able to name letters and words presented to the right but not the left VHF. The authors speculate that this patient lacked the anterior commissure as well as the corpus callosum). The likelihood that these performances reflect the interhemispheric transmission of perceptual information rather than bilateral speech organization is increased by the observation that acallosal subjects can match bilaterally presented patterns and colours (Ettlinger et al, 1972) as well as words (Milner and Jeeves, unpublished data), in vision, and shapes and other qualities in touch (Ettlinger et al, 1974).

In order to try to identify the channel(s) for such putative information transmission, it is clearly helpful to seek to define its limits, in both touch and vision. An interesting attempt to establish such qualitative limits has been put forward by Martin (1985). He has argued that the route for interhemispheric integration in the acallosal brain is indeed the anterior commissure, and that consequently one would expect from nonhuman primate anatomical studies (e.g. Jouandet and Gazzaniga, 1979; Pandya and Rosene, 1985) that it would be limited to carrying information processed in temporal association cortex. (There is also a minor projection from caudal orbitofrontal cortex through the anterior commissure: Barbas and Pandya, 1984. However this would not be expected to participate in perceptual integration.) Behavioural and physiological research on monkeys tends to suggest that, at least in vision, this region of cortex emphasises pattern or object analysis at the expense of spatial qualities, in contrast to posterior parietal cortex, which may be relatively specialized for spatial analysis (Ungerleider and Mishkin, 1982). Hence it would follow that the anterior commissure might transmit pattern information, but not information about spatial location.

In support of this proposal, Martin (1985) reported a series of elegant tests carried out upon an acallosal patient, GB: it was clearly demonstrated that this subject was well able to identify words, shapes, and letters in either VHF, but was very poor indeed at reporting the spatial location of stimuli in his right VHF, despite good performance on the left. This pattern of performance would be expected if there was a strong lateralization of cognitive functions in GB, with spatial processing being dependent upon the right parietal region, in association with an interhemispheric transmission failure for such processing.

An independent report by Meerwaldt (1983) could be interpreted in very similar fashion. His acallosal patient was tested using the Rod-Orientation Test of De Renzi et al (1971), in which a subject must set a moveable hinged rod such as to match a model, both with respect to its orientation in a vertical plane and in the horizontal plane. This patient, when tested in the tactile modality performed very poorly with the right hand only; performance with the left hand was within the normal range. Once more this pattern of results could be explained if there was a right-parietal dependence for performing the task (as is indicated by other work: De Renzi et al, 1971; Meerwaldt and Harskamp, 1982) along with an associated failure of interhemispheric integration. Again the anatomy of the anterior-commissure would be consistent with this interpretation; its cells of origin in the monkey are distributed throughout the temporal lobe (Jouandet and Gazzaniga, 1979), including the multisensory region of superior temporal sulcus (see Desimone and Gross, 1979).

Because of the potential importance of these ideas, we have attempted to establish their generality by carrying out similar tests on our two patients KC and BF. First we considered the visual modality, and used a test similar to one devised by Martin (1985). It involves the simultaneous tachistoscopic presentation of two letters each in a different location within the same VHF, followed by a prolonged presentation of a 5 x 9 matrix of numbers each referring to a different possible location on the screen. The subject is asked to identify each letter along with its place of occurrence. We used a tachistoscopic exposure time of 150 msec for KC and BF, and in order to roughly equalise performance levels, lower exposure times (50-100 msec) for six normal controls. We used a projection system, enabling us to allow subject KC to point to locations on the matrix rather than to identify them by number, which she found difficult. The overall angular size of the stimulus matrix was 18°, i.e. extending 9° into the left and right VHFs, giving a mean lateral eccentricity of the stimuli of 5°.

Our results are shown in Table 1, along with those for Martin's patient GB. It is apparent that we found no deficit in the right VHF for spatial localization performance in either of our patients. Nonetheless there is the expected tendency for letter identification to be superior in the right VHF in our

Table 1. Performance of two acallosal patients in the Martin task

Subject	Spatial Localisation[a]		Letter Identification[a]	
	Left VHF	Right VHF	Left VHF	Right VHF
KC	88	80	55	60
BF	61	70	27	41
Normal Controls (N = 6)	80 (58–89)	84 (73–92)	42 (26–57)	47 (25–52)
GB[b] (Martin, 1985)	70	29	44	50

[a] Performance scores are given as percent correct.
[b] Scores estimated from published graph.

patients (it reaches the .05 level of significance on a chi-square test for BF), and their accuracy on probe trials where letters were presented on the vertical midline of the display reached 83% and 71% respectively.   Since this is markedly better than performance off-centre, it is consistent with accurate central fixation by our patients, which concurrent visual monitoring confirmed.

We next examined the generalizability of Meerwaldt's finding. Each of our two acallosal patients, along with five normal controls, was tested with right and left hand using five different rod settings (cf De Renzi et al, 1971). The results are shown in Table 2 together with Meerwaldt's data. Our controls, like Meerwaldt's, showed no hand asymmetries on any of the three measures: horizontal error, vertical error, or time taken. Patient BF also showed no evidence for a hand asymmetry. KC did show smaller horizontal errors with her left hand than with her right, but it is debatable whether the right hand performance can be regarded as impaired, given the short times taken. (We also tested KC and BF on the visual version of De Renzi's task, and like Meerwaldt, we found no difference between the hands.) Again therefore we have not been able to clearly replicate the published observations of a right-sided spatial impairment, although KC's data are, up to a point, consistent with Meerwaldt's report.

Table 2. Performance of two acallosal patients in the
Rod–Orientation task

Hand:		Horizontal Error[a]		Vertical Error[a]		Time Taken[b]	
		Left	Right	Left	Right	Left	Right
Subject							
KC		12.8	21.9	7.7	10.0	24.8	27.7
BF		7.6	8.6	7.6	8.2	53.1	54.4
Normal Controls	Mean	10.2	10.2	5.4	4.6	42.9	38.6
(Range	Min	3.6	3.6	4.2	9.8	35.2	28.7
N = 5)	Max	14.6	15.4	6.8	9.4	59.4	49.3
Meerwaldt's patient (1983)		7.0	28.6	5.8	18.0	83.0	123.0

a Error measured in degrees
b Time measured in seconds

Although our tachistoscopic findings indicate that Martin's observations on GB are not generalizable to all acallosals, they do not rule out his main contention, namely that the anterior commissure is unable to carry visuospatial information. It is possible that in our two patients there is not the normal degree of lateralization for spatial functions, so that either hemisphere can perform the task without the need for communication with the other. Such bilateralization might likewise enable our two subjects to perform De Renzi's tactile task satisfactorily with either hand. We decided to use a spatial matching task, in both modalities, in order to investigate this possibility.

In the visual modality, we assessed spatial matching in KC (and four normal controls) using a test taken from Holtzman et al (1981). Two 3 x 3 matrices of squares were presented on a TV monitor, one at either side of a central fixation point. An 'x' would appear in one of the squares on each side, following a warning signal, for a 150 msec duration (50 msec for controls) and the subject was required to press either a "same" or "different" button according to whether the spatial locations did or did not match in absolute terms (e.g. bottom left square on one side matching bottom left on the other side). In a control task, only the central square of each matrix was used, but 9 different digits could appear, the subject again being required to make a same/different judgement. Martin's hypothesis would predict normal performance on digit matching, but an impairment on spatial matching.

As shown in Table 3, KC performed clearly above chance (p < 0.01) on both spatial and digit matching. Although, unlike any of the controls, she did (significantly) less well on the former task, an inspection of the data revealed that this could be attributed to an increased incidence of "mirror-image" errors, in which symmetrical spatial locations were responded to as "same" instead of "different". Normal subjects also sometimes made such errors. It seems likely that they occurred as a result of cognitive failures; certainly their occurrence would not reflect a spatial transmission failure. [KC's performance on spatial matching is consistent with Holtzman et al's (1981) report of above-chance performance on the same task in one of their two callosal-section patients whose anterior commissure was spared.]

A parallel prediction of poor intermanual matching of tactile orientation follows from Martin's hypothesis insofar as it can account for Meerwaldt's data. We therefore tested KC on an intermanual version of the Rod-Orientation task in which either the left hand felt the model and the right hand attempted to reproduce it, or vice-versa. Table 4 shows that if the two hands tested alone are averaged, KC performed slightly less accurately (though more quickly) than the controls. When intermanual testing was done, KC performed hardly any worse, unlike the controls. Accordingly it is clear that even though she had shown a slight

Table 3. Interfield matching in vision

Subject	Spatial Matching[a]	Digit Matching[a]
KC	77	93
Normal Controls (N = 4)	91.1 (83-96)	84.2 (78-87)

[a] Performance expressed as percent correct

Table 4. Intermanual performance on the Rod-Orientation task

Subject		Unimanual[a]			Intermanual		
		HE[b]	VE[b]	T	HE[b]	VE[b]	T
KC		17.4	8.9	26.3	20.2	9.5	22.1
Normal Controls	Mean	10.2	5.0	40.8	14.4	8.9	63.8
(Range	Min	7.2	3.9	32.0	9.1	6.2	32.1
N = 5)	Max	14.6	7.0	54.4	17.7	11.7	82.4

[a] Mean of left and right-hand data from Table 2
[b] HE = horizontal error; VE = vertical error

inferiority with the right hand, this can not be accounted for in terms of an interhemispheric transmission failure.

In summary, it can be seen that we have no clear evidence thus far for a qualitative limitation on interhemispheric integration of perceptual information in our acallosal patients. Consequently Martin's argument which led to the implication of the anterior commissure in such integration cannot be sustained on our data.

Motor Integration

The organization of skilled action is dependent upon the continuous use of feedforward as well as feedback information about component movements. It has been known for many years (Jeeves, 1965) that acallosal patients are handicapped in their performance of even simple skilled activities when rapid execution is required. There have been broadly two reasons proposed for this. One is that 'motor' information cannot be adequately distributed between the hemispheres (Milner and Jeeves, 1979; Jeeves, 1986); it may be assumed that the extensive connections between motor and premotor cortical areas which normally pass through the rostral half of the body of the corpus callosum (Pandya et al, 1971; Pandya and Rosene, 1985; de la Coste et al, 1985) are required for this. The other is that the final motor output pathways are compromised, as the result of competition between contralateral and ipsilateral control systems, the latter having become functionally strengthened through disinhibition during ontogeny (Dennis, 1976). We shall return to this second hypothesis in a later section.

The interhemispheric integration of motor control is well exemplified by rapid bimanual skills such as that acquired by the normal subjects of Preilowski (1972). He used an X-Y plotter in which subjects had to guide a pen between parallel straight lines marked on a piece of paper. One crank handle, turned anticlockwise by the left hand, moved the pen in the upward direction (Y-coordinate) and the other, turned anticlockwise by the right hand, moved it leftwards (negative X-coordinate). Complete cerebral commissurotomy evidently prevented subjects from acquiring this skill except in the simplest line-orientations; normal subjects became very quick at all orientations, and the "automaticity" of their skill was demonstrated on trials where visual feedback was removed halfway through. This bimanual skill seems likely to require the continuous monitoring, perhaps by a dominant left hemisphere, of corollary information on the other hemisphere's outputs.

We have trained our two acallosals on this task over nine sessions, and they have performed far better than the complete commissurotomy patients; however, on bimanual trials they never reached the speed and accuracy levels attained by adult or even 10 year-old controls (Jeeves et al, 1986). Furthermore, it seems that the performance of both KC and BF depended critically upon visual feedback. Thus when this was excluded halfway through test trials, their 112.5° and 157.5° pen lines drifted towards 135°, i.e. towards the angle produced by synchronous rotation of the two handles. As Table 5 shows, these deviation errors were more pronounced in KC, although she tended to perform the task visually at a higher speed than BF when visual feedback was available. Interestingly, these results closely resemble those reported by Preilowski (1972) for the two partially commissurotomized patients DM and NF. In these patients, although the anterior commissure and the anterior two-thirds of the corpus callosum were sectioned, the splenium remained intact. This permitted good performance on a number of tests of perceptual integration (Gordon et al, 1971), and it may be assumed that visually-guided performance on the X-Y plotter task was made possible through the mediation of these intact fibres. In the case of KC and BF, it seems reasonable to attribute this visual role to the anterior commissure. However in both DM and NF and our two patients, evidently no purely motor integration was possible.

We conclude that "motor" fibres, probably in the rostral body of the corpus callosum are normally critical for such integration, and that their absence can not be compensated for through other commissures. It thus seems likely that in callosal agenesis the cells which normally give rise to such callosal fibres do not get re-routed through the anterior commissure; it is conceivable that

Table 5. Performance on the X-Y plotter task (Sessions 8 and 9)

Subject		Mean Time on Normal Bimanual Trials[a]			Deviation on 'Blind' Testing[b,c]		
		135°	112.5°	157.5°	135°	112.5°	157.5°
KC		4.73	5.82	5.04	−0.40	+15.55	−6.40
BF		6.35	7.32	6.12	+1.85	+8.90	−1.45
Normal Controls	Mean	3.52	4.15	3.93	+0.36	+0.32	−0.14
(Range	Min	2.58	3.06	3.32	−0.90	−1.88	−7.20
N=10)	Max	4.72	5.40	5.06	+1.35	+1.50	+3.55

[a] Time in seconds required to complete a 15.5 cm line
[b] Mean deviation in degrees following exclusion of visual feedback
[c] A positive value indicates an error in the anticlockwise direction

at least some of them instead contribute to the longitudinal bundle of Probst (Stefanko and Schenk, 1979) and terminate intrahemispherically.

INTERHEMISPHERIC PROCESSES IN ONTOGENY

Development of Hemisphere Specialization

There are known to be morphological asymmetries in the cerebral hemispheres already at birth (Wada et al, 1975; Witelson and Pallie, 1973). Given that there exists a statistical relationship between measures of the brain's morphological and functional asymmetries at maturity (Ratcliff et al, 1980), it is reasonable to suppose that the basic neural substrate for the adult specializations already exists asymmetrically at birth. There is disagreement, however, as to whether there is sharply asymmetrical development after birth, or whether there is a period in childhood before the corpus callosum becomes fully functional during which parallel development of (e.g.) language mechanisms takes place. The former idea (Kinsbourne and Hiscock, 1977; Segalowitz, 1983) derives support from various studies which have found tachistocopic and dichotic asymmetries in young children (although there are some exceptions: e.g. Reynolds and Jeeves, 1978) and from recent analyses which indicate that aphasia-producing lesions in children show a left-side predominance much as in adults (Carter et al, 1982).

On the other hand, the idea that following the completion of myelogenesis in the corpus callosum (around the age of 10) a relatively "strong" lateralized system is able to progressively suppress its weaker partner in the contralateral hemisphere so as to produce the adult asymmetry, has its attractions and numerous adherents (Moscovitch, 1973, 1977; Selnes, 1974; Galin, 1977; Davidson, 1978; Denenberg, 1981). An early bilateral development would have adaptive advantages by providing a fail-safe in the event of brain damage in childhood; it is known that recovery from early aphasia is far more frequent than from adult aphasia (e.g. Lenneberg, 1967).

Clearly a failure of hemispheric specialization in acallosal brains, with the development of bilateral control of speech and other functions, would provide strong support for the second idea. Indeed some of the proponent theorists have argued that such a failure has been demonstrated (e.g. Denenberg, 1981). However, as has been argued elsewhere (Milner and Jeeves, 1979, 1981; Chiarello, 1980; Milner, 1983; Jeeves, 1986) no such demonstration has yet been achieved. There are a number of lines of evidence for functional asymmetries in callosal agenesis.

(i) The amytal test. This is the most direct source of information, in the absence of data on aphasia following unilateral lesions in acallosals. Three bilaterally-tested acallosals have come to our notice. The first patient was tested by B. Milner (personal communication; and cited by Gazzaniga, 1970), and was found to have speech control vested only in the right hemisphere. The other two have shown evidence for bilateral control of speech (Gott and Saul, 1978; McGlone, 1985), an arrangement apparently found in only a small proportion of the general population (Rasmussen and Milner, 1977). However among left-handers the estimated incidence rises to 15%, and among left-handers who suffered early brain damage to 19% (Rasmussen and Milner, 1977). In point of fact all three of the amytal-tested acallosal patients were left-handed, and most acallosal patients show evidence of at least mild hemispheric pathology during childhood. But in any event, as we have pointed out elsewhere (Milner, 1983), only one case of lateralized speech control in callosal agenesis is needed to disprove the generalization that callosal processes are a necessary condition for such lateralization. B. Milner's patient, if totally acallosal, would seem to provide that instantiating case.

(ii) VHF recognition tests. Many observers have confirmed that most acallosal subjects show a right VHF advantage for word recognition; we provide some results on KC and BF in Table 6 to add to this body of data. They were presented with high-frequency 4-letter words taken from the norms of Kucera and Francis (1967), exposed singly for 150 msec to the left or right VHF at a minimum eccentricity of 2°. The right VHF advantage reached significance for KC on a Fisher exact test (p < .05). It may be recalled that KC and BF also showed a right VHF advantage on letter recognition in Martin's task; in that instance BF's asymmetry reached significance. In a different type of VHF study, we measured simple vocal RTs to lateralized light flashes in KC. She showed a highly significant right VHF advantage; yet no overall VHF asymmetry was apparent in manual RT testing (Milner, 1982). Thus indirect tests of both linguistic recognition and production clearly indicate clear moderate-to-strong asymmetries.

(iii) Dichotic testing. As summarised by Chiarello (1980), the usual picture in callosal agenesis is for small but reliable dichotic asymmetries. We have tested KC, on traditional tasks of digit recall and CV-syllable reproduction, and on dichotic monitoring (using the technique of Geffen and Caudrey, 1981). She has shown a consistent right-ear advantage (REA) on all methods of testing. On the monitoring task she was tested on three occasions separated by several months. Her results show a clear and consistent REA in terms of both correct identification and of response latencies. Successive testing gave the following results on the three occasions (1) percentage of correct right ear 69%, percentage correct left ear 51%; (2) right ear 77%, left ear 60%; (3) right ear 65%, left ear 51%. This is evidence consistent with a normal degree of asymmetry in at least some acallosal brains; but that does not, of course, imply a normal mechanism for left-ear performance in agenesis, since that appears to depend substantially on the corpus callosum (data reviewed by Geffen and Quinn, 1984). In the acallosal patient, left ear information could reach the left hemisphere through ipsilateral pathways, or

Table 6. Tachistoscopic recognition of words

Subject	Left VHF[a]	Right VHF[a]
KC	4/12	9/12
BF	3/12	6/12

[a] Scores are number of correct identifications

perhaps via the right hemisphere and the anterior commissure.

(iv) Motor control. A survey of the literature shows that most acallosal patients are described as right-handed (Milner and Jeeves, 1979). A more constrained survey by Chiarello (1980) gave the same result. Although a greater proportion than normal are described as non-right-handed, this would be expected in a neurological population. We have carried out a series of manual performance tests with KC (using the tests and questionnaire summarized by Provins et al, 1982). In every case her asymmetry favoured the right hand, and to an extent near the population average.

In bimanual testing, it is interesting that in the two partial-sectioned patients of Preilowski (1972), an asymmetry emerged in the test probes given at the end of training in which vision was occluded. At 112.5°, where in effect the patient had to inhibit his right hand, performance deteriorated more than at 157.5°, where he had to inhibit his left hand. Preilowski interpreted this asymmetry as reflecting the left hemisphere dominance for motor coordination clearly evident from lesion studies (e.g. Kimura, 1979). To the extent that this finding reflects the normal motor specialization of the left hemisphere, our finding of the same asymmetry in KC and BF (see Table 5) supports the view that their brains have this same specialization.

Discussion. The evidence we have reviewed tends to point towards the conclusion that the corpus callosum is not essential during childhood for lateralization to develop. Accordingly it favours the contrary view that no inhibitory interaction between the hemisphere is required during ontogeny. However there is an alternative interpretation. If indeed it is the case as some anatomists have argued (Bossy, 1970) that some "callosal" fibres are re-routed and pass instead through the anterior commissure in acallosal brains, then it is conceivable that these fibres could mediate the interhemispheric modulation hypothesized by Moscovitch (1977) and others.

Clearly it will be of great interest to be able to use NMR imaging in future years to determine the presence or absence (and if present, the size) of the anterior commissure in acallosal individuals. Very different predictions would be made by the two viewpoints summarised here: if commissural interaction is necessary for lateralization, one would predict decreased specialization in patients lacking the anterior commissure, but if it is not, then one would expect increased behavioural asymmetries in such patients. It is perhaps notable that there are isolated claims of grossly enlarged VHF or hand asymmetries in individual acallosal patients (Kretschmer, 1968; Field et al, 1978; Donoso and Santander, 1982; Meerwaldt, 1983): it would be of great interest to know whether such patients are in that minority of acallosals that lack the anterior commissure (Lemire et al, 1975).

Development of Fine Motor Control

It seems clear from the work of Kuypers and his colleagues (e.g. Brinkman and Kuypers, 1973; Kuypers, 1978) that although both ipsilateral and contralateral motor control can be exerted by either hemisphere of the normal brain, only the corticospinal system (which is purely contralateral) provides the fine control of the fingers needed for precise grasping and manipulation. Recent work also indicates that the preparatory shaping of the hand in reaching for an object is similarly dependent on this system (Jeannerod, 1986). Dennis (1976) has argued that in normal development, the ipsilateral control system is held in check through inhibitory inputs from the contralateral system in the opposite hemisphere. Although the corticospinal projections do not establish their effective synapses until infancy (e.g. Kuypers, 1962) they would be well developed by the time that callosal maturation is well-advanced. By maturity, it is assumed that the crude ipsilateral system will always be over-ruled by the more effective contralateral system, whenever fine control is called for.

As mentioned earlier, callosal agenesis tends to be characterized by slow and clumsy performance even of unimanual acts; Dennis explains this as a consequence of interference from a strong (disinhibited) ipsilateral motor system. The hypothesis is strengthened by her finding of an increased incidence of unintended synkinesic finger movements in two acallosal patients, although it should be noted that gross synkinesias were absent in most of the patients in an earlier study (Ettlinger et al, 1972). If Dennis is correct in interpreting her data as reflecting the disruptive activity of ipsilateral motor control in her acallosal patients, it seems likely that this must be attributed to an abnormal over-development of such control rather than a simple disinhibition at maturity. If the latter were the case, then surgical patients would behave in the same way as Dennis's acallosals, for which we know of no evidence. Certainly Brinkman and Kuypers (1973) found that their split-brain monkeys, when permitted to use both eyes (thus providing visual information to both hemispheres) were perfectly well able to use their contralateral control system for precision-grasp actions. (This is not to say that split-brain patients cannot learn to exploit ipsilateral motor pathways; such patients recover some ability to write, and to obey commands, with the left hand. However this appears not to interfere with or to supercede the contralateral control pathways).

We have used a lateralized test of prehension with KC and BF. Subjects were required to fixate a central point, and following the illumination of an elongated object to the left or the right by a single strobe flash lasting 150 msecs, they had to reach out and grasp it. Their behaviour was videotaped. If there were complete disconnection, then as with the split-brain monkeys of Brinkman and Kuypers, one would expect poor performance only with the crossed combinations of VHF and hand, whilst for example, left hand reaching for a left VHF object should be performed well. However since lateralized visual information can apparently reach both hemispheres, and if fine motor control is chronically subject to ipsilateral interference, one would expect impairment under all conditions. This is what we have found. In KC particularly, poorly-oriented hand formation was frequently apparent, as were fumbling attempts to grasp. Jeannerod (1986) has described the pattern of finger grip formation during natural prehension movements in normal subjects with the help of a quantified film technique. We have adapted this technique to study videotapes of the acallosal KC and normals making similar prehension movements. The change in grip size over time was plotted as in Jeannerod's (1986) Fig. 2. A preliminary analysis showed that the pattern of results for our normals was similar to Jeannerod's. The results for KC showed a characteristically different and flatter function when plotting grip size against time. The grip size was not reduced to match the size of the object prior to contact as was the case with normals. On the other hand, simple reaching towards point sources (aiming) is essentially normal in both patients, as shown in control tests. This would be expected, since gross arm movements (mediated by proximal muscles) would presumably not require corticospinal control. The dissociation is shown in the latency measures summarised in Table 7. We would predict in contrast that only crossed VHF/hand grasping should be impaired in commissurotomized patients if they were to be tested in this manner. Furthermore, since according to the ontogenetic theory these patients should not suffer undue interference from the ipsilateral control system, grasping for centrally presented objects should be unimpaired. It should be noted however that rapid skilled action is impaired in commissurotomized patients (Zaidel and Sperry, 1977) as in acallosal subjects, despite there having presumably been a normal ontogenetic growth of contralateral dominance in motor control. It is possible that in the years following the surgery some reversal of this dominance might take place to permit useful outputs to both hands from lateralised brain mechanisms (e.g. as argued by Gazzaniga, 1970). If so, signs of ipsilateral interference in tasks like ours should become apparent over time.

We conclude that in our patients (although at present we only have quantitative data for KC) there is a generalized disruptive influence of the ipsilateral motor system, even when the contralateral hemisphere is directly (and therefore earlier in time) given the necessary visual information to guide action. Clearly this supports Dennis's hypothesis.

Table 7.   Reaching and grasping performance in K.C.

Subject	Grasp Latency[a]		Reaching Latency[a]
	Crossed	Uncrossed	All Conditions
KC	1205	1193	780
Normal Controls (N = 6)	650 (325–838)	648 (269–775)	790 (553–990)

[a] Times are given in milliseconds

CONCLUSIONS

Some of the earliest studies of acallosal patients (Jeeves and Rajalakshmi, 1964; Jeeves, 1965) indicated deficits on perceptual motor tasks performed under speed stress. These findings have been replicated by other workers (see Chiarello, 1980). The majority of the studies reviewed in this chapter can be brought together by asking how they help to understand the poor performance of acallosals on such perceptual motor tasks. On the one hand, some studies increase our understanding of the ability of acallosals to process sensory inputs whether visual, tactile or auditory and to integrate them in readiness for an appropriate motor output. On the other hand, we have reported studies which directly investigated possible sources of difficulty in the motor output side of perceptual motor tasks.

As regards the input side, we saw that on tasks where a simple manual response was required to a lateralised point source of light the acallosal deficit observed could be attributed to using an alternative sensory interhemispheric pathway rather than the normal callosal one. Thus when the intensity of the stimulus was varied interhemispheric transmission time co-varied suggesting that when this task is done by acallosals we are measuring interhemispheric sensory transfer time and not motor transfer as is presumed to be the case in normals. The possible alternative pathways to consider are the anterior commissure or subcortical pathways. Sergent (1986) has presented evidence from studies designed to examine the capacity of commissurotomized patients to process information simultaneously received by each hemisphere and to produce a unified response after bilateral stimulation presentation, which she believes "suggest that subcortical structures subserve interhemispheric integration . . .". In support of her view she cites earlier work on animals (Trevarthen, 1965) as well as work by Trevarthen and Sperry (1973) and Holtzman (1984) on callosotomized subjects.

However, for the tasks we have administered the subcortical pathways seem unlikely candidates since when patients with surgically disconnected commissures including both the corpus callosum and the anterior commissure were investigated on similar tasks (Sergent & Myers, 1986) it was found (a) that their interhemispheric transmission times were more greatly lengthened than were those of the acallosals, and (b) that there was no variation as a function of stimulus intensity. If subcortical pathways mediated the transfer of visual sensory information in our acallosal patients, then they should act similarly in both of these respects in the sectioned patients.

As regards the transfer of spatial and pattern information it is clear that acallosals can integrate such information and a priori the most likely route for the visual pattern information would be the anterior commissure, connecting as it does parts of the temporal association cortex. Martin (1985) presented data supporting his view that pattern but not spatial information was transmitted and his results gained support from Meerwaldt's (1983) study. However, our own studies on two acallosal patients indicated that there was no marked deficit in the transfer of spatial information as measured by Martin's and Meerwaldt's procedures, although one of the patients did produce a pattern of results partially similar to Meerwaldt's acallosal patient.

Our studies requiring spatial matching in both the visual and haptic modalities gave no evidence of deficits. Before concluding too firmly that there are no deficits on the transfer of spatial information as tested by De Renzi's rod orientation test we need to exclude the possibility that patients could have used information from more proximal musculature. There has been a hint that such a strategy may have been attempted by some adult patients in our more recent studies. On balance, however, we believe that the anterior commissure remains the most likely candidate for the transfer of spatial as well as sensory information, and it may do this in two ways. First, the visually driven cells in the inferotemporal cortex (a subset of which project through the anterior commissure) have well-defined if rather large receptive fields and thus may together be able to code for location. (Indeed many cells in the posterior-parietal cortex in the monkey, thought to have a role in spatial perception, have larger receptive fields.) It is also apparent that lesions of inferotemporal cortex produce deficits on at least one "spatial" task, namely the reversal version of the landmark task (Pohl, 1973; Ungerleider and Brody, 1977). Therefore even in normal monkeys the anterior commissure may be capable of carrying information about location. Secondly, there may be a wider field of origin of cells contributing to the anterior commissure in the acallosal brain than is apparent in the normal brain. In some cases the commissure is larger than normal, and this may be because it contains fibres which would normally be destined for the corpus callosum (Bossy, 1970). It may also be the case that the acallosal brain preserves some of the exuberant inputs which it receives early in foetal life but which normally regress as the corpus callosum develops (Auroux and Roussel, 1967). In either case it is possible that parietal regions in at least some acallosal brains do have interconnections through the anterior commissure which could mediate visuo-spatial integration. This would make them analogous to the brains of marsupials, such as the opossum, in which widespread regions of the cerebral hemispheres are connected, in the absence of a corpus callosum, through the anterior commissure (Granger et al, 1985).

Turning now to the output side; there is evidence of two factors which could contribute to the poor performance observed on perceptuo-motor skills under speed stress. First, in the study of bimanual coordination using the X-Y plotter, it was evident that the acallosals whilst better than split-brain patients did not achieve the level of performance attained by normals. At the same time it was clear that the skilled performance of acallosals after extended practice is critically dependent on visual feedback throughout the task. This indicates that no purely motor integration through the existing midline structures is possible in the acallosal. In normals it is fibres in the rostral part of the callosum that are presumed to be critical for such integration. The second factor involves the continuing competition between the crossed and uncrossed motor outputs which result in less than optimal performance. It is assumed that in normal subjects inhibitory processes operating through the callosum prevent competition by the ipsilateral uncrossed pathways with the crossed pathways during ontogeny. Supporting evidence for this came from studies of grasping and reaching behaviour in one acallosal. Simple reaching and pointing behaviour was not affected whilst grasping behaviour was.

Our observations on the integration of both perceptual input and motor output may be combined to explain the observed behaviour of acallosals. First, it would seem that the difficulty with perceptual motor integration is due mainly to the motor output side rather than being at the level of perceptual integration: however exceptions to this may arise where rapid sensory feedback is at a premium. In those rare patients where there is a major deficit in perceptual integration (Donoso and Santander, 1982) it may be that both the anterior commissure and the corpus callosum are absent: in such cases more dramatic sensorimotor deficits would also be expected. In the normal subject the evidence indicates that to achieve highly skilled performance on bimanual tasks motor feedforward processes are crucial and these probably occur through the rostral parts of the corpus callosum. One might expect that in the early years of life where these pathways are not fully myelinated there may be deficits as compared with performance at and after maturity.

A major difficulty running through this research is uncertainty as to the status of the anterior commissure (specifically the pars interhemispherica) in any given acallosal patient. Evidence from CT scans is insufficient to determine whether the commissure is enlarged, or even whether it is present. However, on the assumption that it is present in most totally acallosal patients (Lemire et al, 1975), we would suggest that it probably provides the key to interpreting the performance of such patients on tests of interhemispheric integration.

With the rapid improvement of imaging techniques some of these unresolved issues are potentially resolvable. Studies using NMR imaging should make it possible not only to focus on the role of the anterior commissure, but also to study deficits associated with disease of, or damage to, specific parts of the corpus callosum.

REFERENCES

Auroux, M. and Roussel, C. (1967). Relations Neocorticales Assurees par la Commissure Blanche Anterieure Chez le Foetus Humain. C.R. Assoc. Anat., 137, 152-159.

Barbas, H. and Pandya, D.N. (1984). Topography of commissural fibers of the prefrontal cortex in the rhesus monkey. Exp. Brain Res., 55, 187-191.

Berlucchi, G. (1978). Interhemispheric Integration of Simple Visuomotor Responses. In Cerebral Correlates of Conscious Experience. (eds. P.A. Buser and A. Rougeul-Buser). North Holland, Amsterdam.

Bossy, J.G. (1970). Morphological Study of a Case of Complete, Isolated and Asymptomatic Agenesis of the Corpus Callosum. Arch. D'Anat., D'Histol. et D'Embryol., 53, 289-340.

Brinkman, J. and Kuypers, H.G.J.M. (1973). Cerebral Control of Contralateral and Ipsilateral Arm, Hand and Finger Movements in the Split-Brain Rhesus Monkey. Brain, 96, 653-674.

Carter, R.L., Hohenegger, M.K. and Satz, P. (1982). Aphasia and Speech Organization in Children. Science, 218, 797-799.

Chiarello, C. (1980). A House Divided? Cognitive Functioning with Callosal Agenesis. Brain and Language, 11, 128-158.

Davidson, R.J. (1978). Lateral Specialization in the Human Brain: Speculations Concerning its Origins and Development (Commentary). Behavioral Brain Sciences, 2, 291.

Denenberg, V.H. (1981). Hemispheric Laterality in Animals and the Effects of Early Experience. Behavioral Brain Sciences, 4, 1-49.

Dennis, M. (1976). Impaired Sensory and Motor Differentiation with Corpus Callosum Agenesis: A Lack of Callosal Inhibition During Ontogeny? Neuropsychologia, 14, 455-469.

De Renzi, E., Faglioni, P. and Scotti, G. (1971). Judgement of Spatial Orientation in Patients with Focal Brain Damage. J. Neurol. Neurosurg. Psychiat., 34, 489-495.

Desimone, R. and Gross, C.G. (1979). Visual Areas in the Temporal Cortex of the Macaque. Brain Res., 178, 363-380.

Donoso, A. and Santander, M. (1982). Hemialexia y Afasia Hemianoptica en Agenesia del Cuerpo Calloso. Rev. Chil. Neuropsiquiat, 20, 137-144.

Ettlinger, G. (1977). Agenesis of the Corpus Callosum. In Handbook of Clinical Neurology, Vol. 30, Part 1, "Congenital Malformations of the Brain and Skull". North Holland, Amsterdam.

Ettlinger, G., Blakemore, C.B., Milner, A.D. and Wilson, J. (1972). Agenesis of the Corpus Callosum: A Behavioural Investigation. Brain, 95, 327-346.

Ettlinger, G., Blakemore, C.B., Milner, A.D. and Wilson, J. (1974). Agenesis of the Corpus Callosum: A Further Behavioural Investigation. Brain, 97, 225-234.

Field, M., Ashton, R. and White, K. (1978). Agenesis of the Corpus Callosum: Report of 2 Pre-School Children and Review of the Literature. Devel. Med. Child Neurol., 20, 47-61.

Galin, D. (1977). Lateral Specialization and Psychiatric Issues: Speculations on Development and the Evolution of Consciousness. Ann. N. Y. Acad. Sci., 299, 397-411.

Gazzaniga, M.S. (1970). The Bisected Brain. Appleton Century Crofts, New York.

Geffen, G. and Caudrey, D. (1981). Reliability and Validity of the Dichotic Monitoring Test for Language Laterality. Neuropsychologia, 19, 413-423.

Geffen, G. and Quinn, K. (1984). Hemispheric Specialization and Ear Advantages in Processing Speech. Psychol. Bull., 96, 273-291.

Gordon, H.W., Bogen, J.E. and Sperry. R.W. (1971). Absence of Deconnexion Syndrome in 2 Patients with Partial Section of the Neocommissures. Brain, 94, 327-336.

Gott, P.S. and Saul, R.E. (1978). Agenesis of the Corpus Callosum: Limits of Functional Compensation. Neurol., 28, 1272-1279.

Granger, E.M., Masterton, R.B. and Glendenning, K.K. (1985). Origin of Interhemispheric Fibers in Acallosal Opossum (with a Comparison to Callosal Origins in Rat). J. Comp. Neurol., 241, 82-98.

Hamilton, C.R. and Vermeire, B.A. (1986). Localization of Visual Functions with Partially Split-Brain Monkeys. In Two Hemispheres - One Brain. (eds. F. Lepore, M. Ptito and H.H. Jasper). Alan R. Liss, New York.

Holtzman, J.D. (1984). Interactions between Cortical and Subcortical Visual Areas: Evidence from Human Commissurotomy Patients. Vision Res., 24, 801-813.

Holtzman, J.D., Sidtis, J.J., Volpe, B.T., Wilson, D.H. and Gazzaniga, M.S. (1981). Dissociation of Spatial Information for Stimulus Localization and the Control of Attention. Brain, 104, 861-872.

Jeannerod, M. (1986). The Formation of Finger Grip during Prehension, a Cortically Mediated Visuomotor Pattern. Behav. Brain Res., 19, 99-116.

Jeeves, M.A. (1965). Psychological Studies of Three Cases of Congenital Agenesis of the Corpus Callosum. In Functions of the Corpus Callosum, (ed. E. G. Ettlinger). Churchill, London.

Jeeves, M.A. (1969). A Comparison of Interhemispheric Transmission Times in Acallosals and Normals. Psychon. Sci., 16, 245-246.

Jeeves, M.A. (1986). Callosal Agenesis: Neuronal and Developmental Adaptations. In Two Hemispheres - One Brain. (eds. F. Lepore, M. Ptito and H.H. Jasper). Alan R. Liss, New York.

Jeeves, M.A. and Rajalakshmi, R. (1964). Psychological Studies of a Case of Congenital Agenesis of the Corpus Callosum. Neuropsychologia, 2, 247-252.

Jeeves, M.A., Silver, P.H. and Jacobson, I. (1986). Bimanual Coordination in Callosal Agenesis and Partial Commissurotomy (in preparation).

438

Jouandet, M.L. and Gazzaniga, M.S. (1979). Cortical Field of Origin of the anterior commissure of the Rhesus Monkey. Exp. Neurol., 66, 381-397.

Kimura, D. (1979). Neuromotor Mechanisms in the Evolution of Human Communication. In Neurobiology of Social Communication in Primates. (eds. H.D. Steklis and M.J. Raleigh). Academic Press, New York.

Kinsbourne, M. and Hiscock, M. (1977). Does Cerebral Dominance Develop? In Language Development and Neurological Theory. (eds. S.J. Segalowitz and F.A. Gruber). Academic Press, New York.

Kretschmer, H. (1968). Zur Klinik des Balkensyndroms. Arch. Psychiat. Z. ges. Neurol., 211, 250-265.

Kucera, H. and Francis, W.N. (1967). Computational Analysis of Present-Day American English. Brown University Press, Providence RI.

Kuypers, H.G.J.M. (1962). Cortico-Spinal Connections: Postnatal Development in the Rhesus Monkey. Science, 138, 678-680.

Kuypers, H.G.J.M. (1978). From Motor Control to Conscious Experience. In Cerebral Correlates of Conscious Experience. (eds. P.A. Buser and A. Rougeul-Buser). North Holland, Amsterdam.

LaCoste, M.G. de, Kirkpatrick, J.B. and Ross, E.D. (1985). Topography of the Human Corpus Callosum. J. Neuropath. and Eexperimental Neurology, 44, 578-591.

Lemire, R.J., Loeser, J.D., Leech, R.W. and Alvord, E.C. (1975). Normal and Abnormal Development of the Human Nervous System. Harper & Row, Hagerstown, MD.

Lenneberg, E. (1967). Biological Foundations of Language. Wiley, New York.

Lines, C.R., Rugg, M.D. and Milner, A.D. (1984). The Effect of Stimulus Intensity on Visual Evoked Potential Estimates of Interhemispheric Transmission Time. Exp. Brain Res., 57, 89-98.

McGlone, J. (1985). Congenital Absence of the Corpus Callosum: Neuropsychological and Sodium Amytal Findings. Paper presented at INS Annual Meeting, San Diego, CA.

Martin, A. (1985). A Qualitative Limitation on Visual Transfer via the Anterior Commissure. Brain, 108, 43-63.

Meerwaldt, J.D. (1983). Disturbances of Spatial Perception in a Patient with Agenesis of the Corpus Callosum. Neuropsychologia,

<u>21</u>, 161-165.

Meerwaldt, J.D. and Harskamp, F. van (1982). Spatial Disorientation in Right-Hemisphere Infarction. J. Neurol. Neurosurg. Psychiat., <u>45</u>, 586-590.

Milner, A.D. (1982). Simple Reaction Times to Lateralized Visual Stimuli in a Case of Callosal Agenesis. Neurospychologia, <u>20</u>, 411-419.

Milner, A.D. (1983). Neuropsychological Studies of Callosal Agenesis. Psychological Medicine, <u>13</u>, 721-725.

Milner, A.D. and Jeeves, M.A. (1979). A Review of Behavioural Studies of Agenesis of the Corpus Callosum. In <u>Structure and Function of Cerebral Commissures</u>. (eds. I. Steele Russell, M.W. van Hof and G. Berlucchi). Macmillan, London.

Milner, A.D. and Jeeves, M.A. (1981). The Functions of the Corpus Callosum in Infancy and Adulthood (Commentary). Behav. Brain Sci., <u>4</u>, 30-31.

Milner, A.D. and Lines, C.R. (1982). Interhemispheric Pathways in Simple Reaction Time to Lateralized Light Flash. Neuropsychologia, <u>20</u>, 171-179.

Milner, A.D., Jeeves, M.A., Silver, P.H., Lines, C.R. and Wilson, J. (1985). Reaction times to Lateralized Visual Stimuli in Callosal Agenesis: Stimulus and Response Factors. Neuropsychologia, <u>23</u>, 323-331.

Milner, A.D., Milne, A.B. and Mackenzie, A. (1986). Simple Reaction Times to Lateralized Stimuli Using Finger-Thumb Apposition (in preparation).

Moscovitch, M. (1973). Language and the Cerebral Hemispheres: Reaction-Time Studies and their Implications for Models of Cerebral Dominance. In <u>Communication and Affect: Language and Thought</u>. (eds. P. Pliner, T. Alloway and L. Krames). Academic Press, New York.

Moscovitch, M. (1977). Development of Lateralization of Language Functions and its Relation to Cognitive and Linguistic Development: A Review. In <u>Language Development and Neurological Theory</u>. (eds. S. Segalowitz and F. Gruber). Academic Press, New York.

Pandya, D.N. and Rosene, D.L. (1985) Some observations on trajectories and topography of commissural fibers. In <u>Epilepsy and the Corpus Callosum</u> (ed. A.G. Reeves). Plenum Press, New York.

Pandya, D.N., Karol, E.A. and Heilbronn, D. (1971). The Topographical Distribution of Interhemispheric Projections in the Corpus Callosum of Rhesus Monkey. Brain Res., 32, 31-43.

Poffenberger, A.T. (1912). Reaction Time to Retinal Stimulation with Special Reference to the Time Lost in Conduction through Nerve Centers. Archs. Psychol. (N.Y.), 3, 1-73.

Preilowski, B.F.B. (1972). Possible Contributions of the Anterior Forebrain Commissures to Bilateral Motor Coordination. Neuropsychologia, 10, 267-277.

Provins, K.A., Milner, A.D. and Kerr, P. (1982). Asymmetry of Manual Preference and Performance. Perceptual and Motor Skills, 54, 179-194.

Rasmussen, T. and Milner, B. (1977). The Role of Early Left-Brain Injury in Determining Laterality of Cerebral Speech Functions. Ann. NY Acad. Sci., 299, 355-369.

Ratcliff, G., Dila, C., Taylor, L. and Milner, B. (1980). The Morphological Asymmetry of the Hemispheres and Cerebral Dominance for Speech: A Possible Relationship. Brain and Language, 11, 87-98.

Reynolds, D.M. and Jeeves, M.A. (1977). Further studies of tactile perception and motor coordination in agenesis of the corpus callosum. Cortex, 13, 257-272.

Reynolds, D.M. and Jeeves, M.A. (1978). A Developmental Study of Hemisphere Specialization for Recognition of Faces in Normal Subjects. Cortex, 14, 511-520.

Rizzolatti, G. (1979). Interfield Differences in RTs to Lateralized Visual Stimuli in Normal Subjects. In Structure and Function of Cerebral Commissures. (eds. I. Steele Russell, M.W. van Hof and G. Berlucchi). Macmillan, London.

Rugg, M.D., Lines, C.R. and Milner, A.D. (1984). Visual Evoked Potentials to Lateralized Visual Stimuli and the Measurement of Interhemispheric Transmission Time. Neuropsychologia, 22, 215-225.

Rugg, M.D., Lines, C.R. and Milner, A.D. (1985). Further Investigation of Visual Evoked Potentials elicited by Lateralized Stimuli: Effects of Stimulus Eccentricity and Reference Site. Electroencephalography and clinical Neurophysiology, 62, 81-87.

Rugg, M.D., Milner, A.D. and Lines, C.R. (1985). Visual Evoked Potentials to Lateralised Stimuli in Two Cases of Callosal Agenesis. Journal of Neurology, Neurosurgery and Psychiatry, 48,

367-373.

Segalowitz, S.J. (1983). On the Requirements of a Developmental Theory of Lateralization. In Language Functions and Brain Organization. (ed. S.J. Segalowitz). Academic Press, New York.

Selnes, O.D. (1974). The Corpus Callosum: Some Anatomical and Functional Considerations with Special Reference to Language. Brain and Language, 1, 111-139.

Sergent, J. (1986). Subcortical Coordination of Hemisphere Activity in Commissurotomized Patients. Brain, 109, 357-369.

Sergent, J. and Myers, J.J. (1985). Manual, Blowing, and Verbal, Simple Reactions to Lateralized Flashes of Light in Commissurotomized Patients. Perc. and Psychophysics, 37, 571-578.

Stefanko, S.Z. and Schenk, V.W.D. (1979). Anatomical Aspects of the Agenesis of the Corpus Callosum in Man. In Structure and Function of Cerebral Commissures. (eds. I. Steele Russell, M.W. van Hof and G. Berlucchi). Macmillan, London.

Trevarthen, C. (1965). Functional Interactions between the Cerebral Hemispheres of the Split-Brain Monkey. In Functions of the Corpus Callosum. (ed. E.G. Ettlinger). Churchill, London.

Trevarthen, C. and Sperry, R.W. (1973). Perceptual Unity of the Ambient Visual Field in Human Commissurotomy Patients. Brain, 96, 547-570.

Ungerleider, L.G. and Mishkin, M. (1982). Two Cortical Visual Systems. In Analysis of Visual Behavior. (eds. D.J. Ingle, Goodale, M.A. and Mansfield, R.J.W.). MIT Press, Cambridge MA.

Wada, J.A., Clarke, R. and Hamm, A. (1975). Cerebral Hemispheric Asymmetry in Humans: Cortical Speech Zones in 100 Adult and 100 Infant Brains. Arch. Neurol., 32, 239-246.

Witelson, S.F. and Pallie, W. (1973). Left Hemisphere Specialization for Language in the Newborn. Neuroanatomical Evidence of Asymmetry. Brain, 96, 641-646.

Zaidel, D. and Sperry, R.W. (1977). Some Long-Term Motor Effects of Cerebral Commissurotomy in Man. Neuropsychologia, 15, 193-204.

26
Hemispheric Functions Evaluated by Measurements of the Regional Cerebral Blood Flow

Jarl Risberg

INTRODUCTION

Modern non-invasive methodology for measurement of regional cerebral blood flow (rCBF) has made it possible to study healthy normal subjects as well as any patient group of interest. The evaluation of rCBF provides information about the functional level of the cerebral cortex. This interpretation of blood flow changes as mirroring functional and metabolic changes is today supported by ample evidence (Raichle et al., 1976). In the present paper a review will be given of some studies which have focused on the question of hemispheric differences in function.

MEASUREMENT OF THE REGIONAL CEREBRAL BLOOD FLOW

The first rCBF-technique involved intra-arterial injection of 133-xenon. This restricted the measurements to selected patient groups being measured in only one hemisphere. During the last ten years the intra-carotid injection has been replaced by inhalation or intra-venous injection of the tracer. Our rCBF-technique is based on inhalation of 133-Xe (90 MBq/l) during one minute followed by 10 min of normal air breathing as suggested by Obrist et al. (1975). Part of the xenon gas dissolves in arterial blood and arrives to the brain. The inert and freely diffusible tracer emits gamma rays, which are recorded by multiple scintillation detectors covering both hemispheres. The arrival and disappearance of 133-Xe is recorded and

Acknowledgements: Supported by the Swedish Medical Research Council (Proj. nr 4969). Special thanks to Helena Fernö, B.A., for aid in the preparation of the manuscript.

forms the basis for calculation of flow values (in ml/100g/min) according to principles and computer algorithms developed by Obrist et al. (1975), Risberg et al. (1975A) and Risberg (1980). Several different flow measures can be derived from the recordings. The most useful ones for description of changes during mental activation are called f_1 (grey matter blood flow) and ISI (Initial Slope Index), both indicating the blood flow of the cortical grey matter. The absorption of the gamma photons in brain tissue limits the influence from deeper brain structures. The method thus gives information restricted to superficial cortical areas.

Until now commercially available equipments have at the most had 32 detectors covering both hemispheres. This gives a rather poor spatial resolution of 3-4 cm. We have recently been involved in the development of a recording system with 254 mini-detectors (the "Cortexplorer", by Scan. Detectronic Inc., Hadsund, Denmark), which has a spatial resolution of about 1 cm. An example from our first experiences with this system will be presented later (Fig 2).

THE RCBF ACTIVATION RESPONSE

The fact that the blood flow level of a cortical area varies with its functional level was discovered about 20 years ago (Ingvar and Risberg, 1967) using the intra-arterial injection technique. A verbal memory and attention task (digit-span backwards) caused flow increases in temporal and frontal areas of the measured left hemisphere. Since then numerous studies have demonstrated rCBF-increases and redistributions during sensory stimulation, cognition and motor performance (for a review see Risberg, 1986). Several of the findings have later been confirmed by measurements of local oxygen or glucose metabolism using positron emission tomography (Mazziotta et al.,1983). Modern rCBF-techniques allow simultaneous bilateral recordings, which makes it possible to specifically address the problem of hemispheric asymmetries of the activation response. Some studies of verbal, spatial and emotional functions will now be reviewed in more detail.

Verbal Functions

When a "resting" baseline recording is made in an awake normal subject laying comfortably with closed eyes in a silent laboratory, the results will show very similar flow levels and regional patterns in the two hemispheres (Risberg et al., 1975B, Prohovnik et al., 1980). A very well documented finding is the hyperfrontality of the pattern with frontal flow values being 10-20% higher than the hemispheric mean, and with lower flow levels postcentrally (Wilkinson et al., 1969, Ingvar, 1979, Prohovnik et al., 1980). One of the factors suggested to cause this frontal elevation is verbally mediated planning and behaviour control activities. A blood flow increase which is partly due to verbal processes might thus be present already during "resting".

The first attempt to induce hemispheric blood flow asymmetries by giving mental tasks to normal subjects was made by Risberg et al. (1975B). Healthy right-handed normal young volunteers were given a special version of the Miller analogies test. Significant asymmetries with the left hemisphere being higher that the right, especially in temporo-parietal and frontal regions, were seen in 12 subjects, who were motivated for an optimal performance by a monetary reward. A less motivated group of 12 subjects, who got a fixed payment irrespective of performance, showed only tendencies to asymmetries. An important observation was that the major part of the rCBF-activation was symmetrical (reward group: rt 12%, lt 15% increase) indicating the importance also of the right hemisphere for different aspects of verbal functions. These findings were later replicated by Gur and Reivich (1980) using the same test for activation and a similar rCBF-technique. The engagement also of the right hemisphere in different forms of verbal activity has been a common finding in the rCBF litterature. As a matter of fact Ryding et al. (1985) found greater right than left hemisphere activation during automatic speech (repetitive recital of the week-days). The left hemisphere activation seems, however, to dominate when the speech is more creative like in a word fluency test or a test of verbal creative thinking. This was found in a recently completed study by Warkentin et al. (1986). Subjects were healthy right-handed volunteers of both sexes. The instruction for the word fluency test was to tell the investigator as many words as possible beginning with a given letter. The target letter was changed every minute. The creative thinking test consisted of a list of common objects (like a brick), one object being given every minute with the instruction to mention as many ways as possible of using the object. The answers were given aloud via a microphone in the face mask. The results obtained by a 32 detectors instrument are shown in Figure 1. Significant increases from resting baseline were seen in left frontal areas during both tests with an additional elevation in a left parietal region during the creative thinking test. An additional interesting observation (not illustrated) was that this left parietal increase was significant only in the males, while the left frontal basal (Broca) increase was more marked in the females. This finding might indicate that males and females use somewhat different cognitive strategies (spatial-verbal) when performing the creative thinking test.

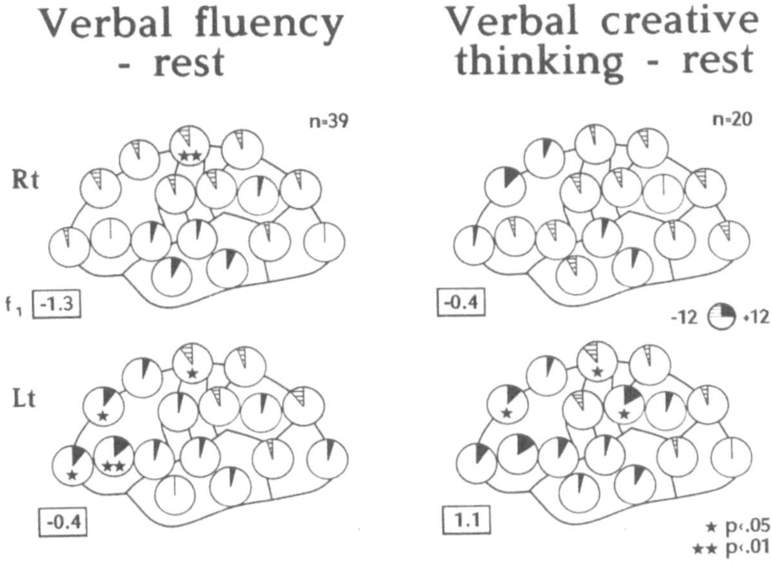

Verbal fluency - rest

Verbal creative thinking - rest

Figure 1. Changes of rCBF during a word fluency test and a test of verbal creative thinking in normal subjects. Changes of the hemispheric means are shown in the boxes. Regional changes from rest to activation are shown as clock symbols with black indicating increases. (From Warkentin et al. 1986).

As mentioned earlier, we have recently made the first recordings with our new high-resolution 254 detectors Cortexplorer instrument. The assumption motivating the development of this equipment was that the flow changes, which occur when the brain is engaged in any higher mental function, are multi-focal and strictly localized. This assumption is strongly supported by earlier results using the intra-arterial injection technique and a high-resolution instrument (Lassen et al. 1977). Many details of these complex patterns are likely to be lost when equipment with poor spatial resolution is used. Results from measurements in a neurologically normal patient with normal speech functions are shown in Figure 2. The recording during rest (upper part of figure) was made with closed eyes and during silence. The second measurement was made during verbal creative thinking according to the procedure described

446

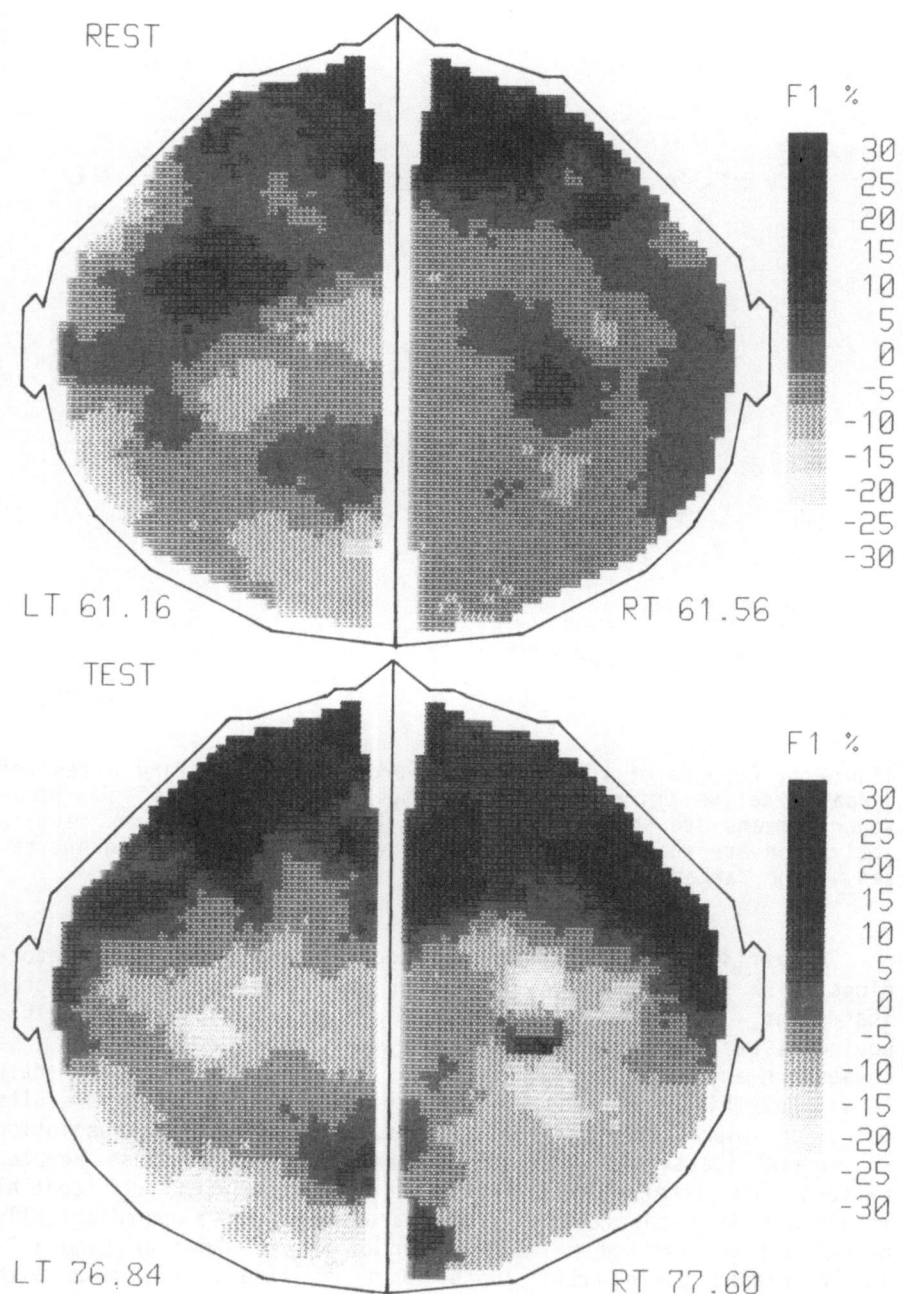

Figure 2. High-resolution rCBF in a normal subject (see text).

above. All superficial cortical areas are shown in grey tones indicating % of the mean of the two hemispheres. It is evident that a much more detailed pattern of changes is seen than with the old equipment (Fig. 1). A left frontal increase is clearly seen together with temporal increases, which are most pronounced on the right side. What is likely to be the right supplementary motor area also shows an increase and tendencies to elevations are seen in left parietal and right occipital areas. The earlier findings are thus confirmed in this single case with the addition of several other focal changes and asymmetries, the interpretation of which has to await confirmation in an extended series of measurements.

Spatial Functions

It has been more difficult to reliably induce hemispheric asymmetries by spatial tasks. We have for example been unable to demonstrate any asymmetries during puzzle tests (Risberg and Hagstadius, 1983). Such tasks induce symmetric, mainly parietal, increases. The group motivated by a monetary reward in the study by Risberg et al. (1975B) did, however, show significantly higher right than left hemisphere increases while silently solving the Street test of perceptual closure. As for the verbal test it was found that the major part of the rCBF increase was bilateral. This is not surprizing since the incomplete figures of the test represent objects which can be verbalized. In a replication of this study by Gur and Reivich (1980), right advantage during the Street test was seen only in the most skilled in a group of normal right-handed subjects. The authors speculated that the cortical organization of spatial abilities is more subject to individual variations than that of verbal functions. The latter might be more "hard-wired" to the left hemisphere.

Somato-Sensory Stimulation

Tactile stimulation has been investigated by bilateral rCBF-technique in only one study (Risberg and Prohovnik, 1983). The stimuli were nine aluminum pins arranged in a 2cm x 2cm square. One to nine of the pins touched the palms of both hands of the subject in a random order throughout the rCBF-measurement. Two different conditions were created by instruction: sensation, in which the subject was encouraged to ignore the stimuli, and attention, in which the subject was asked to attend to the stimulation of one of the hands and respond by saying "one" only when a single pin touched the palm. Subjects were 10 healthy right-handed young male volunteers. The sensation condition was accompanied by parietal and

448

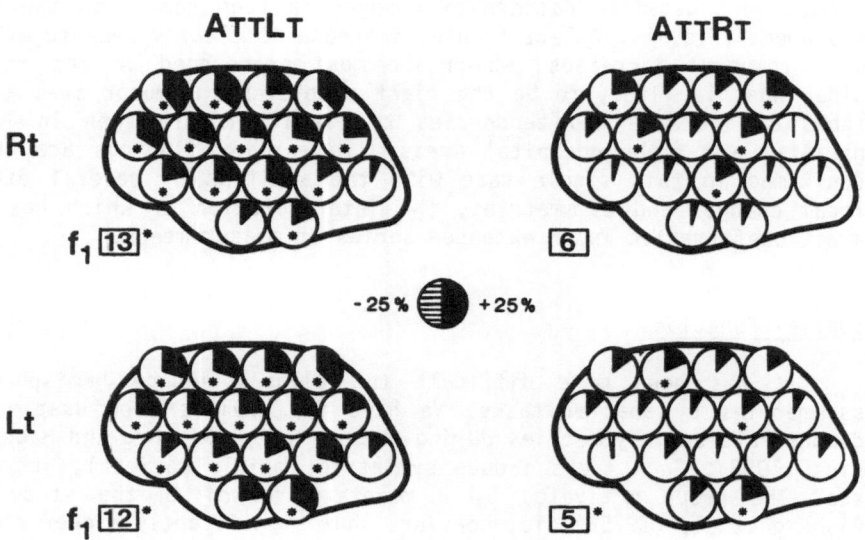

Figure 3. Changes of rCBF from rest to attention to tactile stimulation of the left (AttLt) and right (AttRt) hand in 10 normal subjects. Changes of the hemispheric means are shown in the boxes and regional changes are depicted as clock symbols. The asterisks indicate significant increases (p<0.05) by Anova and Scheffé test. (From Risberg and Prohovnik, 1983).

pre-frontal flow increases contralateral to the side of stimulation. As is shown in Figure 3, attention to the right hand increased flow in the same areas as during sensation but with a larger amplitude, especially on the right side. Attention to the left hand caused much larger flow increases than attention to the right hand. The former condition was accompanied by large generalized flow increases in both hemispheres with a tendency to somewhat larger changes on the right side. These findings are interesting to relate to the results of Halsey et al. (1979), who compared rCBF changes during movements of the right and left hand. They found larger flow increases in contralateral motor areas during left hand movements in line with the present findings. Attention to the left hand seems thus to be more activating that attention to the right hand. These results might be interpreted as supporting the hypothesis of right hemisphere dominance for attentional mechanisms (Heilman and van den Abell, 1979).

Anxiety

We have recently completed the first systematic study of the effects of an increased level of anxiety on bilateral rCBF (Johanson et al., 1986). Twelve patients, 8 women and 4 men, diagnosed as anxiety neurotics participated in the study. Before the rCBF measurement the patients were subjected to personality testing by means of the Meta-Contrast Technique (MCT; Smith and Nyman, 1961). This technique is based on an interpretation of the answers given by the patient following subliminal presentation of a threatening picture. The method is thought to activate perceptual defences, which are analogous to the defence systems described in the psychoanalytic litterature. The MCT-testing might also elucidate signs of anxiety. In this study the anxiety responses from the MCT-session were used as anxiety provoking stimuli during an rCBF recording, which followed a resting baseline measurement. During the anxiety provocation the experimenter asked the patient to freely associate on the basis of anxiety-answers given during the MCT-session. The patient was asked to remain silent with closed eyes during the rCBF recording. Observations of the behaviour of the patient were made and the provocation of anxiety was adjusted to the reactions of the patient.

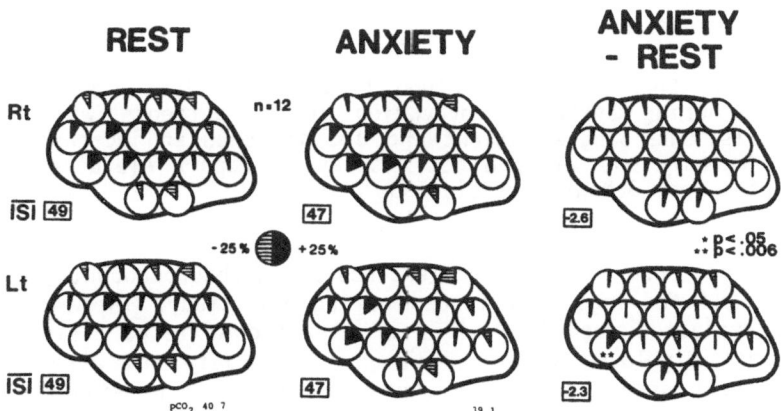

Figure 4. rCBF in 12 patients with anxiety neurosis during rest and during increased anxiety (see text). Regional values are shown as % (clock symbols) of the hemispheric mean in the left and middle part of the figure. Changes from rest to anxiety are shown to the right. Note the significant (by t-test) increase in the left fronto-temporal area. (From Johanson et al., 1986).

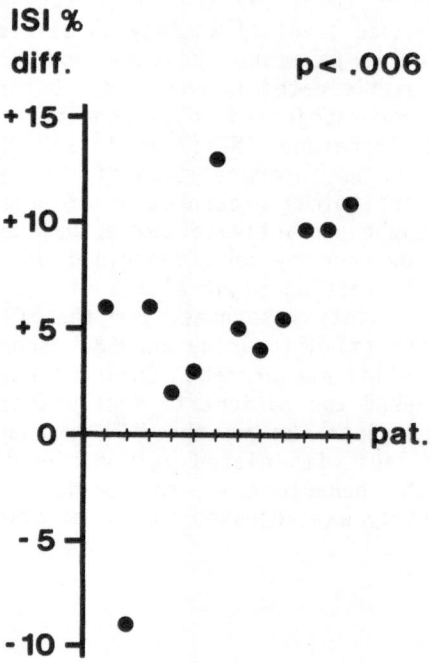

ISI %
diff.   **p< .006**

Figure 5. Changes of the blood flow of the left fronto-temporal region during anxiety provocation in 12 patients with anxiety neurosis. The single patient who did not show an increase of flow was the only one who did not display or report elevated anxiety. (Adapted from Johanson et al. 1986).

After the measurement the patient was asked about thoughts, associations and feelings experienced during the study. These data were then evaluated regarding signs of elevated anxiety. Eleven of the patients displayed vegetative signs of anxiety and also reported that their associations and thoughts had been very anxiety-loaded. The remaining patient did not show or report any increased anxiety. Figur 4 shows the results for the total group. The flow level was about the same during rest and during anxiety but regionally a focal increase was seen in a left fronto-temporal region (p<0.006). Figure 5 shows the individual flow increases in this region. All but one patient had an increased flow level in this area during anxiety-provocation. The non-responding patient was the one, who did not show or report increased anxiety.

The results thus suggest a linkage between elevated anxiety and increased neuronal activity in the left fronto-temporal region. It is likely that several other areas, especially limbic ones, also are activated, but they are not recorded by the present rCBF-technique. It is, however, interesting that the changes noted are in an area which is closely connected to the limbic system. The linkage between anxiety and the frontal lobes has also been demonstrated in a series of rCBF studies in patients submitted to stereotactic psychosurgery due to severe anxiety (Kullberg and Risberg, 1978). A reduction of frontal flow levels were seen after the operation parallel to a reduction of symptoms of anxiety. It is also interesting to relate the results to findings in neuropsychological studies of trait-anxious individuals indicating increased left hemisphere activation (Tucker et al., 1978). Stress has, however, been linked to right hemisphere activation (Tucker et al., 1977), which might explain why Reivich et al. (1983) found right frontal increases in individuals who experienced stress-related anxiety while being subjected to a PET-study.

FINAL REMARKS

Some examples of the use of bilateral rCBF-measurements in the study of hemispheric asymmetries of function have been given. The present method offers a mapping of the tonic level of neuronal activity in superficial cortical areas. It is important to realize that the blood flow level is a quantitative measure of the neuronal activity and not a qualitative. (The loosing chess player might have higher rCBF than the winner!) It is a matter of debate to what extent an rCBF increase might be caused by inhibitory neuronal activity. The generally good agreement between well-established theories of the cortical organization of mental functions and the rCBF-results does, however, speak in favour of a dominating influence of exitatory activity. It has also been shown that global inhibition caused by a GABA agonist causes a very marked global decrease of rCBF (Roland and Friberg, 1983).

To link the different subcomponents of a multi-focal rCBF activation response to the different subcomponents of a mental process is far from easy. Some increases might be specifically related to the nature of the mental task while other changes might be caused by attentional or emotional functions. One of our earlier studies showed diffuse rCBF changes when the task was new to the subject and more restricted and focal changes following habituation and training (Risberg et al., 1977). Careful control of the mental setting of the measurement situation and the use of a design with several repeated measurements, including sensory-motor control tasks, aid in the interpretation of the results. A more exact localization of the activated cortical regions by the use of more resolute instruments like the "Cortexplorer" will also make it easier to draw conclusions about brain-behaviour relations. The rCBF method will most likely be a valuable tool also in the future exploration of the dual brain.

REFERENCES

Gur, R.C. and Reivich, M. (1980). Cognitive Task Effects on Hemispheric Activation. Brain Lang., 9, 78-92.
Halsey, J.H., Blauenstein, U.W., Wilson, E.M and Wills, E.L. (1979). Regional Cerebral Blood Flow Comparison of Right and Left Hand Movement. Neurology, 29, 21-28.
Heilman, K.M. and van den Abell, T. (1979). Right Hemisphere Dominance for Mediating Cerebral Activation. Neuropsychologia, 17, 315-321.
Ingvar, D.H. (1979). "Hyperfrontal" Distribution of the Cerebral Grey Matter Flow in Resting Wakefulness: On the Anatomy of the Conscious State. Acta Neurol. Scand., 60, 12-25.
Ingvar, D.H and Risberg, J. (1967). Increase of Regional Cerebral Blood Flow during Mental Effort in Normals and in Patients with Focal Brain Disorders. Exp. Brain Res., 3, 195-211.
Johanson, A.M., Risberg, J., Silfverskiöld, P. and Smith, G. (1986). Regional Changes of Cerebral Blood Flow during Increased Anxiety in Patients with Anxiety Neurosis. In The Roots of Perception. (eds. U. Hentschel, G. Smith and J.G. Graguns). North Holland, Amsterdam.
Kullberg, G. and Risberg, J. (1978). Changes in Regional Cerebral Blood Flow following Stereotactic Psychosurgery. Appl. Neurophysiol., 41, 79-85.
Lassen, N.A., Roland, P.E., Larsen, B., Melamed, E. and Soh, K. (1977). Mapping of Human Cerebral Functions: A Study of the Regional Cerebral Blood Flow Pattern during Rest, its Reproducibility and the Activation seen during Basic Sensory and Motor Functions. Acta Neurol. Scand., 56, suppl. 64, 262-263.
Mazziotta, J.C., Phelps, M.E. and Halgren, E. (1983). Local Cerebral Glucose Metabolic Response to Audiovisual Stimulation and Deprivation: Studies in Human Subjects with Positron CT. Human Neurobiol., 2, 11-23.
Obrist, W.D., Thompson, H.K., Wang, H.S. and Wilkinson, W.E. (1975). Regional Cerebral Blood Flow Estimated by 133-Xenon Inhalation. Stroke, 6, 245-256.
Prohovnik, I., Håkansson, K. and Risberg, J. (1980). Observations on the Functional Significance of Regional Cerebral Blood Flow in Resting Normal Subjects. Neuropsychologia, 18, 203-217.
Raichle, M.E., Grubb, R.L., Gado, M.H., Eichling, J.O. and Ter-Pogossian, M.M. (1976). Correlation between Regional Cerebral Blood Flow and Oxidative Metabolism. Arch. Neurol., 33, 523-526.

Reivich, M., Gur, R. and Alavi, A. (1983). Positron Emission Tomographic Studies of Sensory Stimuli, Cognitive Processes and Anxiety. Human Neurobiol., 2, 25-33.

Risberg, J. (1980). Regional Cerebral Blood Flow Measurements by 133 Xe-Inhalation: Methodology and Applications in Neuropsychology and Psychiatry, Brain Lang., 9, 9-34.

Risberg, J. (1986). Regional Cerebral Blood Flow. In Experimental Techniques in Human Neuropsychology. (ed. J.H. Hannay). Oxford University Press, New York.

Risberg, J., Ali, Z., Wilson, E.M., Wills, E.L. and Halsey, J.H. (1975A). Regional Cerebral Blood Flow by 133-Xenon Inhalation. Preliminary Evaluation of an Initial Slope Index in Patients with Unstable Flow Compartments. Stroke, 6, 142-148.

Risberg, J. and Hagstadius, S. (1983). Effects on the Regional Cerebral Blood Flow of Long-Term Exposure to Organic Solvents. Acta Psychiatr. Scand., 67, suppl. 303, 92-99.

Risberg, J., Halsey, J.H., Wills, E.L. and Wilson, E.M. (1975B). Hemispheric Specialization in Normal Man Studied by Bilateral Measurements of the Regional Cerebral Blood Flow. Brain, 98, 511-524.

Risberg, J., Maximilian, V.A. and Prohovnik, I. (1977). Changes of Cortical Activity Patterns during Habituation to a Reasoning Test. A Study with the 133-Xe Inhalation Technique for Measurement of the Regional Cerebral Blood Flow. Neuropsychologia, 15, 793-798.

Risberg, J. and Prohovnik, I. (1983). Cortical Processing of Visual and Tactile Stimuli Studied by Non-Invasive rCBF Measurements. Human Neurobiol., 2, 5-10.

Roland, P.E. and Friberg, L. (1983). Are Cortical rCBF Increases during Brain Work in Man due to Synaptic Exitation or Inhibition? J. Cereb. Blood Flow Metabol., 3, suppl. 1, 244-245.

Ryding, E., Brådvik, B. and Ingvar, D.H. (1985). Simultaneous Bilateral Changes in Regional Cerebral Blood Flow (rCBF) from Verbal and Nonverbal Vocal Activations. J. Cereb. Blood Flow Metabol., 5, suppl. 1, 207-208.

Smith, G.J-W. and Nyman, G.E. (1961). A Serial Tachistoscopic Experiment and its Clinical Application. Psychological Issues, 52, International Universities Press, New York.

Tucker, D.M., Antes, J.R., Stenslie, C.E. and Barnhardt, T.M. (1978). Anxiety and Lateral Cerebral Function. J. Abn. Psychol., 87, 380-383.

Tucker, D.M., Roth, R.S., Arneson, B.A. and Buchingham, V. (1977). Right Hemisphere Activation during Stress. Neuropsychologia, 15, 697-700.

Warkentin, S., Risberg, J., Nilsson, A., Rodriguez, G., Karlson, S. and Flekkøy, K. (1986). Cortical Changes of Regional Cerebral Blood Flow during Verbal Production and Verbal Creative Thinking. Paper presented at the Third Nordic Meeting in Neuropsychology, Bergen, Norway, August 12-15, 1986.

Wilkinson, I.M., Bull, J.W., Boulay, G.H. du, Marshall, J., Ross Russell, R.W. and Symon, L. (1969). Regional Cerebral Blood Flow in the Normal Cerebral Hemisphere. J. Neurol. Neurosurg. Psychiat., 32, 367-378.

27

Interhemispheric Interaction: Models, Paradigms and Recent Findings

Joseph B. Hellige

There is now a great deal of evidence that information processing differences between the left and right cerebral hemispheres
can have important consequences for information processing in neurologically normal humans. Until recently, the emphasis has been
on learning as much as possible about the processing abilities and
propensities of each hemisphere and not so much on the manner in
which the two hemispheres interact to form an integrated processing
system. As humans look around their world, hear sounds and touch
objects it is the case that most information is presented to both
hemispheres and it is often the case that both hemispheres are
capable of generating a behavioral response to the stimulation. We
have only limited knowledge about the nature of interhemispheric
interactions in such situations. With these things in mind, the
present chapter examines some of the ways that interhemispheric
interaction might be studied by comparing performance in situations
involving only one hemisphere to performance in similar situations
involving both hemispheres. Specifically, consideration will be
given to dual-task studies of interference, to studies demanding
cross-hemispheric integration of information and to investigation
of the effects of simultaneous presentation of the same stimulus
information to both hemispheres.

DUAL-TASK INTERFERENCE

One question about interhemispheric interaction has to do with
the extent to which the two hemispheres in neurologically normal
individuals can function as separate processing systems with attentional resources that are at least somewhat independent of each
other. Dual-task interference paradigms have been useful in studying this particular aspect of interaction. For example, a number
of findings from different laboratories and using a variety of
experimental tasks lead to the conclusion that two concurrently
performed tasks interfere with each other more if they require
processes from the same cerebral hemisphere than if the processing

can be spread more evenly across the two hemispheres. The existence of laterality effects in dual-task experiments has led to a re-evaluation of the traditional single-resource model of attention in which all resource-demanding processes are proposed to compete for resources from a single undifferentiated pool. Instead, such results suggest that the two cerebral hemispheres function as somewhat independent processing systems, each with a limited capacity that is separate from the capacity of the other hemisphere. For discussion of the relative merits of specific alternatives to the single-resource model see Friedman & Polson (1981), Hellige (1985) and Kinsbourne and Hiscock (1983).

The fact that two logically unrelated tasks interfere with each other more when they require processing resources from the same cerebral hemisphere raises the possibility that a single task may be performed most efficiently when the different processing components can be spread equitably across the two hemispheres. Consistent with this possibility, Green (1984) reports several choice reaction-time tasks in which responses are faster when stimuli occur in the visual field contralateral to the hand generating responses; that is, when the hemisphere that must program the response output is not the same hemisphere that must initiate stimulus processing. Perhaps one advantage of functional cerebral asymmetries is the reduction of maladaptive interaction between simultaneous tasks and between processing components within what we generally regard as a single task.

CROSS-HEMISPHERIC INTEGRATION

A somewhat different question about interhemispheric interaction has to do with the consequences for information processing when the hemispheres are required to share information. Consider, for example, a task that requires neurologically normal subjects to indicate whether two simultaneously presented letters are same or different. If both letters are presented to the same visual field (same hemisphere), then the task does not require any collaboration of the two hemispheres. However, if one letter is presented to each visual field (different hemispheres) then some cross-hemisphere integration of information is required. A comparison of within-field and across-field conditions can, therefore, a shed light on the nature of interhemispheric interaction for situations in which the hemispheres are forced to collaborate.

From the dual-task studies and the results reported by Green (1984), one might expect better performance in the across-field condition because it is in this condition that the perceptual processing load would seem to be spread more evenly across the two cerebral hemispheres. In fact, such an across-field advantage has been reported for a variety of stimuli when the presentation of the two stimuli is simultaneous (e.g., Davis & Schmit, 1971 and review by Banich, 1986). However, when the two stimuli to be compared are presented successively, the more typical result has been better performance in the within-field condition than in the between-field condition (e.g., Dimond, Gibson & Gazzaniga, 1972), suggesting that in some situations placing the entire processing load on one hemisphere may be more efficient than demanding interhemispheric interaction.

That simultaneous versus successive presentation of the stimuli can be a relevant factor is suggested by letter matching experiments conducted by Ron Webster and myself. Our experiments required subjects to make a button pressing response to indicate whether two letters of different case had the same name. On some trials the two letters were presented to the same visual field and on other trials the two letters appeared in opposite visual fields. Orthogonal to this, the letters were separated in time by 100 msec or 500 msec (Experiment 1) and by 0 msec or 500 msec (Experiment 2). Reaction time (RT) was shorter with a 500 msec interval between letters than with a 0 or 100 msec interval. Of particular interest was the finding in both experiments that this decrease in RT as the interval between letters became longer was significantly greater for the within-field than for the across-field condition. Put another way, an across-field advantage tended to be larger with a 0 msec or 100 msec interval than with a 500 msec interval. One possible interpretation of this is that any advantage of spreading the processing load more evenly across the two hemispheres is greatest in those conditions that make the heaviest demands--in this case, when the letters are presented simultaneously or so close together in time that the perceptual processing of one letter is not complete before presentation of the other letter.

Recent work has begun to introduce elegant refinements into the comparison of within- and across-field presentations in an attempt to be more precise about exactly what information is passed from one hemisphere to the other and when it is passed (Banich, 1986; Liederman, Merola and Martinez, 1985). A description of this work is beyond the scope of this chapter, but it appears that the specific information passed depends on a variety of task conditions including the type of stimulus material, simultaneous versus successive presentation of stimuli and level of practice.

SIMULTANEOUS PRESENTATION TO BOTH HEMISPHERES

As noted earlier, in the natural environment it is often the case that the same relevant stimulus information is presented to both hemispheres simultaneously or nearly so and it is important to discover the nature of interhemispheric interaction in these cases. There are several plausible possibilities that can be considered most easily after description of an experimental paradigm that can address questions about this aspect of interhemispheric interaction. The critical features of the paradigm will be discussed in the context of a tachistoscopic visual half-field experiment, although in principal the logic extends to other stimulus modalities.

This paradigm uses tasks for which both cerebral hemispheres have competence, but for which there is some evidence of a qualitative difference in the nature of processing depending on which hemisphere (visual field) is stimulated. The operational definition of a qualitative difference in processing is that there is an interaction of hemisphere-of-presentation and some experimentally manipulated task variable (cf., Hellige, 1983). In addition to presenting the critical stimulus information unilaterally, trials are included on which the same stimulus information is presented simultaneously to both visual fields (referred to hereafter as BILATERAL trials). It is then possible to compare qualitative aspects of information processing when both hemispheres have simultaneous access to the same stimulus information with the processing found on each of the unilateral trials. In this way, it is possible to obtain information about which of the two bilaterally presented stimuli dominates performance during the various experimental conditions and, thereby, to differentiate various models of interhemispheric interaction. Consider the following examples.

Figure 1 shows four different patterns of hypothetical results from a visual half-field experiment of the type just described. Specifically, the experiment is presumed to include three different visual field conditions (LVF-RH, RVF-LH and BILATERAL trials) and two levels of an independent variable chosen because it interacts with left versus right visual field (the levels being A1 and A2). In addition, it is assumed that both hemispheres have equivalent access to response output and that only the first response that is made is recorded. For the present discussion, performance is assumed to be RT of correct responses so that lower values reflect better performance. Note that the results for right visual field - left hemisphere (RVF-LH) and left visual field - right hemisphere (LVF-RH) trials are identical in all four panels, with there being better performance in the A1 condition than in the A2 condition on RVF-LH trials and no difference between A1 and A2 performance on LVF-RH trials. The results for BILATERAL trials in the four panels illustrate four models of interhemispheric interaction. Although other models could also be considered, these four all have high a priori plausibility and span a sufficient range of outcomes for illustrative purposes.

According to the Best Performance Model the performance on BILATERAL trials for each experimental condition is equal to the better of the two unilateral trials for that condition. Consider one way that this pattern of performance could come about with RT as the performance measure. Suppose that on BILATERAL trials each hemisphere processes completely the stimulus with which it is stimulated and that the two hemispheres do so completely independently of each other. Under this assumption of complete independence, it is reasonable to suppose that as soon as one hemisphere finishes pro-

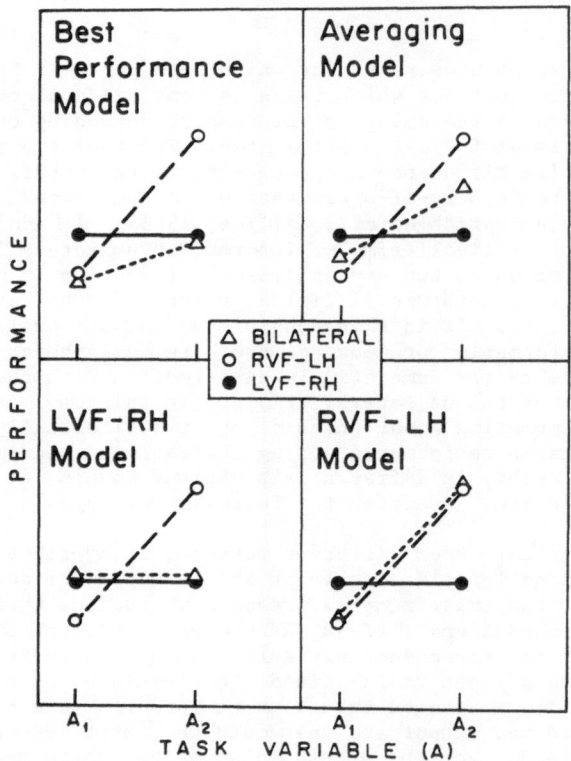

Figure 1. Hypothetical results from an experiment including three
visual field conditions. See text for explanation.

cessing, it will produce a response. Thus, in a situation such as
described where both hemispheres have equal access to response output and only the first response is recorded, RT on BILATERAL trials
should equal the faster of the two unilateral RTs for each experimental condition. Given the notion considered earlier that the two
hemispheres are at least somewhat independent in terms of their processing resources, it is plausible to consider that they may also
operate independently in the sense just outlined.

According to the Averaging Model the performance on BILATERAL
trials for each experimental condition is equal to the average of
the two unilateral trials for that condition. There are several
ways in which this could come about. For example, suppose that on
each BILATERAL trial one of the two cerebral hemispheres completely
dominates processing, so that the response is equivalent to one of
the unilateral RTs. If the hemisphere that dominates varies from
trial to trial in a way such that each hemisphere dominates half the

time, then the mean for BILATERAL trials in each condition would be equal to the average of the two unilateral RTs for that condition. The same thing could also appear in the mean results for a group if for half the subjects in the group the left hemisphere dominated all of the BILATERAL trials and for the other half of the subjects the right hemisphere dominated all of the BILATERAL trials. Thus, when the Averaging Model provides the best description of group data it is necessary to examine individual subject scores to differentiate these possibilities.

The notion that when faced with simultaneous bilateral presentation of the same stimulus information one hemisphere or the other might consistently dominate processing is related to the concept of "metacontrol" as discussed by Levy and Trevarthen (1976). Metacontrol refers to the neural mechanisms that determine which hemisphere will attempt to control cognitive operations. Based on their research with four commissurotomy patients, Levy and Trevarthen conclude that when both hemispheres were presented with information, one frequently asserted control. Furthermore, the controlling hemisphere was not always the one most effective for the task being performed, varied with task demand and from subject to subject within one task. In addition, several investigators have proposed that an experimental task can induce arousal asymmetries between the two cerebral hemispheres, with the consequence that attention is directed more readily to the side of space contralateral to the more activated hemisphere (e.g., Kinsbournce, 1975). It is reasonable to assume that, in the present paradigm, on BILATERAL trials the stimulus that is presented to the more aroused hemisphere would dominate performance. Thus, models proposing the dominance for one hemisphere on BILATERAL trials have a priori plausibility.

Concepts such as metacontrol and arousal asymmetries underly the two remaining models outlined in Figure 1. According to the LVF-RH model the right hemisphere dominates processing consistently for all BILATERAL trials and for all subjects and even for conditions like A1 where performance would be better if the left hemisphere were allowed to dominate. A similar situation is seen for the RVF-LH model where it is the left hemisphere that dominates performance consistently. Thus, the experimental paradigm that has been described has the potential to provide converging information about the nature of interhemispheric interaction. In order to illustrate the use of the paradigm I will describe briefly the results of an experiment completed recently by Jon Jonsson, Chikashi Michimata and myself (Hellige et al., 1986).

In our experiment, right-handed subjects performed a comparison task using schematic face stimuli employed in previous laterality experiments by Sergent (1982, 1984). This stimulus set consists of 16 faces generated by all the combinations of four features (hair, eyes, mouth or jaw), each feature having two different exemplars. On each trial, subjects were exposed for one sec to a face in the center of a viewing screen. When this face disappeared, a second face was presented for 250 msec to either the LVF-RH, RVF-LH or bilaterally (that is, the same probe face to both visual fields). The task of the subject was to indicate whether or not the second face was identical to the first (centered) face. The faces were either identical (SAME trials) or differed on only one of the four features (DIFFERENT trials). Subjects made their responses by pressing two of four microswitches arranged so that the index finger of each hand was used to make one type of response (SAME or DIFFERENT, counterbalanced across subjects) while the middle finger of each hand was used to make opposite responses. An advantage of this arrangement is that it allows both hemispheres equivalent access to both responses (Hellige & Sergent, 1986).

This particular task was chosen because previous research (Sergent, 1982, 1984) indicates that both cerebral hemispheres have competence to perform the task (e.g., there is no overall advantage for one visual field of the other) but the hemispheres perform the face comparison in qualitatively different ways. Specifically, when the two faces differed on only one feature, Sergent found that the RT of correct responses depended on which of the four features differed and, more importantly, found that the pattern of feature-location effects was different on LVF-RH and RVF-LH trials. The interesting question addressed by our experiment is what feature location pattern will emerge on BILATERAL trials, using the models considered in Figure 1 as a guideline.

Figure 2 shows the RT of correct responses on DIFFERENT trials as a function of the location of the feature difference for each of the three visual field conditions. Consider first the two unilateral conditions. As Figure 2 indicates, RT depended on the location of the feature difference, but of greater importance for present purposes, the feature location pattern was significantly different for RVF-LH and LVF-RH trials. This interaction can be seen most clearly by examining the inner features, eyes and mouth. Note that RT increased as the feature difference was moved from the eyes to the mouth, with the increase being significantly greater on RVF-LH trials than on LVF-RH trials. As a result of this, there was a trend toward an RVF-LH advantage when the eyes differed but a trend toward an LVF-RH advantage when the mouth differed. Given that the pattern of feature location effects differs for the two unilateral conditions, it is possible to consider the similarity of the pattern on BILATERAL trials to each of the unilateral patterns.

Inspection of Figure 2 suggests that the pattern of feature location effects on BILATERAL trials is identical to the pattern shown on RVF-LH trials, but with the RT on BILATERAL trials increased by a constant amount (approximately 24 msec). As a consequence of this, the BILATERAL pattern differs from the LVF-RH pattern in the same way that the two unilateral patterns differ from each other. These observations are corroborated by statistical analyses showing a significant LVF-RH versus BILATERAL by Feature Location interaction ($p < .01$) but no RVF-LH versus BILATERAL by Feature

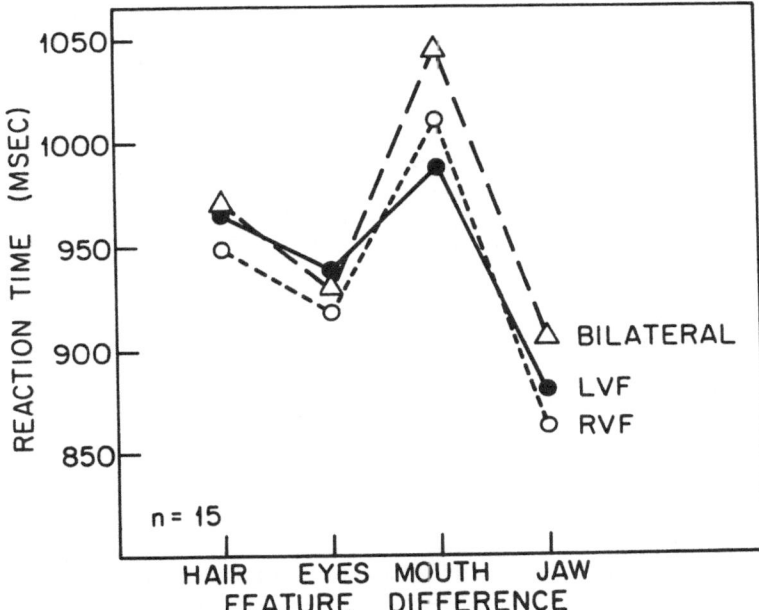

Figure 2.   Reaction time of correct responses on DIFFERENT trials as
a function of the feature difference (hair, eyes, mouth
and jaw) for each of the three visual field conditions.
From Hellige et al. (1986).

Location interaction.

The results shown in Figure 2 are clearly inconsistent with
the Best Performance Model, as the RT on BILATERAL trials is significantly longer than on unilateral trials. This finding is inconsistent with the hypothesis that each hemisphere processes the probe
stimulus in complete independence of the other hemisphere with the
faster of the two producing the response. One possible alternative
is that each hemisphere began processing the probe stimulus but at
some point in the information processing sequence there is a comparison of the processing to that point. Such a comparison operation
would presumably occur regardless of the location of feature difference and would take time. If it were the case that for most subjects only the RVF-LH stimulus continued to be processed after the
comparison, this type of explanation would be consistent with the
finding that RT on BILATERAL trials is equal to RT on RVF-LH trials
plus a constant. Another possibility is based on the claims by
Holtzman and Gazzaniga (1982) that the two hemispheres have access
to a common pool of processing resources (i.e., the hemispheres are

not completely independent in terms of resources) and that these resources must be distributed between the two hemispheres when the subject is faced with demanding bilateral stimulation. Perhaps it is a need to divide limited resources between the hemispheres that causes a constant decrement in performance on BILATERAL trials.

Strictly speaking, the finding that performance was worse for BILATERAL presentation than for unilateral presentation is also inconsistent with the other three models outlined Figure 1, none of which predict such a main effect of visual field. However, it is important to note that the pattern of feature location effects is identical on BILATERAL trials and RVF-LH trials and different for LVF-RH trials. In this sense, the data clearly favor the RVF-LH Model over either the LVF-RH Model or the Averaging Model. In order to determine the extent to which this was the case for individual subjects, Hellige et al. (1986) determined for each subject which of these three models best described the pattern of feature location effects on BILATERAL trials (ignoring any main effect of visual field). The RVF-LH Model provided the best fit to the RT data for 10 of 15 subjects, with the LVF-RH Model and Averaging Model providing the best fit to the RT data for 2 and 3 subjects, respectively. Similar individual difference results were reported for the error data. Thus, of the four models considered in Figure 1, the data from most subjects in this experiment are best described by the RVF-LH Model, with the addition of a constant increase in RT (and percentage of errors) on BILATERAL trials compared with RVF-LH trials.

It is interesting to note that in the group data shown in Figure 2 the RVF-LH stimulus dominates performance even when such domination leads to poorer performance. For example, when the feature difference is moved from the eyes to the mouth the increase in RT is greater for RVF-LH trials than for LVF-RH trials. The corresponding increase on BILATERAL trials matches the increase on RVF-LH trials despite the fact that the LVF-RH stimulus is also present and could lead to better performance. This dissociation of the relative efficiency of LVF-RH and RVF-LH performance and the visual field that dominates performance in the BILATERAL situation is similar to findings dealing with metacontrol in split-brain patients (Levy & Trevarthen, 1976) and suggests that the concept of metacontrol extends to the intact brain. Additional research using the BILATERAL paradigm is necessary to determine the extent to which the nature of interhemispheric interaction is influenced by specific task demands, handedness, and so forth.

Although the BILATERAL data from most of the subjects in the Hellige et al. (1986) experiment were described better by the RVF-LH Model than by the other models outlined in Figure 1, there were substantial individual differences in the overall feature location pattern and some interesting deviations from the majority pattern. Some illustration of these individual differences is provided in Figure 3, which shows both error rates and RT for each of the experimental conditions for three different subjects. Subject 15

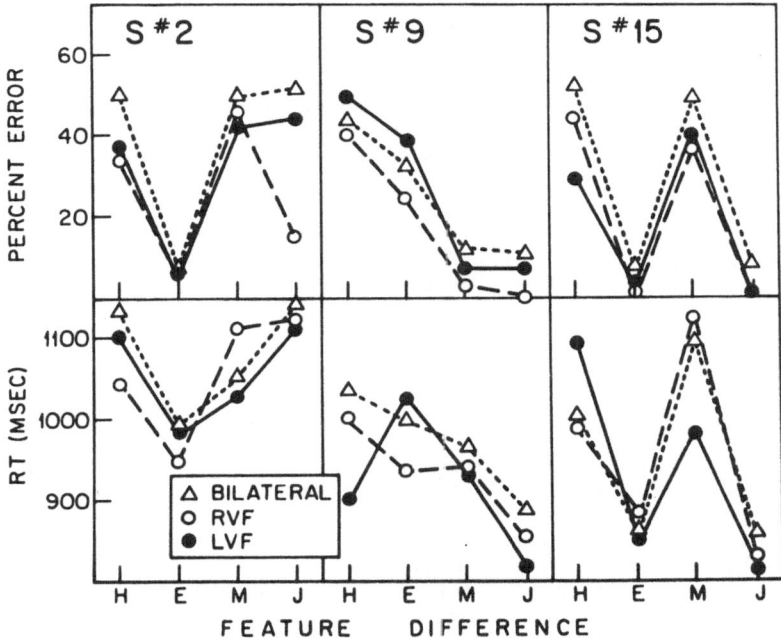

Figure 3. Percent error and RT data from three individual subjects. See text for discussion.

shows RT performance similar to the group data shown in Figure 2, both in terms of the main effect of feature location and because the RT pattern for BILATERAL trials is identical to the pattern for RVF-LH trials and different from the pattern for LVF-RH trials. However, for this particular subject the RT is the same for BI-LATERAL and RVF-LH trials; that is, there is no constant disadvantage for BILATERAL presentation in the RT data (but note such a disadvantage for percentage of errors). Subject 9 shows a somewhat different main effect of feature location, but is similar to Subject 15 in the sense that the feature location pattern on BILATERAL trials is the same as on RVF-LH trials and different from that seen on LVF-RH trials. In contrast to Subject 15, for Subject 9 both the RT and percentage of errors on BILATERAL trials were consistently greater than on RVF-LH trials. Subject 2 is, in some ways, the subject most deviant from the majority pattern. Subject 2 was the only subject in the experiment for whom the LVF-RH Model provided the best description of BILATERAL results for both RT and percentage of errors. Note that the main effect of feature location is some-what different from that of each of the other two subjects and that the pattern of feature location effects follows closely the LVF-RH

464

pattern, with BILATERAL performance well-described as equal to LVF-RH performance plus some constant increase in RT and errors. Thus, one might argue that for this subject it is the right hemisphere that asserts metacontrol for this task.

This brief illustration of individual differences indicates that the specific way in which the two hemispheres interact in a situation where both receive relevant stimulus information may vary from subject to subject, even among right-handers. Furthermore, such variation may be independent of any individual differences in the superiority of one hemisphere or the other for the task. It remains for future research to examine the stability of such individual differences across tasks and to examine whether such differences across tasks are at all predictive of individual differences in :her aspects of cognitive processing.

CONCLUDING COMMENTS

In our enthusiasm for studying information processing differences between the left and right cerebral hemispheres we have tended to put aside questions about how the two hemispheres interact with each other to provide an efficient and unified single information processing system. However, in the neurologically intact organism the hemispheres undoubtedly interact in a variety of ways that are likely to depend on such things as task demands and individual differences. The paradigms reviewed in this chapter have begun to indicate the range of such interactions. Although much remains to be learned, it has already become clear that cognitive processing in the natural environment is likely to be influenced at least as much by the nature of interhemispheric interaction as by hemispheric asymmetries.

Acknowledgement

Preparation of this chapter was facilitated by a grant from the National Science Foundation to the author (BNS-8317655).

References

Banich, M.T. (1986) Models of interhemispheric interaction: Implications for resource allocation between the cerebral hemispheres. Submitted for publication.

Davis, R. & Schmit, V. (1971) Timing the transfer of information between the hemispheres in man. Acta Psychologica, 35, 335-346

Dimond, S.J., Gibson, A.R., & Gazzaniga, M.S. (1972) Cross field and within field integration of visual information. Neuropsychologia, 10, 379-381.

Friedman, A. & Polson, M.C. (1981) Hemispheres as independent resource systems: Limited capacity processing and cerebral specialization. J. exp. Psychol., Hum. Percept. Perform., 7, 1031-1058.

Green, J. (1984) Effects of intrahemispheric interference on reaction times to lateral stimuli. J. exp. Psychol., Hum. Percept. Perform., 10, 292-306.

Hellige, J.B (1983) Hemisphere X task interaction and the study of laterality. In Cerebral Hemisphere Asymmetry: Method, Theory and Application. (ed. J. Hellige). Praeger, New York.

Hellige, J.B. (1985) Hemisphere-specific priming and interference: Issues in conceptualization. Presented at International Neuropsychological Society, San Diego, CA.

Hellige, J.B., Jonsson, J.E., & Michimata, C. (1986) Processing from LVF, RVF and BILATERAL presentations: Examinations of metacontrol and interhemispheric interaction. Submitted for publication

Hellige, J.B. & Sergent, J. (1986) Role of task factors in visual field asymmetries. Brain Cog., 5, 200-222.

Holtzman, J.D. & Gazzaniga, M.S. (1982) Dual task interactions due exclusively to limits in processing resources. Science, 218, 1325-1327.

Kinsbourne, M. (1975) The mechanism of hemispheric control of the lateral gradient of attention. In Attention and Performance V. (eds. P. Rabbitt & S. Dornic), Academic Press, New York.

Kinsbourne, M. & Hiscock, M. (1983) Asymmetries of dual-task performance. In Cerebral Hemisphere Asymmetry: Method, Theory and Application. (ed., J. Hellige), Praeger, New York.

Levy. J. & Trevarthen, C. (1976) Metacontrol of hemispheric function in human split-brain patients. J. exp. Psychol., Hum. Percept. Perform., 2, 299-312.

Liederman, J., Merola, J., & Martinez, S. (1985) Interhemispheric collaboration in response to simultaneous bilateral input. Neuropsychologia, 23, 673-684.

Sergent, J. (1982) About face: Left-hemisphere involvement in processing physiognomies. J. exp. Psychol., Hum. Percept. Perform., 8, 253-272.

28
Neuroanatomical Aspects of Hemisphere Specialization in Humans

Sandra F. Witelson* and Debra L. Kigar**

NEUROANATOMICAL ASYMMETRIES

Anatomical asymmetries between the right and left hemispheres have been documented in the human brain since the last century. Much of the early work focussed on the Sylvian (lateral) fissure and the observation that this fissure was longer on the left more often than on the right side (e.g., Cunningham, 1892; Eberstaller, 1890). More recent anatomists pursued this morphological difference and looked within the Sylvian fossa in order to measure the horizontal surface of the temporal lobe which lies posterior to the first transverse gyrus (Heschl's gyrus, the primary auditory receiving area) (e.g., Pfeiffer, 1936; von Economo & Horn, 1930). Early on, the area of this region, called the planum temporale, was found to be larger on the left side (see Figure 1). Much of the current interest in neuroanatomical asymmetry was rekindled by the report of Geschwind and Levitsky (1968) who observed that the linear extent of the lateral edge of the planum temporale was greater on the left side in 65 percent of 100 brain specimens. Other more recent studies (e.g., Wada, Clarke & Hamm, 1975; Witelson & Pallie, 1973) have confirmed this finding. Several reviews of asymmetry in the planum temporale in adults as well as in children are available (see e.g., Witelson, 1977; 1983).

The finding of asymmetry in the planum has proved quite consistent in the past twenty years, but defining this region is not as simple as was originally thought (see Witelson, 1977, for further discussion). As can be seen in Figure 1, the Sylvian fissure does not always follow a single course, but sometimes branches posteriorly. In such cases it is often ambiguous which posterior ramus is the main one, which could make a difference in gross measurement of the planum. Moreover, since the Sylvian

---

*   Departments of Psychiatry, Psychology, and Neurosciences,
         McMaster University, Hamilton, Ontario, Canada.
**   Department of Psychiatry, McMaster University

fissure sometimes turns sharply upward, any studies that expose the planum by a dissection in the plane of the horizontal axis of the Sylvian fissure could exclude the posterior part of the planum situated within the ascending part of the fissure. Because the

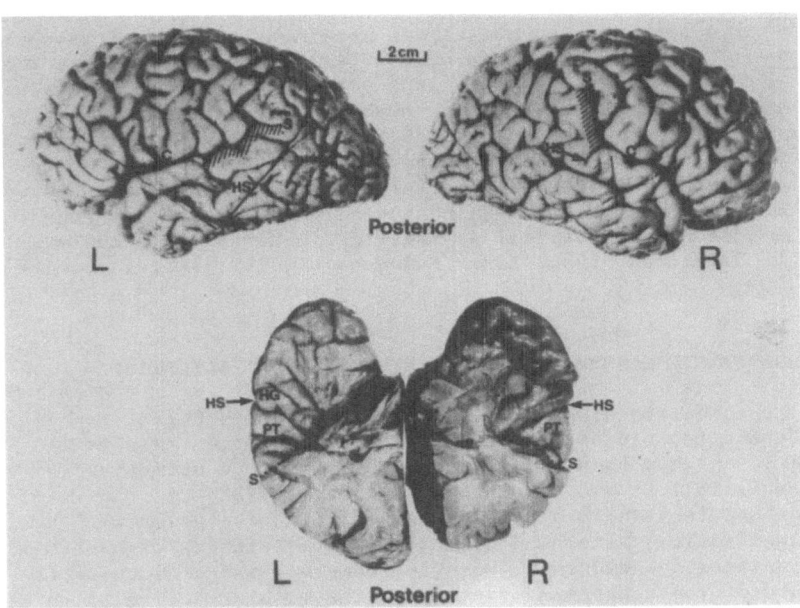

Figure 1. Lateral views of the two hemispheres and the exposed superior surface of each temporal lobe of a right-handed woman show the typical asymmetries of a sharper upswing of the Sylvian fissure on the right side, a larger posterior parietal operculum on the left side, and a larger planum temporale on the left side (519mm^2 versus 386mm^2). Each planum temporale was exposed by dissection which followed the curve of the Sylvian fissure to its end. Note the bifurcation in the posterior part of the Sylvian fissures. Note also that the anterior end of the right temporal block is tilted upward so that the planum is photographed in a horizontal plane. Abbreviations: S, posterior end of the Sylvian fissure; C, inferior tip of the central sulcus; HS, Heschl sulcus; HG, Heschl gyrus; PT, planum temporale. Dotted lines indicate the posterior boundary of the planum; hatched lines indicate the lateral edge of the planum. The parietal operculum is between C and S, above the Sylvian fissure. (Figure reprinted from Witelson, 1987a, with permission.)

Sylvian fissure curves sharply upward on the right more often than on the left side, such dissection could lead to a systematic bias in measuring the planum. Such difficulties are inherent in measuring gyral regions in the human cortex. Histological analyses of right and left planum regions may be useful in resolving such issues. Some histological work has been reported (e.g., Galaburda, Sanides & Geschwind, 1978) and other is in progress (e.g., Witelson & Colonnier, 1983).

It is possible to investigate some neuroanatomical asymmetries in vivo. With the use of carotid arteriograms, the different angulation of the Sylvian fissure in the temporoparietal region can be observed. Associated with this anatomical asymmetry is a larger postcentral operculum on the left than the right side (see Figure 1). CT scans and skull X-rays have also been used to compare the breadths of the frontal and occipital lobes of both hemispheres and asymmetries have been found in these dimensions also. Several reviews of neuroanatomical asymmetries in general are now available (e.g., Galaburda, 1984; LeMay & Geschwind, 1978; Witelson, 1977; 1983).

NEUROANATOMICAL ASYMMETRY AND HEMISPHERE SPECIALIZATION

The functional specialization of the hemispheres in mediating different aspects of cognition, such as language and spatial perception, has been well documented. Numerous extensive reviews are available (e.g., Bradshaw & Nettleton, 1983, of studies of normal adults; Heilman & Valenstein, 1985, of studies with brain-damaged adults; Witelson, 1985b; 1987b, of studies with normal and brain-damaged children). However, numerous issues such as the nature of the functional dichotomy, the absolute or relative nature of this dichotomy, and the relationship of functional asymmetry and level of cognitive ability are still unresolved (see Witelson, 1987a).

A further issue concerns whether the neuroanatomical asymmetries are a substrate of functional brain lateralization. Many researchers have assumed too enthusiastically that the anatomical and functional asymmetries are correlated. Research findings related to this issue are only beginning to accumulate.

Individuals vary in the direction and degree of functional brain lateralization. There is also variation in neuroanatomical asymmetry. One approach to evaluating whether the asymmetries are related is to study whether variation in one is correlated with variation in the other. Such work may serve an additional purpose. The various neuropsychological approaches to the study of functional asymmetry have methodological and theoretical limitations which make it difficult to resolve some of the issues pertaining to functional asymmetry (see Witelson, 1983). The study of possible neuroanatomical correlates may provide independent biological variables to help delineate some psychological issues. One example will be described later concerning the classification of hand preference.

To date, a small body of evidence is available supporting the association of some anatomical asymmetries with patterns of speech lateralization. For example, some data show an association of hand preference with frontal and occipital lobe breadths and, in a few cases, with planum asymmetry. Speech lateralization, as determined by amytal testing, was found to be associated with asymmetrical morphology in the Sylvian fissure and parietal operculum region. In a few cases, ear asymmetry on dichotic listening was found to be associated with planum asymmetry. This work relating anatomical and functional asymmetries has been reviewed in more detail elsewhere (Witelson, 1980; 1983).

ANATOMICAL STUDIES OF THE HUMAN CORPUS CALLOSUM

With the recent interest in right-left anatomical asymmetries there has been a concurrent surge of studies examining the gross anatomy and morphology of the corpus callosum, which is the main fiber tract connecting the two hemispheres. Clearly, the callosum is essential for interhemispheric functioning. Individuals in whom the corpus callosum is neurosurgically sectioned or interfered with by disease, making stimulation directed to one hemisphere inaccessible to the other, may behave as if they have two separate sets of perceptions and memories. The corpus callosum may also play a role in the manifestation, and possibly even in the maintenance, of the functional specialization of the hemispheres (Witelson, 1985b). Thus it is reasonable to consider whether the anatomy of the corpus callosum may have some relationship to hemisphere specialization.

There is considerable variation in callosal anatomy which is consistent across different studies (see Witelson & Kigar, 1987). There are also individual differences in patterns of hemisphere specialization, which may be indexed to some degree by behavioral measures such as hand preference (e.g., see Bryden, 1982). Thus the hypothesis arises that left handers, who as a group have greater bilateral functional representation than right handers, might have a callosal anatomy that allows for greater inter- hemispheric connectivity. Greater bihemispheric representation of cognitive functions has also been reported for females compared to males (e.g., see Bryden, 1982). If this proves to be the case, perhaps callosal anatomy may also vary between the sexes.

Some of the earliest anatomical studies of the human corpus callosum appeared at the turn of the century (see Witelson & Kigar, 1987, for review). The early studies (e.g., Bean, 1906; Mall, 1909) examined the size of the callosum in relation to intelligence, race and sex. Within the last decade, work has focussed on the size and shape of the callosum and its various subdivisions in relation to variables such as hand preference and sex which have relevance for neuropsychological considerations. Most of the work has been based on measurements obtained from postmortem study of brain specimens. Recently some work examining psychiatric patients compared to control groups of normal indivi- duals has provided normative data using in vivo measurements from magnetic resonance imaging (MRI) scans.

The studies to date have dealt with measurements of the total area of the callosum along its midsagittal (mid-longitudinal) axis, and of the area and width (midsagittal height) of its various sub- divisions. The anatomical features and dimensions measured in the available studies are shown in Figure 2. A summary of the main findings concerning callosal anatomy in normal adults is presented in Table 1. Some reports are available only in abstract form and these are listed separately at the bottom of the table.

HAND PREFERENCE AND CALLOSAL ANATOMY

Almost all the studies of the anatomy of the corpus callosum involve postmortem examination of the brain. Accordingly, infor- mation about hand preference was not usually available. Not only was writing hand not known, but tested measures of hand preference or proficiency were certainly unavailable. Witelson and colleagues have been studying a group of cognitively normal individuals composed of seriously ill hospital out-patients with non-neural metastatic disease who have consented to neuropsychological testing and, in the event of death, to autopsies for research purposes (for further details, see Witelson, 1983). One behavioral measure was a test of observed hand preference based on the twelve-item hand preference questionnaire of Annett (1967). There are several possible classification schemes for hand preference (see e.g., Bryden, 1982). Following Annett's (1972) model, the subjects in Witelson's studies were classified as consistent right handers (having 100% right hand preference on the battery of manual tasks), consistent left handers (100% left hand preference) and mixed handers (less than 100% preference for either right or left hand). Mixed handers thus included both right and left hand writers. Consistent left hand preference occurs in only about 4 percent of the population and none were available in the sample of 42 cases.

In the first report of this group, a subgroup of 15 individ- uals with mixed hand preference was found to have a larger callosal area by about 11 percent compared to a group of 27 individuals with consistent right hand preference (Witelson, 1985a). The anterior and posterior halves were also larger in area in the mixed handed group. These differences continued to be observed even when cerebrum weight and sex were controlled for. The larger callosum

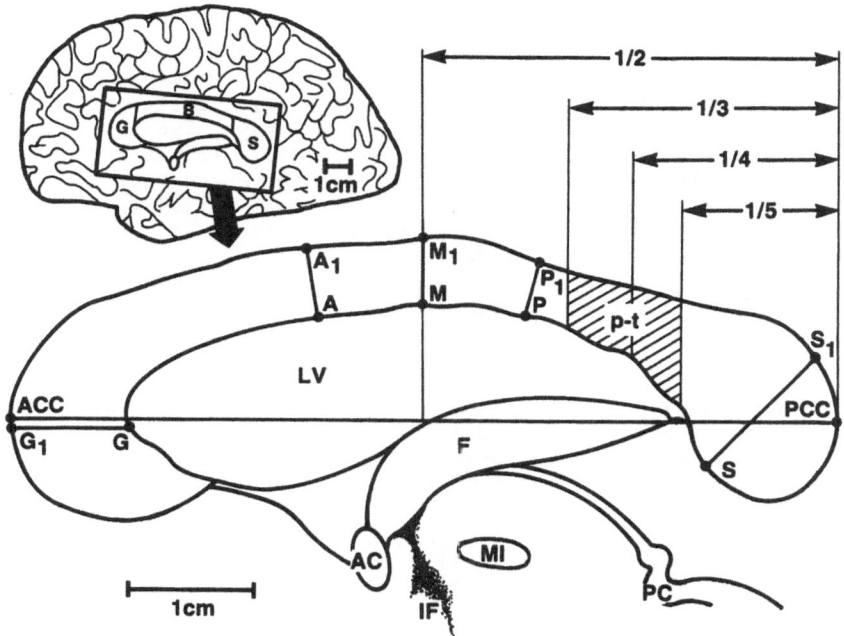

Figure 2. The corpus callosum of an adult human is shown in midsagittal view. The various measures and subdivisions referenced in Tables 1 and 2 are indicated. Abbreviations: G, genu; B, body; S, splenium; ACC and PCC, anteriormost and posteriormost points of the corpus callosum, respectively, which form the line of maximal length used to obtain the callosal subdivisions (the anterior and posterior halves, and the posterior third, quartile, and fifth). The midposterior region (cross-hatched area), labelled the parietotemporal region (p-t), is defined as the posterior third minus the posterior fifth region; GG_1, maximal width of the genu; SS_1, maximal dorsoventral width of the splenium; AA_1, a width of the anterior part of the body of the callosum; MM_1, midbody width; PP_1, a width of the posterior part of the body; LV, lateral ventricle; F, fornix; IF, interventricular foramen; MI, massa intermedia; AC, anterior commissure; PC, posterior commissure. (Reprinted from Witelson & Kigar, 1987, with permission.)

of the mixed handers is compatible with the greater bilateral functional representation reported for left versus right handers.

In a subsequent report, further callosal subdivisions were measured. It was found that the posterior part of the body or trunk of the callosum, labelled the parietotemporal region (see Figure 2) was particularly large in mixed handers compared to right handers (by 19 percent) (Witelson, 1986). This result is of considerable interest with regard to the possible anatomical substrate of hemisphere specialization as this callosal region has been found to house fibers that connect the parietotemporal cortex of the two hemispheres in monkeys (e.g., Cipolloni & Pandya, 1985) and in humans (de Lacoste, Kirkpatrick & Ross, 1985). In humans, these are the cortical regions crucial for language, praxis (learned motoric functions) and visuospatial functions, all of which tend to be asymmetrically represented in the parietotemporal regions of the hemispheres. The larger parietotemporal callosal region in the group of mixed handers may be related to greater bihemispheric representation of these cognitive functions.

The splenium is the bulbous region at the posterior end of the callosum, but there are no clear anatomical landmarks to define its anterior boundary. It is roughly the posterior fifth region of the callosum and this geometric definition has been used in most studies. There is considerable variation in the size and shape of the splenium (see Witelson & Kigar, 1987). It has been demonstrated to house fibers which cross the hemispheres from the occipital and anterior temporal regions relevant to visual functions in monkeys (e.g., Rockland & Pandya, 1986) as well as in humans as evidenced in clinical cases of callosal sectioning.

The findings for the splenium are unique, in that in contrast to the rest of the callosum, it did not differ in size or in the extent of variation between the two hand groups (Witelson, 1985a; 1986). This was so despite the fact that this bulbous region of the callosum shows considerable individual variation in shape.

The splenial region also differs from the rest of the callosum in other anatomical and functional respects. The neural developmental course is different for the splenium. After birth, the splenium undergoes the greatest relative increase in overall size compared to the rest of the callosum, tripling its width compared to its size at birth (Rakic & Yakovlev, 1968). It is also the first region to begin myelination (at about the fourth postnatal month), with callosal myelination progressing anteriorly from the splenium (Yakovlev & Lecours, 1967).

At a physiological level, the posterior region of the callosum may have a different function in interhemispheric transfer of information than the more anterior regions. The splenium appears to transmit mainly sensory information, rather than higher

level processed perceptual information (Sidtis et al, 1981).
Experimental work with monkeys has indicated that splenial functioning may result in unilateral engrams whereas anterior
commissure functioning results in bilateral traces (Doty, Overman &
Negrao, 1979). Consistent with these results are the findings
based on metabolic mapping of monkey visual cortical areas by the
deoxyglucose method. Callosal input to the prestriate visual
cortex which courses through the splenium appears to have sup-
pressive effects compared to the facilitative electrophysiological
influences that the anterior commissure appears to have (Macko &
Mishkin, 1985).

Since patterns of hemisphere specialization are related to
variation in hand preference, and the available data suggest that
corpus callosum size varies with direction of hand preference, it
follows that patterns of hemisphere specialization may be related
to variation in callosal size. If so, greater bilateral representation of hemispheric functions which may require more interhemispheric communication may be associated with a larger callosum.
Whether a larger callosum reflects more fibers, fibers of greater
diameter, or thicker myelin sheaths remains to be determined.

Recent findings in developmental neurobiology suggest that
few, if any, additional callosal fibers cross the midline after
birth. In fact, experimental evidence in different species
suggests that within days after birth there is a period of intense
loss of callosal fibers, and that this period of axonal elimination, part of the early regressive events in neural development,
appears to end with the onset of callosal myelination and rapid
synaptogenesis. An estimate of the period of axonal elimination in
the human callosum is from one to four months of age postnatally
(Innocenti, 1986). This feature of neural development suggests
that the number of fibers in the callosum is established in early
infancy.

If the larger callosum of mixed handers is a reflection of
more fibers and not merely thicker fibers or some other histological feature, then the anatomical difference is not likely the
result of differential experience associated with different hand
preference. The key question then may be not what biological
factor results in a larger callosum in mixed handers, but what
prevents mixed handers from having a smaller callosum, as do right
handers. As argued previously (Witelson, 1985a; 1986), there may
be some mechanism, possibly genetically or epigenetically based,
which results in less axonal elimination, a larger callosum, and
the behavioral manifestation of non consistent-right-handedness.
Along similar lines, within Changeux's (1985) hypothesis of
selective stabilization, it was suggested that the postnatal loss
of fibers in the corpus callosum might be a factor in hemisphere
specialization (p.241).

474

Table 1. RESULTS OF STUDIES OF THE ANATOMY OF THE CORPUS CALLOSUM IN NORMAL ADULTS IN RELATION TO HEMISPHERE SPECIALIZATION[a]

STUDY	MATERIAL	SAMPLE	SUBGROUPS	TOTAL CORPUS CALLOSUM (CC) AREA	ANTERIOR HALF CC AREA	POSTERIOR HALF CC AREA	SPLENIUM WIDTH[b]	SPLENIUM AREA[c]
BEAN[d] 1906	POSTMORTEM mid-sagittal drawings	67 CAUCASIAN(C) (includes 26 cases from Retzius, 1900) X̄ = 52 yr	53 Male	691	368[f]	301[f]	g	173[h]
			14 Female	634	314*	289		164
		80 NEGRO (N) X̄ = 44 yr	52 Male	635	321	316		186
			28 Female	571*	286*	286*		173
				*M;F: C>N	*Males: C>N			*Males: C<N
MALL[d] 1909	POSTMORTEM mid-sagittal drawings	33 CAUCASIAN	30 Male	612				150
			3 Female	i				
		70 NEGRO	50 Male	575				143
			20 Female	548				137
DE LACOSTE-UTAMSING & HOLLOWAY 1982	POSTMORTEM mid-sagittal	N = 14 X̄ = 37 yr[j]	9 Male	704.3			11.4	186.1
		X̄ = 28 yr[j]	5 Female	708.3			16.4*	218.3

Table 1 continued

STUDY	MATERIAL	SAMPLE	SUBGROUPS	TOTAL CORPUS CALLOSUM (CC) AREA	ANTERIOR HALF CC AREA	POSTERIOR HALF CC AREA	SPLENIUM	
							WIDTH[b]	AREA[c]
HOLLOWAY & DE LACOSTE 1986	POSTMORTEM mid-sagittal	N = 16 CAUCASIAN AND NEGRO						
		$\bar{X}$ = 64 yr	8 Male	618.4			10.3	
		$\bar{X}$ = 73 yr	8 Female	744.4			13.3*	
WITELSON 1985a	POSTMORTEM mid-sagittal	N = 42 CAUCASIAN	Consistent Right Handers (CRH)					
		$\bar{X}$ = 48 yr	7 Male	672.1	367.4	304.6		180.7
		$\bar{X}$ = 51 yr	20 Female	654.8	345.1	309.7		172.4
			Mixed Handers (MH)					
		$\bar{X}$ = 49 yr	5 Male	800.6	423.2	377.4		202.4
		$\bar{X}$ = 49 yr	10 Female	697.1	376.6	320.5		178.4
				*CRH<MH	*CRH<MH	*CRH<MH		

Table 1 continued

STUDY	MATERIAL	SAMPLE	SUBGROUPS	TOTAL CORPUS CALLOSUM (CC) AREA	ANTERIOR HALF CC AREA	POSTERIOR HALF CC AREA	SPLENIUM	
							WIDTH[b]	AREA[c]
WITELSON 1986	POSTMORTEM mid-sagittal	N = 42 CAUCASIAN $\bar{X}$ = 50 yr	27 CRH / 7M & 20F	659.4	350.9	308.4		174.5
		$\bar{X}$ = 49 yr	15 MH / 5M & 10F	731.6*	392.1*	339.5* p-t-k 62.3(CRH) 74.1*(MH)		186.4
NASRALLAH ANDREASEN COFFMAN OLSON DUNN EHRHARDT & CHAPMAN 1986	MRI[m] mid-sagittal	N = 41 NORMAL VOLUNTEERS $\bar{X}$ = 28 yr (control group for schizophrenics)	Right Handers / 11 Male	705[n]				
			10 Female	605				
			Left Handers / 10 Male	660				
			10 Female	No values or comparisons given.				

477

Table 1 continued

STUDY	MATERIAL	SAMPLE	SUBGROUPS	TOTAL CORPUS CALLOSUM (CC) AREA	ANTERIOR HALF CC AREA	POSTERIOR HALF CC AREA	SPLENIUM WIDTH[b]	SPLENIUM AREA[c]
ABSTRACTS:								
DE LACOSTE-UTAMSING HOLLOWAY KIRKPATRICK & ROSS 1981	POSTMORTEM mid-sagittal	N = 28	15 Male / 13 Female	Total Group 662.2			Min/Max 9/14 / 14/18*	*F>M
DEMETER RINGO & DOTY 1985	POSTMORTEM mid-sagittal	N = 31	19 Male / 12 Female				No sex difference	No sex difference
NASRALLAH ANDREASEN OLSON COFFMAN COFFMAN DUNN & EHRHARDT 1985	MRI mid-sagittal	N = 41 NORMAL VOLUNTEERS (control group for schizophrenics)	Right Handers 11 Male / 10 Female; Left Handers 10 Male & 10 Female			% of CC Area 33.4p / 36.3*		

No values or comparisons given for left handers.

Table 1 continued

STUDY	MATERIAL	SAMPLE	SUBGROUPS	TOTAL CORPUS CALLOSUM (CC) AREA	ANTERIOR HALF CC AREA	POSTERIOR HALF CC AREA	SPLENIUM WIDTH[b]	SPLENIUM AREA[c]
CLARKE KRAFTSIK INNOCENTI & VAN DER LOOS 1986	POSTMORTEM & MAGNETIC RESONANCE IMAGING (MRI) combined, mid-sagittal	N = 58	32 Male 26 Female	Min/Max 480/930 490/760	% of CC Area No sex difference		*F>Mq	% of CC Area 25.6 27.9*
								Absolute area[r] No sex difference

Footnotes

* Indicates a statistically significant difference: either between the starred value and the one directly above it; or between the group comparisons as indicated. No asterisk indicates no statistical difference. Level of significance used was .05.

a All measurements are mean values in mm or mm^2. The anatomical locations are schematically represented in Figure 2.

b Maximal dorsoventral width of the splenium.

c Splenial area is defined as the posterior fifth region of the corpus callosum based on a linear subdivision as shown in Figure 2.

Table 1 continued

d The statistical analyses were done by the present authors based on raw data presented in the original reports in Tables I, VI, VII (Bean, 1906) and pages 28-31 (Mall, 1909). Mall did not report age, but in the Bean report, those cases 16 yr of age or less were excluded. Independent t-tests (two-tailed) were done for all callosal area measures presented and also for brain weight between sex and race subgroups. All significant results are indicated.

e Race is given only when reported.

f Ns for the anterior and posterior half areas are different than for the total corpus callosum: 38 Caucasian males, 8 Caucasian females, 54 Negro males, 25 Negro females.

g Blank indicates no information available.

h Ns for the splenial area are different than for the other areas: 53 Caucasian males, 14 Caucasian females, 52 Negro males, 25 Negro females.

i The female sample is too small for valid mean scores, and the total callosal area of one case was atypically small ($270mm^2$).

j Mean ages obtained from de Lacoste-Utamsing (June, 1983), personal communication.

k These are measurements of the mid region of the posterior half of the corpus callosum (labelled the parietotemporal region) as shown in Figure 2.

m In contrast to postmortem material, midsagittal MRI scans may not be aligned with the true midsagittal plane. Scans also represent the maximal area over a 3-dimensional cut of the callosum.

n These numbers are based on a magnification of the MRI scans using a factor which only gave approximately true values.

p Splenial area in this study was defined as the posterior quartile region based on a linear subdivision of the corpus callosum as shown in Figure 2.

q Females had a greater splenial width relative to width of the posterior callosal body.

r Personal communication, Clarke et al (June, 1986).

480

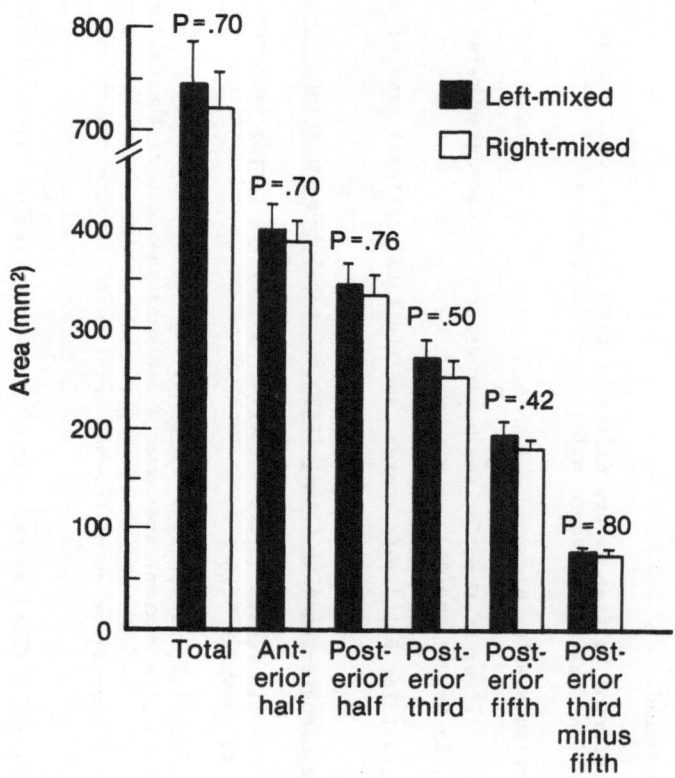

Figure 3. Mean area measurements (±SEM) of the corpus callosum regions for the right mixed (N = 9) and left mixed (N = 6) subgroups. Two-tailed t-tests (df = 13) were used. (Reprinted from Witelson, 1986, with permission.)

Neuroanatomical Key for Hand Preference Classification

The size of callosal regions other than the splenium was found to vary with direction of hand preference: consistent right hand versus mixed hand preference. Further statistical analyses were done to determine whether callosal size also varied with degree or magnitude of hand preference (scores could vary from +12 to −12). Since the 27 consistent right handers had almost identical hand preference scores, correlations were calculated only for the 15 mixed handers, which included individuals of both sexes. To rule out differing brain size between the sexes partial correlations were done. The partial correlation with cerebrum weight as the covariate for total callosal area and hand score, which reflects both direction and degree of hand preference, was r = −.03 (df = 12, p = .92). The partial correlation for absolute hand score, which reflects only degree of hand preference, was r = .26 (df = 12, p = .40). Therefore, no evidence was obtained that callosal size is associated with degree of hand preference (see Witelson, 1985a; 1986). These anatomical findings support the biological validity of a simple functional dichotomy of hand preference: consistent right versus mixed hand preference.

Further analyses supported the observation that within the mixed handers, callosal anatomy is not differentiated by laterality of hand preference. The mixed handed group was subdivided into two subgroups according to writing hand. No differences in callosal anatomy were observed for these subgroups (see Figure 3). Further details are presented elsewhere (Witelson, 1986).

The question remains whether laterality (consistent right versus non consistent right) or consistency (right or left) versus nonconsistency (mixed) is the key factor associated with callosal size. In the former case, consistent left handers would be expected to show a callosal anatomy similar to mixed handers; in the latter, consistent left handers would be expected to be similar to consistent right handers. No consistent left handers were available for postmortem study, but MRI scans of the midsagittal view of the callosum were obtained for two normal men having consistent left hand preference. The results of these scans suggested that they were similar in callosal anatomy to the mixed handed group (Witelson, 1985a). Such preliminary results suggest that laterality is the key factor and support a dichotomy between right and non right hand preference in the classification of handedness. Further work is needed concerning consistent left handers.

Only one other study has reported data on callosal size in right and left handed individuals. This was an MRI investigation of schizophrenics which examined normal male volunteers as controls (Nasrallah et al, 1986). In this study the left handers had a smaller mean value for total callosal area than did right handers. However, no statistical analyses were presented comparing the scores of the right and left handed groups; nor were the raw data given so that no statistical analyses could be done by the present authors. No data were given to indicate whether the groups were comparable in chronological age. Total cerebral area in the mid-sagittal plane could be calculated for each group from the information given in Tables 2 and 3 of the Nasrallah et al (1986) report. The left handed subgroup had a smaller cerebral area which might have confounded any group differences in callosal size, but no statistical analysis was done. Finally, the definition of right and left handed individuals was not specified. The criteria of right handed probably involved the usual definition of right handedness, namely, the use of the right hand in writing. Such a classification would likely include some mixed handers in the right handed group and could mask possible differences between consistent right handers and others.

The in vivo imaging techniques currently available (e.g., MRI scans) offer an easier route to studying brain, and specifically callosal anatomy in relation to hand preference or other indices of hemisphere specialization. However, such technology leads to anatomical measures that may necessarily be less precise than direct postmortem measurements. It must also be noted that if hand preference is defined simply as the hand preferred for writing, then right handed groups will inevitably contain some mixed handers (as defined in Witelson's studies) and differences between consistent right handers and others may be masked.

SEX AND CALLOSAL ANATOMY

In the studies of callosal anatomy, sex was usually known, since sex can be ascertained readily from medical records even in postmortem studies (see Table 1). Although some of the early studies were primarily interested in callosal anatomy in relation to race, heredity and intellectual ability, sex was also docu-mented, as for example in the work of Bean (1906) and Mall (1909) who studied large groups of brain specimens. Several recent studies have focussed directly on possible sex differences in callosal size (de Lacoste-Utamsing & Holloway, 1982; Holloway & de Lacoste, 1986; and others having only preliminary reports: Clarke, Kraftsik, Innocenti & van der Loos, 1986; Demeter, Ringo & Doty, 1985). Similar work is underway in a few other laboratories but remains unpublished to date. Other recent reports have also recorded and evaluated sex as a possible factor in callosal size although their main focus was not sex differences: Witelson (1985a) in a study of the possible relationship between hand preference and callosal anatomy, and Nasrallah et al (1985; 1986) in a study of callosal anatomy in schizophrenia.

The sex difference in brain size (as measured by weight or volume) is well documented. The male brain is significantly larger than the female brain by about 10 to 15% (e.g., Holloway, 1980). This fact presents a complication in evaluating any difference between the sexes in the size of parts of the brain, such as the callosum. It might be expected that since the total brain is larger in males, the callosum might also be larger in absolute size in males compared to females. For this reason, it may be most appropriate to consider the callosum relative to brain size and some studies have done this.

Furthermore, in the large studies of Bean and of Mall, brain weight was different in the two racial groups (Negro and Caucasian). Possible factors such as nutrition, experience, body size, or autopsy and fixation procedures that could be relevant remain undetermined and are not discussed.

No statistical analyses were given in these reports, but the raw data were reported, enabling the present authors to determine whether the brain weights differed significantly between the races and whether the races could be combined for the purpose of sex comparisons. The brain weights proved to be significantly different between the races, at least for males. (Bean report: 1304gm versus 1216gm for Caucasian and Negro males, respectively, $t = 3.34$, df = 108, $p = .001$; 1105gm versus 1068gm for Caucasian and Negro females, respectively, $t = .98$, df = 38, $p = .33$. Mall report: 1421gm versus 1312gm for Caucasian and Negro males, respectively, $t = 3.11$, df = 50, $p = .003$; no comparison was possible for females as only three Caucasian females were in the sample, with only one having a recorded brain weight.)

The findings concerning possible sex differences in callosal anatomy will be presented according to various callosal subdivisions and for absolute as well as relative size (see Table 1 for further details). For the Bean and Mall studies, the statistical comparisons reported were done by the present authors. A review of some of these results is presented in an earlier report (Witelson & Kigar, 1987).

Total Callosal Area

No sex difference was present in absolute size in Bean's sample of Caucasians ($t = 1.73$, df = 65, $p = .09$). Mall's raw data were also available, but there were only three female Caucasian brains, and one had an atypically small callosum of $270mm^2$. Males had a significantly greater total callosal area than females in Bean's Negro samples ($t = 2.75$, df = 78, $p = .008$), but not in Mall's ($t = 1.26$, df = 68, $p = .21$). The discrepant results in callosal sex differences in the Negro samples could be related to the variation in brain size. In Bean's study there was a particularly large difference in brain weight between Negro males and females (1216gm versus 1068gm, $t = 5.06$, df = 83, $p < .00001$). An analysis of callosal size relative to brain weight might attenuate this difference. Mall found a much smaller difference, although still statistically significant, between the sexes of his Negro samples (1312gm versus 1222gm, t = 2.25, df = 47, p = .03).

Two recent studies involving much smaller samples found no statistical difference in total callosal area between the sexes (de Lacoste-Utamsing & Holloway, 1982; Holloway & de Lacoste, 1986). The female groups tended to have larger mean values than did the males and much emphasis has been placed on this statistically non-significant variation by these authors and others. In the earlier study (1982) it was concluded that the total corpus callosum "was greater in females relative to brain weight" (p.1432), on the basis of the finding that brain weight differed significantly as expected, but callosal area did not. However, no statistical analysis of any ratio score was done.

In the later paper (1986) a ratio score of total callosal area relative to brain weight was considered and found to be significantly greater for females. However, there appear to be some unusual features of the samples of eight male and eight female adult brains studied in the Holloway and de Lacoste (1986) report. The male group did not have a significantly larger brain weight, as is usually the case. This is also the only report in which the size of the mean callosal area of the males was not greater than that of the females. Also surprising is the low absolute mean value of $618mm^2$ for total callosal area in the male group, considerably less than most other male groups. The female sample used as a matched comparison group, had a mean callosal area of $744mm^2$, higher than most other female groups, in spite of their much older chronological age of 73 years. Such factors suggest that these samples may not be representative of the population.

In another study involving 12 male and 30 female brains (Witelson, 1985a), sex differences were evaluated both for absolute total callosal area and for area with the variable of cerebrum weight factored out (weight of the hemispheres with the hindbrain removed), using an analysis of covariance. In neither case was a significant sex difference observed. One further study (Clarke et al, 1986) also did not observe any sex difference in total callosal area. An additional study of schizophrenia (Nasrallah et al, 1986) included normal male and female groups as controls. Unfortunately no statistical analyses of the normal groups were reported, but the mean value of the males exceeded that of the females.

In summary, the evidence does not support a sex difference in total callosal area, either in absolute size or in proportion to brain size. Only one study (Bean, 1906) observed a significant difference in absolute callosal size between the sexes. Males were observed to have a larger callosum, in Negroes only, but the Negro males and females also differed markedly in brain weight. Only one

study observed a significant difference in relative callosal size.
Females were found to have a larger callosum (Holloway & de
Lacoste, 1986), but the absolute values of callosal size as well as
of brain weight in these samples were not very comparable to those
of other studies.

Anterior and Posterior Halves

Some of the above studies also measured callosal subdivision.
In Bean's study, within each racial group males had a significantly
greater absolute area for the anterior half of the callosum than
did females (Caucasians: t = 2.14, df = 44, p = .04; Negroes: t =
2.5, df = 77, p = .02).

For the posterior half, in Bean's report no sex difference in
absolute size was found for the Caucasian group (t = .52, df = 44,
p = .60), but in the Negro group males had a significantly greater
posterior half region compared to females (t = 2.36, df = 77,
p = .02). However the variation in brain size between the sexes
makes these comparisons of absolute size only a preliminary
analysis.

In contrast, in Witelson's (1985a) study, analyses for sex
differences within each hand group, showed no difference for either
absolute area or for area corrected for brain size for the anterior
or posterior halves of the callosum. Clarke et al (1986) also
observed no sex difference in the anterior half proportional to
total callosal area. However, in Witelson's study, for the
posterior half region, an analysis of covariance with cerebrum
weight as the covariate, indicated an interaction between the
factors of sex and hand preference which almost reached statistical
significance (p = .08). This analysis suggested that among mixed
handers, males may have a proportionately larger posterior region
relative to brain weight than do females, but that this sex
difference is not present among right handers.

The posterior quartile region of the corpus callosum was
measured from MRI scans of normal males and females, used as
controls in a study of schizophrenia (Nasrallah et al, 1985). The
mean area of this region relative to total callosal area was
reported to be larger in the female than male group. However, no
data were presented in this abstract, nor in their subsequent 1986
report.

In sum, these results raise the possibility that the size of
some posterior region proportional to the anterior region of the
corpus callosum may be different between the sexes.

Splenium

Maximal width. Several of the studies measured the posterior bulbous region (splenium) of the corpus callosum. One measure used by five of the studies (see Table 1) is the maximal dorsoventral width of the splenium (see Figure 2). The maximal splenial width has been defined as the maximal perpendicular width between two parallel lines drawn as tangents to the dorsal and to the ventral splenial surfaces. In three reports (de Lacoste-Utamsing et al, 1981; de Lacoste-Utamsing & Holloway, 1982; Holloway & de Lacoste, 1986), splenial width was found to be significantly greater in females than males. In the report by Clarke et al (1986), involving larger samples (32 males and 26 females), maximal width of the splenium relative to the width of the body of the callosum was measured. This value was reported as significantly greater in the female group, although no mean values were given in the abstract. In contrast, Demeter et al (1985) reported no sex difference in splenial width, but no values were given in the abstract.

Absolute area. Seven studies in Table 1 measured the absolute area of the splenium, in all cases defined as the posterior fifth region of the callosum. Not one study found a significant sex difference on this measure. (Analyses by the present authors, for Bean: Caucasians; $t = .85$, $df = 65$, $p = .40$; Negroes; $t = 1.61$, $df = 75$, $p = .11$; for Mall: Negroes; $t = 1.22$, $df = 68$, $p = .23$). De Lacoste-Utamsing and Holloway (1982) reported their findings as statistically significant although the difference was significant only at the .08 level.

Four of the seven relevant studies presented the mean values and in all cases, except the de Lacoste-Utamsing and Holloway (1982) report, the value for the male group tended to be larger. The results of Bean (1906), Mall (1909), Witelson (1985a), Demeter et al (1985), and Clarke et al (1986) clearly indicate that females do not have an absolutely larger posterior fifth region compared to males. In the de Lacoste-Utamsing and Holloway (1982) study, the female group had the larger score. It is difficult to account for this discrepancy. The female group of the de Lacoste-Utamsing and Holloway (1982) study is very small ($n = 5$) and one possibility is that for some reason it is an atypical sample. Moreover, it is not clear if and how the 9 male and 5 female cases in the (1982) report were selected from the 15 male and 13 female cases of an earlier report (de Lacoste-Utamsing et al, 1981). The minimum and maximum values for splenial width in the female groups were the same in both these reports, yet the mean callosal area of the total group in the 1981 report was considerably less than that of either sex group in the 1982 report.

No area measurements of the posterior fifth region were presented in the Holloway and de Lacoste (1986) report, although it was described as a replication of their earlier study in which splenial area was a key variable.

TABLE 2. ABSOLUTE AND RELATIVE SIZE OF THE POSTERIOR FIFTH REGION OF THE CALLOSUM IN MALES AND FEMALES IN DIFFERENT STUDIES

STUDY	POSTERIOR FIFTH REGION (mm^2)				AREA PROPORTIONAL TO TOTAL CALLOSAL AREA	
	ABSOLUTE AREA					
	MALE		FEMALE		MALE	FEMALE
	N	$\bar{X}$	N	$\bar{X}$		
BEAN, 1906[a] Caucasian	53	173	14	164	.25	.26
Negro	52	186	25	173	.29	.30
MALL, 1909 Caucasian	30	150	3[b]	—	.25	—
Negro	50	143	20	137	.25	.25
DE LACOSTE-UTAMSING et al, 1981 Abstract	15	—[c]	13	—	—	—
DE LACOSTE-UTAMSING & HOLLOWAY, 1982	9	186	5	218	.26	.31

TABLE 2 continued

STUDY	POSTERIOR FIFTH REGION (mm²)					
	ABSOLUTE AREA				AREA PROPORTIONAL TO TOTAL CALLOSAL AREA	
	MALE		FEMALE		MALE	FEMALE
	N	X̄	N	X̄		
HOLLOWAY & DE LACOSTE 1986	8	—c	8	—	—	—
WITELSON, 1985a Consistent Right Handers	7	181	20	172	.27	.26
Mixed Handers	5	202	10	178	.25	.26
CLARKE et al 1986 Abstract	32	—d	26	—	.26	.28

a Ratio scores are based on Ns for which both total and posterior fifth areas were available.
b The female sample is too small for valid mean scores, and the total callosal area of one case was atypically small (270 mm²).
c No measurements reported.
d No absolute values reported.

Relative area. One approach to controlling for the possible contribution of differences in brain size between the sexes is to consider the relative size of one brain part to another, for example the posterior fifth callosal region to the total callosal area. Such a ratio score also allows some comparison between different studies as, to some extent, it controls for baseline differences due to factors such as fixation methods, body height and nutritional history.

Table 2 includes those studies reviewed in Table 1 in which the posterior fifth area was measured and presents the results both in terms of absolute size when available and as a ratio score. As indicated previously, the absolute area of the posterior fifth region is not significantly different between the sexes but the values are consistently smaller in the female than in the male group in each study, except the de Lacoste-Utamsing and Holloway (1982) report. The mean ratio scores of the posterior fifth area to total callosal area tend to be quite similar between sex groups in each study, except for the male and female groups (.26 and .31, respectively) in the above report. However, it is noted that none of these ratio scores were subjected to statistical analyses by the various authors nor by the present authors.

The source of the discrepancy between the de Lacoste-Utamsing and Holloway (1982) report and the other reports is not obvious. In fact it may not even be a statistically significant discrepancy. Unfortunately the 1986 replication study by Holloway and de Lacoste did not consider area of the posterior fifth region. Bean's (1906) Negro male and female groups are the only other groups with ratio scores as high as that of the female group in the de Lacoste-Utamsing and Holloway (1982) report. Mall's (1909) Negro samples had ratio scores similar to the majority of other studies. If callosal morphology should prove to be different in some respects between Negroes and Caucasians, one possible explanation for the relatively larger splenium in the de Lacoste-Utamsing and Holloway (1982) female group may be that it included a high proportion of Negro women. Their report did not specify the racial origin of their sample, but did indicate that the brains were obtained from the Dallas Forensic Institute, Texas. In their subsequent 1986 report, the brain specimens were in fact reported to be from both Caucasian and Negro individuals, but the distribution of each in each sex group was not reported.

In sum, the evidence does not support a sex difference in the absolute or even relative size of the area of the posterior fifth region. However, there is a tenuous suggestion of possible sex differences in the posterior regions of the callosum, reflected perhaps in the maximal width of the splenium or the proportional size of some posterior regions.

490

SUMMARY

A body of data is beginning to accumulate which indicates that
variation in brain structure is associated with patterns of hemisphere lateralization of function. As mentioned briefly above,
some work has shown that right-left asymmetries in the parietotemporal cortical regions, such as the orientation of the Sylvian
fissure, the extent of the postcentral parietal operculum, or the
size of the planum temporale, have some association with hand
preference and speech lateralization as determined by amytal
testing (see Witelson, 1983, for a review). There is little
evidence to date that there are any sex differences in these neuroanatomical asymmetries (e.g., see Witelson, 1983, pp. 106-108).

The size of the corpus callosum has been found to be associated with hand preference -- specifically, to be smaller in
individuals with consistent right hand preference than in individuals with all other hand preference patterns, including mixed right
handers, mixed left handers, and consistent left handers. Since
hand preference is correlated to some degree with the lateralization of speech representation, such findings suggest that callosal
size may be related to patterns of functional lateralization. Of
particular interest is the observation that the segment of the
callosum which appears to differ most between different hand
preference groups is the posterior part of the body of the
callosum, which is involved in interhemispheric transmission
between cortical regions known to be involved in those cognitive
functions having lateralized representation.

Sex differences in callosal anatomy are neither marked nor
reliable. In general, differences in patterns of hemisphere
specialization found to be related to sex may not have any associated differences in gross anatomy. Differences related to sex, if
they exist, may be at a histological, neurochemical or neuro-
endocrinological level.

Needless to say, such results are just the first steps in
delineating the possible neuroanatomical substrate of hemisphere
specialization. New technologies such as magnetic resonance
imaging offer efficient opportunities to study further the
correlation of some neuroanatomical and psychological variables.
Anatomical features like the size of the planum temporale are
difficult to obtain from computerized scans, while others, such as
the size of the callosum, can be more readily measured
unambigu-ously.

If hand preference is to be used as a behavioral index of
hemisphere specialization, the study of consistent right handers as
a homogeneous group may be worthwhile. The finding that callosal
size varies with a particular dichotomous classification of hand
preference suggests that such a model of behavioral variation may

have biological validity and may be of general use in other neuropsychological research. Such work illustrates how neuroanatomical
analysis may be a key to the study of some psychological functions.

Anatomical asymmetries exist soon after birth (e.g., Witelson,
1983) as does hemisphere functional specialization (e.g., Witelson,
1985b; 1987b). It remains to be determined whether functional
asymmetry in infancy varies with anatomical asymmetry as is now
beginning to be documented for adults. Clearly this is a difficult
issue to investigate.

The finding of anatomical variation early in life in dimensions documented to be associated with functional asymmmetry at
maturity suggests that the morphological bases of hemisphere
specialization may be genetically or epigenetically determined. In
a developmental neurobiological model, such as Edelman's recent
proposal (Edelman & Finkel, 1984), the network of neuronal groups
and higher level neuronal maps may be differentially established in
different individuals by genetic and subsequent molecular influences, with left handers having more bilateral circuits involving
callosal fibers than right handers.

One of the crucial issues for psychology is what significance
the anatomical or functional asymmetries might have for level of
cognitive ability. To date there is little evidence to support an
association. A few cognitive differences have been noted between
right and left handers (see Witelson, 1986). A recent study based
on the finding of a larger callosum in mixed than in consistent
right handers, is an illustration of how neuroanatomical findings
may lead to fertile psychological hypotheses. It was predicted
that mixed handers, possibly having greater interhemispheric
transfer of information, would perform better than consistent right
handers on tasks requiring interhemispheric transmission of information for their completion. The mixed handers indeed were better
on a group of such tasks (Potter & Graves, 1987). Since there is
also some evidence of sex differences in functional lateralization
and level of performance on some cognitive tasks, the interesting
possibility of some correlated sex differences in anatomy remains.

The finding of associations between behavioral measures and
neuroanatomical measures supports the general hypothesis that the
documented neuroanatomical asymmetries may in fact be related to
functional asymmetry. Further work is needed in order to elucidate
the nature of the relationship of neuroanatomical measures with
more direct neuropsychological indices of hemisphere
specialization, such as performance on lateral perceptual tests,
electrophysiological measures, or deficits associated with
localized lesions.

492

ACKNOWLEDGEMENT

Preparation of this paper was supported in part by U.S.
NIH-NINCDS Contract N01-NS-6-2344 and NINCDS Grant R01-NS-18954
awarded to SFW. Diane Clews produced the manuscript.

REFERENCES

Annett, M. (1967). The Binomial Distribution of Right, Mixed and
   Left Handedness. Quart. J. Exper. Psychol., 19, 327-333.
Annett, M. (1972). The Distribution of Manual Asymmetry. Brit.
   J. Psychol., 63, 343-358.
Bean, R. B. (1906). Some Racial Peculiarities of the Negro
   Brain. Am. J. Anatomy, 5, 353-432.
Bradshaw, J. L. and Nettleton, N. C. (1983). Human Cerebral
   Asymmetry. Prentice-Hall, New Jersey.
Bryden, M. P. (1982). Laterality: Functional Asymmetry in the
   Intact Brain. Academic Press, New York.
Changeux, J-P. (1985). Neuronal Man. The Biology of Mind,
   Translated by Dr. L. Garey. Pantheon Books, New York.
Cipolloni, P. B. and Pandya, D. D. (1985). Topography and
   Trajectories of Commissural Fibers of the Superior Temporal
   Region in the Rhesus Monkey. Brain Res., 57, 381-389.
Clarke, S., Kraftsik, R., Innocenti, G. M. and van der Loos, H.
   (1986). Sexual Dimorphism and Development of the Human Corpus
   Callosum. Neurosc. Letters, 26, S299, Abstract.
Cunningham, D. J. (1892). Contribution to the Surface Anatomy of
   the Cerebral Hemispheres. Royal Irish Academy, Dublin.
De Lacoste-Utamsing, C. and Holloway, R. L. (1982). Sexual
   Dimorphism in the Human Corpus Callosum. Science, 216,
   1431-1432.
De Lacoste-Utamsing, C., Holloway, R. L., Kirkpatrick, J. B. and
   Ross, E. D. (1981). Anatomical and Quantitative Aspects of
   the Human Corpus Callosum. Soc. Neurosc., 7, Abstract No.
   127.5.
De Lacoste, M. C., Kirkpatrick, J. B. and Ross, E. D. (1985).
   Topography of the Human Corpus Callosum. J. Neuropath.
   Exper. Neurol., 44, 578-591.
Demeter, S., Ringo, J. and Doty, R. W. (1985). Sexual Dimorphism
   in the Human Corpus Callosum? Soc. Neurosc., 11, Abstract
   No. 254.12.
Doty, R. W., Overman, W. H. and Negrão, N. (1979). Role of
   Forebrain Commissures in Hemispheric Specialization and Memory
   in Macaques. In Structure and Function of Cerebral
   Commissures. (eds. I. S. Russell, M. W. Van Hof and G.
   Berlucchi). Univ. Park Press, Baltimore.
Eberstaller, O. (1890). Das Stirnhirn. Urban and Schwartzenberg,
   Wien and Leipzig.

Edelman, G. M. and Finkel, L. H. (1984). Neuronal Group Selection in the Cerebral Cortex. In Dynamic Aspects of Neocortical Function. (eds. G. M. Edelman, W. E. Gall and W. M. Cowan). Wiley, New York.

Galaburda, A. M. (1984). Anatomical Asymmetries. In Cerebral Dominance: The Biological Foundations. (eds. N. Geschwind and A. M. Galaburda). Harvard Univ. Press, Cambridge, Mass.

Galaburda, A. M., Sanides, F. and Geschwind, N. (1978). Cytoarchitectonic Left-right Asymmetries in the Temporal Speech Region. Arch. Neurol., 35, 812-817.

Geschwind, N. and Levitsky, W. (1968). Human Brain: Left-right Asymmetries in Temporal Speech Region. Science, 161, 186-187.

Heilman, K. M. and Valenstein, E. (Eds.) (1985). Clinical Neuropsychology. 2nd Edition. Oxford Univ. Press, New York.

Holloway, R. L. (1980). Within-species Brain-body Weight Variability: A Reexamination of the Danish Data and Other Primate Species. Am. J. Physical Anthropol., 53, 109-121.

Holloway, R. L. and De Lacoste, M. C. (1986). Sexual Dimorphism in the Human Corpus Callosum: an Extension and Replication Study. Human Neurobiology, 5, 87-91.

Innocenti, G. M. (1986). The General Organization of Callosal Connections. In Cerebral Cortex. (eds. E. G. Jones and A. A. Peters). Plenum Press, New York.

LeMay, M. and Geschwind, N. (1978). Asymmetries of the Human Cerebral Hemispheres. In Language Acquisition and Language Breakdown: Parallels and Divergencies. (eds. A. Caramazza and E. B. Zurif) Johns Hopkins University Press, Baltimore.

Macko, K. A. and Mishkin, M. (1985). Metabolic Mapping of Higher-order Visual Areas in the Monkey. In Brain Imaging and Brain Function. (ed. L. Sokoloff). Raven Press, New York.

Mall, F. P. (1909). On Several Anatomical Characters of the Human Brain, said to Vary According to Race and Sex, with Special Reference to the Weight of the Frontal Lobe. Am. J. Anatomy, 9, 1-32.

Nasrallah, H. A., Andreasen, N. C., Coffman, J. A., Olson, S. C., Dunn, V. D., Ehrhardt, C. and Chapman, S. M. (1986). A Controlled Magnetic Resonance Imaging Study of Corpus Callosum Thickness in Schizophrenia. Biolog. Psychiat., 21, 272-282.

Nasrallah, H. A., Andreasen, N. C., Olson, S. C., Coffman, J. A. Coffman, C. E., Dunn, V. D. and Ehrhardt, J. C. (1985). Absence of Sexual Dimorphism of the Corpus Callosum in Schizophrenia: A Magnetic Resonance Imaging Study. Soc. Neurosc., 11, Abstract No. 382.8.

Pfeifer, R. A. (1936). Pathologie der Horstrählung und der Corticalen Hörsphäre. In Handbuch der Neurologie, Vol.6. (eds. O. Bumke and O. Foerster). Springer, Berlin.

Potter, S. and Graves, R. (1987). Interhemispheric Transfer in Consistent Right-handers and Mixed-handers. J. Clin. Exper. Neuropsychol., 9, Abstract 24.

494

Rakic, P. and Yakovlev, P. I. (1968). Development of the Corpus
     Callosum and Cavum Septi in Man. J. Comp. Neurology, 132,
     45-72.
Retzius, G. (1900). Ueber das Hirngewicht der Schweden.
     Biologische Untersuchungen, 9, 51-68.
Rockland, K. S. and Pandya, D. N. (1986). Topography of Occipital
     Lobe Commissural Connections in the Rhesus Monkey. Brain
     Res., 365, 174-178.
Sidtis, J. J., Volpe, B. T., Holtzman, J. E., Wilson, D. H. and
     Gazzaniga, M. S. (1981). Cognitive Interaction after Staged
     Callosal Section: Evidence for Transfer of Semantic
     Activation. Science, 212, 344-346.
von Economo, C. and Horn, L. (1930). Über Windungsrelief, Maße
     und Rindenarchitektonik der Supratemporalfläche, Ihre
     Individuellen und Ihre Seitenunterschiede. Zeitschrift fur
     die gesamte Neurologie und Psychiatrie, 130, 678-757.
Wada, J. A., Clarke, R. and Hamm, A. (1975). Cerebral Hemisphere
     Asymmetry in Humans. Arch. Neurol., 32, 239-246.
Witelson, S. F. (1977). Anatomic Asymmetry in the Temporal Lobes:
     Its Documentation, Phylogenesis, and Relationship to
     Functional Asymmetry. Ann. N.Y. Acad. Sci., 299, 328-354.
Witelson, S. F. (1980). Neuroanatomical Asymmetry in Left
     Handers: A Review and Implications for Functional Asymmetry.
     In The Neuropsychology of Left Handers. (ed. J. Herron).
     Academic Press, New York.
Witelson, S. F. (1983). Bumps on the Brain: Right-left Anatomic
     Asymmetry as a Key to Functional Asymmetry. In Language
     Functions and Brain Organization. (ed. S. Segalowitz).
     Academic Press, New York.
Witelson, S. F. (1985a). The Brain Connection: The Corpus
     Callosum is Larger in Left Handers. Science, 229, 665-668.
Witelson, S. F. (1985b). On Hemisphere Specialization and
     Cerebral Plasticity from Birth. Mark II. In Hemispheric
     Function and Collaboration in the Child. (ed. C. Best).
     Academic Press, New York.
Witelson, S. F. (1986). Wires of the Mind: Anatomical Variation
     in the Corpus Callosum in Relation to Hemispheric
     Specialization and Integration. In Two Hemispheres - One
     Brain: Functions of the Corpus Callosum. Neurology and
     Neurobiology. (eds. F. Leporé, M. Ptito and H. H. Jasper).
     17, 117-137.
Witelson, S. F. (1987a). Brain Asymmetry, Functional Aspects. In
     Encyclopedia of Neuroscience. (ed. G. Adelman). Birkhauser
     Boston, Cambridge.
Witelson, S. F. (1987b). Neurobiological Aspects of Language and
     Spatial Perception in Children. Child Devel., in press.
Witelson, S. F. and Colonnier, M. (1983). Neuroanatomical
     Asymmetry in the Human Temporal Lobes and Related
     Psychological Characteristics. Final Report, U.S. NIH-NINCDS
     contract N01-NS-6-2344.

Witelson, S. F. and Kigar, D. L. (1987). Individual Differences in the Anatomy of the Corpus Callosum: Sex, Hand Preference, Schizophrenia and Hemisphere Specialization. In Individual Differences in Hemisphere Specialization, NATO ASI Series, Life Sciences. (ed. A. Glass). Plenum Press.

Witelson, S. F. and Pallie, W. (1973). Left Hemisphere Specialization for Language in the Newborn: Neuroanatomical Evidence of Asymmetry. Brain, 96, 641-646.

Yakovlev, P. I. and Lecours, A-R. (1967). The Myelogenetic Cycles of Regional Maturation of the Brain. In Regional Development of the Brain in Early Life. (ed. A. Minkowski). Blackwell Scientific, London.

Index